An Introduction to the Physiology of Hearing

2nd edition

James O. Pickles

Department of Physiology,
University of Birmingham,
Birmingham, England

ACADEMIC PRESS
Harcourt Brace Jovanovich, Publishers
London · San Diego · New York · Boston · Sydney · Tokyo · Toronto

ACADEMIC PRESS LIMITED
24–28 Oval Road
London NW1 7DX

United States Edition published by
ACADEMIC PRESS INC.
San Diego, CA 92101

British Library Cataloguing in Publication Data

Pickles, James O.
An introduction to the physiology
of hearing—2nd ed.
i. Man. Hearing Physiology
I. Title
612.85

ISBN 0–12–554753–6
ISBN 0–12–554754–4 Pbk

Typeset by Profile Information London
Printed in Great Britain by
St Edmundsbury Press Ltd, Bury St Edmunds, Suffolk

Preface to the Second Edition

This book deals with the way that the auditory system processes acoustic signals. Since the first edition appeared, rapid and exciting developments have been made in several areas close to this theme, most notably in the understanding of cochlear function, including cochlear mechanics, hair cell function, and mechanisms of transduction. The sections of the book dealing with these topics have been completely rewritten. Chapter 10, dealing with physiological correlates of sensorineural hearing loss, has been expanded. However, all chapters have been completely revised. Given the rapid expansion of the amount of material available, severe selection has had to be applied, both in the topics presented, and in the references that could be quoted, in order to ensure adequate treatment of the main theme of the book within a reasonable length. A reading scheme is provided (p. viii), so as to guide readers to the sections of the book most appropriate for their interests.

I am indebted to E. de Boer, D. Caird, B.C.J. Moore, S.M. Rosen, I.J. Russell and M.B. Sachs for their comments on sections of an earlier version of the manuscript, and for their help on various points. However, the author takes full responsibility for the interpretations presented here. I would also like to express thanks to my colleagues in the Department of Physiology, namely S.D. Comis and M.P. Osborne, who collaborated in producing micrographs shown in the book, and to T.L. Hayward, who prepared the illustrations. I would also like to thank the authors and publishers who have allowed me to reproduce illustrations from their published work.

April 1988   J.O. Pickles
                                                                            Birmingham

From the Preface to the First Edition

The last 15 years have seen a revolution in auditory physiology, but the new ideas have been slow to gain currency outside specialist circles. Undoubtedly, one of the main reasons for this has been the lack of a general source for non-specialists, and it is hoped that the present book will introduce current thinking to a much wider audience.

While the book is primarily intended as a student text, it is hoped that it will be equally useful to teachers of auditory physiology. It should be particularly useful to those teaching physiology to medical students, because general texts of physiology aimed at medical students commonly contain only a small section on hearing, which is based on material that is 20 or more years out of date. The increasing concern about the extent of hearing loss in the community should increase the attention paid to auditory physiology in the medical curriculum.

The book is written at a level suitable for a degree course on the special senses or as a basis for a range of postgraduate courses. It is organized so as to be accessible to those approaching the subject at a number of levels and with a variety of backgrounds (see Reading plan, p. viii). Only the most elementary knowledge of physiology is assumed, and even such basic concepts as ionic equilibrium potentials are explained where appropriate, so that the book should be accessible to those with only a small background in physiology. The treatment is non-mathematical, and only a few very elementary algebraic equations appear. In Chapter 3, some of the reasoning behind theories of cochlear mechanics is explained verbally for the benefit of those without mathematical training. I have found that for those *with* such training, this approach makes the subject more, rather than less, confusing. I can only apologize to them, and refer them to the several excellent accounts that have already appeared at their own level (e.g. Dallos, 1973a; de Boer, 1980, 1983; Zwislocki, 1965).

I should like to express my thanks to colleagues who read and commented on sections of an earlier version of the manuscript, and in particular to G.R. Bock, S.D. Comis, J.L. Cranford, L.U.E. Kohllöffel, O. Lowenstein, B.C.J. Moore, G.F. Pick, H.F. Ross, I.J. Russell, R.L. Smith and G.K. Yates. I am also grateful to T.L. Hayward for help with the illustrations.

Reading Plan

Chapter 1, on the physics and analysis of sound, contains elementary information which should be read by everyone. Readers who need only a brief introduction to auditory physiology may then read only Chapter 3 on the cochlea. Those whose interests lie in the psychophysical correlates may read Chapters 1 to 4, and then turn to Chapter 9. Readers who are interested in audiological and clinical aspects may read Chapters 1 to 3, part of Chapter 4 (as indicated in the introduction to Chapter 4), and then Chapter 10. Chapter 5, which explores the newer ideas on cochlear physiology, is written at a more advanced level than the rest of the book, and if desired may be omitted without affecting the understanding of the other chapters. Chapters 6, 7 and 8, on the brainstem, cortex and centrifugal pathways should appeal primarily to specialist physiology students, although the latter part of Chapter 7 on the cortex, some parts of Chapter 8 on centrifugal pathways, and Chapter 9 on psychophysical correlates contain information that should be of interest to physiological psychologists.

Contents

To Charlotte

1. *The Physics and Analysis of Sound*

Some of the basic concepts of the physics and analysis of sound, which are necessary for the understanding of the later chapters, are presented here. The relations between the pressure, displacement and velocity of a medium produced by a sound wave are first described, followed by the decibel scale of sound level, and the notion of impedance. Fourier analysis and the idea of linearity are then described.

A. The Nature of Sound

In order to understand the physiology of hearing, a few facts about the physics of sound, and its analysis, are necessary. As an example, Fig. 1.1 shows a tuning fork sending out a sound wave, and shows the distribution of the sound wave at one point in time, plotted over space, and at one point in space, plotted over time. The tuning fork sends out a travelling pressure wave, which is accompanied by a wave of displacement of the air molecules making them vibrate around their mean positions. There are two important variables in such a sound wave. One is its frequency, which is the number of waves to pass any one point in a second, measured in cycles per second, or hertz (Hz). This has the subjective correlate of pitch, sounds of high frequency having high pitch. The other important attribute of the wave is its amplitude or intensity, which is related to the magnitude of the movements produced. This has the subjective correlate of loudness.

If the sound wave is in a free medium, the pressure and velocity of the air vary exactly together, and are said to be in phase. The displacement however lags by a quarter of a cycle. It is important to understand that the pressure variations are around the mean atmospheric pressure. The variations are in fact a very small proportion of the total atmospheric pressure – even a level as high as 140 dB SPL (defined on p. 4), as intense as anything likely to be encountered in everyday life, makes the

1

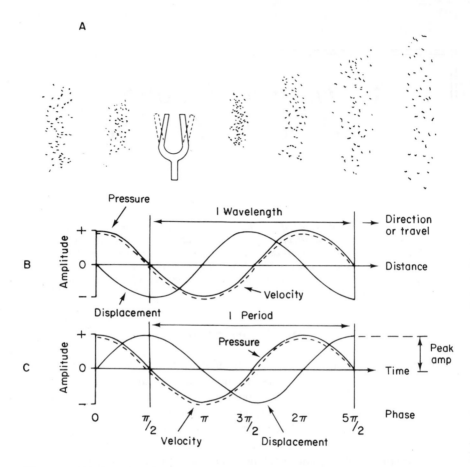

Fig. 1.1   (A) A tuning fork sending out a sound wave. (B) The variation of the pressure, velocity and displacement of the air molecules in a sinusoidal sound wave are seen at one moment in time. The variations are plotted as a function of distance. The pressure and velocity vary together, and the displacement lags by a quarter of a cycle. (C) The same variation is plotted as a function of time, as measured at one point. Because times further in the past are plotted to the right of the figure, the curve of displacement is here plotted to the right, not to the left of the pressure curve, as in part B. The phase increases by 2π (or 360°) in one cycle. The sound wave is defined by its peak amplitude and by its frequency.

pressure vary by only 0.6%. The displacement is also about the mean position, and the sound wave does not cause a net flow of molecules.

The different parameters of the sound wave can easily be related to each other. The peak pressure above atmospheric (p) and the peak velocity of the sinusoid (v) are related by:

$$p = zv \tag{1}$$

where z is a constant of proportionality, called the *impedance*. It is a function of the medium in which the sound is travelling, and will be dealt with later.

The intensity of the sound wave is the amount of power transmitted through a unit area of space. It is a function of the *square* of the peak pressure and, by equation 1, also of the square of the peak velocity. In addition, it depends on the impedance; for a sine wave,

$$\text{Intensity } I = p^2/2z = zv^2/2 \tag{2}$$

In other words, if the intensity of a sound wave is constant, the peak pressure and the peak velocity are constant. They are also independent of the frequency of the sound wave. It is for these reasons that the pressure and velocity will be of most use later.

Unlike the above parameters, the peak *displacement* of the air molecules does vary with frequency, even when the intensity is constant. For constant sound intensity, the peak displacement is inversely proportional to the frequency:

$$d = \frac{1}{2\pi f} \sqrt{\frac{(2I)}{z}} \tag{3}$$

where d is the peak displacement, and f is the frequency. So for sounds of constant intensity, the displacement of the air particles gets smaller as the frequency increases. We can see correlates of this when we see a loudspeaker cone moving. At low frequencies the movement can be seen easily, but at high frequencies the movement is imperceptible, even though the intensities may be comparable.

B. The Decibel Scale

We can measure the intensity of a sound wave by specifying the peak excess pressure in normal physical quantities, e.g. newtons/metre^2, sometimes called pascals. In fact it is often more useful to record the RMS pressure, meaning the square Root of the Mean of the Squared pressure, because such a quantity is related to the energy (actually to the square root of the energy) in the sound wave over all shapes of waveform. For a sinusoidal waveform the RMS pressure is $1/\sqrt{2}$ of the peak pressure. While it is perfectly possible to use a scale of RMS pressure in terms of N/m^2, for the purposes of physiology and psychophysics it turns out to be much more convenient to use an intensity scale in which equal increments roughly correspond to equal increments in sensation, and in which the very large range of intensity used is represented by a rather narrower range of numbers. Such a scale is made by taking the ratio of the sound intensity to a certain reference intensity, and then taking the logarithm of the ratio. If logarithms to the base 10 are taken, the units in the resulting scale, called Bels, are rather large, so the scale is expressed in units 1/10th the size, called decibels, or dB.

$$\text{Number of dB} = 10 \log_{10} \left(\frac{\text{Sound intensity}}{\text{Reference intensity}} \right)$$

Because the intensity varies as the square of the pressure, the scale in dB is 10 times the logarithm of the *square* of the pressure ratio, or 20 times the logarithm of the pressure ratio:

$$\text{Number of dB} = 20 \log_{10} \left(\frac{\text{Sound pressure}}{\text{Reference pressure}} \right)$$

It only now remains to choose a convenient reference pressure. In physiological experiments the investigator commonly takes any reference he finds convenient, such as that, for instance, given by the maximum signal in his sound stimulating system. However, one scale in general use has a reference close to the lowest sound pressure that can be commonly detected by man, namely, $2 \times 10^{-5} N/m^2$ RMS, or 20 μpascals RMS. In air under standard conditions this corresponds to a power of 10^{-12} watts/m². Intensity levels referred to this are known as dB SPL.

$$\text{Intensity level in dB SPL} = 20 \log_{10} \left(\frac{\text{RMS sound pressure}}{2 \times 10^{-5} N/m^2} \right)$$

We are then left with a scale with generally positive values, in which equal intervals have approximately equal physiological significance in all parts of the scale, and in which we rarely have to consider step sizes less than one unit. While we often have to use only positive values, negative values are perfectly possible. They represent sound pressures less than $2 \times 10^{-5} N/m^2$, for which the pressure ratio is less than 1.

C. Impedance

Materials differ in their response to sound; in a tenuous, compressible medium such as air a certain sound pressure will produce greater velocities of movement than in a dense, incompressible medium such as water. The relation between the sound pressure and particle velocity is a property of the medium and was given in equation 1 by Impedance $z = p/v$. For plane waves in an effectively infinite medium the impedance is a characteristic of the medium alone. It is then called the *specific impedance*. In the SI system, z is measured in $(N/m^2)/(m/sec)$, or $N \ sec/m^3$. If z is large, as for a dense, incompressible medium such as water, relatively high pressures are needed to achieve a certain velocity of the molecules. The pressure will be higher than is needed for a medium of low specific impedance, such as air.

The impedance will concern us when we consider the transmission of sounds from the air to the cochlea. Air has a much lower impedance than the cochlear fluids. Let us take, as an example, the transmission of sound from air into a large body of water, such as a lake. The specific impedance of air is about $400 \ N \ sec/m^3$, and that of water $1.5 \times 10^6 \ N \ sec/m^3$, a ratio of 3750 times. In other words, when a sound wave meets a water surface at normal incidence, the pressure variation in the wave is only large enough to displace the water at the boundary by 1/3750 of the displacement of the air near the boundary. However, continuity requires that the displacements of the molecules immediately on both sides of the boundary must be equal. What happens is that much of the incident sound wave is reflected; the pressure at the boundary stays high, but because the reflected wave is travelling in the opposite direction to the incident wave it produces movement of the molecules in the opposite direction. The movements due to the incident and reflected waves therefore substantially cancel, and the net velocity of the air molecules will be small. This leaves a net ratio of pressure to velocity in the air near the boundary which is the same as that of water.

One result of the impedance jump is that much of the incident power is reflected. Where z_1 and z_2 are the specific impedances of the two media, the proportion of the incident power transmitted is $4z_1z_2/(z_1 + z_2)^2$. At the air–water interface this means that only about 0.1% of the incident power is transmitted, corresponding to an attenuation of 30 dB. In a later section (p. 17) we shall see how the middle ear converts a similar attenuation in the ear to the near-perfect transmission estimated as occurring at some frequencies.

Finally, in analysing complex acoustic circuits, it is convenient to use analogies with electrical circuits, for which the analysis is well known. Impedance in an electrical circuit relates the voltage to the rate of movement of charge, and if we are to make an analogy we need a measure of impedance which relates to the amount of medium moved per second. We can therefore define a different acoustic impedance, known as acoustic ohms, which is the pressure to move a unit *volume* of the medium per second. Acoustic ohms will not be used in this book and, where necessary, values will be converted from the literature, which is done by multiplying the number of acoustic ohms by the cross-sectional area of the structure in question.

D. The Analysis of Sound

Figure 1.2 shows a small portion of the pressure waveform of a complex acoustic signal. There is a regularly repeating pattern with two peaks per cycle. The pattern can be approximated by adding together the two sinusoids shown, one at 150 Hz, and the other at 300 Hz. Such an analysis of a complex signal into component sinusoids is known as Fourier analysis. and forms one of the conceptual cornerstones of auditory physiology. The result of a Fourier transformation is to produce the *spectrum* of the sound wave (Fig. 1.2C). The spectrum shows here that, in addition to the main components, there are also smaller components, at 1/15th of the amplitude or less, at 450 Hz and 600 Hz. Such a spectrum tells us the amplitude of each frequency component, and so the energy in each frequency region.

The principles of Fourier analysis can be illustrated most easily by the reverse process of Fourier synthesis, that is, by taking many sinusoids and adding them together to make a complex wave. Fig. 1.3 shows how it is possible to make a good approximation of a sine wave by adding many sinusoids together. If this process were continued indefinitely, it would be possible to make a waveform indistinguishable from a square wave. Fourier analysis is simply the reverse of this – finding the elementary sinusoids, which when added together, will give the required waveform.

Why do we analyse sound waves into sinusoids rather than into other elementary waveforms? One reason is that it is mathematically convenient to do so. Another reason is that sinusoids represent the oscillations of a very broad class of physical systems, so that examples are likely to be found in nature. However, the most compelling reason from our point of view is that the auditory system itself seems to perform a Fourier transform, like that of Fig. 1.2C, although with a more limited resolution. Therefore, sinusoids are not only simple physically, but are simple physiologically. This has a correlate in our own sensations, and a sinusoidal sound wave has a particularly pure timbre. In understanding the physiology of the lower stages of the auditory system, one of our concerns will be with the way in which the system analyses sound into sine waves, and how it handles the frequency and intensity information in them.

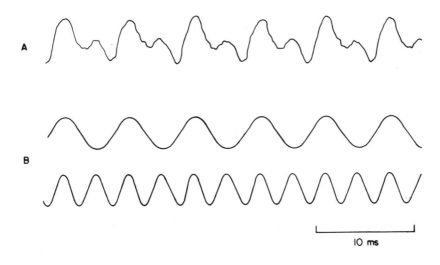

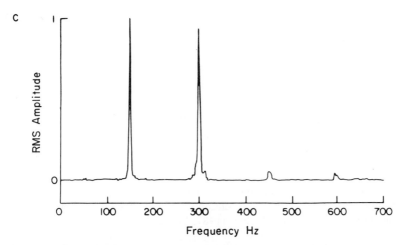

Fig. 1.2 (A) A portion of a complex acoustic waveform. (B) The waveform can be closely approximated by adding together two sine waves. (C) A Fourier analysis of the waveform in A shows that in addition to the main components, there are other smaller ones at higher frequencies. Components at still higher frequencies, responsible for the small high-frequency ripple on the waveform in A, lie outside the frequency range of the analysis, and are not shown.

Figure 1.4 shows some common Fourier transforms. In the most elementary case, a simple sinusoid, which lasts for an infinite time, has a Fourier

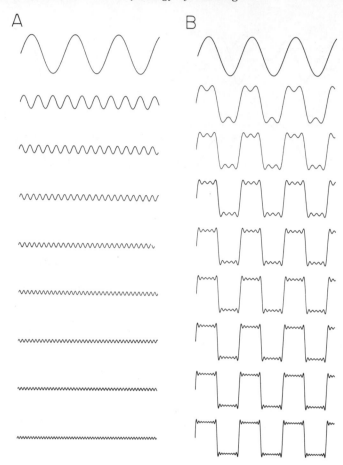

Fig. 1.3 A square wave can be approximated by adding together sinusoids of relative frequencies 1, 3, 5, 7, etc. The column in B shows the effect of successively adding the sinusoids in A. From Pickles (1987, Fig. 2.3).

transform represented by a single line, corresponding to the frequency of the sinusoid (Fig. 1.4A). A wave such as a square wave, similarly lasting for an infinite time, has a spectrum consisting of a series of lines (Fig. 1.4B). But physical signals do not of course last for an infinite time, and the result of shortening the duration of the signal is to broaden each spectral line into a band (Fig. 1.4C). The width of each band turns out to be inversely proportional to the duration of the waveform, and the exact shape of each band is a function of the way the wave is turned on and off. If, for instance, the waveform is turned on and off abruptly, sidelobes appear around each spectral band (Fig. 1.4D).

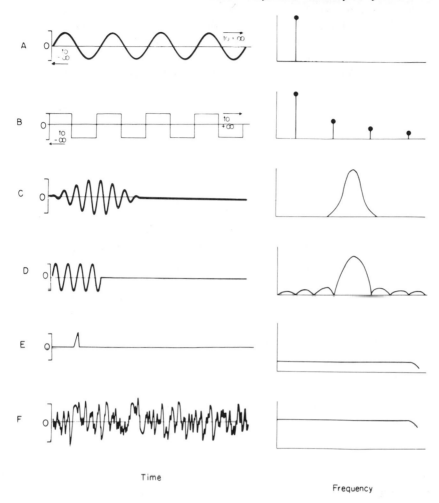

Fig. 1.4   Some waveforms (left) and their Fourier analyses (right). (A) Sine wave. (B) Square wave (in these cases the stimuli last an infinite time, and have line spectra, the components of which are harmonically related). (C) Ramped sine wave. (D) Gated sine wave. (E) Click. (F) White noise.

In the most extreme case, the wave can be turned on for an infinitesimal time, in which case we have a click. The spread of the spectrum will be in inverse proportion to the duration, and so, in the limit, will be infinite. The spectrum of a click therefore covers all frequencies equally. In practice, a click will of course last for a finite time, and this is associated with an upper frequency limit to the spectrum (Fig. 1.4E). Another quite different signal, namely white noise, also contains all frequencies equally (Fig. 1.4F). Although the spectrum determined over short periods shows considerable random variability, the spectrum determined over a long period is flat. It differs from a click in the relative phases of the frequency components, which for white noise are random.

E. Linearity

One concept which we shall meet many times, is that of a *linear* system. In such a system, if the input is changed by a certain factor k, the output is also changed by the same factor k, but is otherwise unaltered. In addition, linear systems satisfy a second criterion, which is that the output to two or more inputs applied at the same time, is the sum of the outputs that would have been obtained if the inputs had both been applied separately.

We can therefore identify a linear system as one in which the amplitude of the output varies in proportion to the amplitude of the input. A linear system also has other properties. For instance, the only Fourier frequency components in the output signal are those contained in the input signal. A linear system never generates new frequency components. Thus it is distinguished from a non-linear system. In a non-linear system, new frequency components are introduced. If a single sinusoid is presented, the new components will be harmonics of the input signal. If two sinusoids are presented, there will, in addition to the harmonics, be intermodulation products produced; that is, Fourier components whose frequency depends on both of the input frequencies. In the auditory system, we shall be concerned with whether certain of the stages act as linear or non-linear systems. The tests used will be based on the properties described above.

F. Summary

1. A sound wave produces compression and rarefaction of the air, the molecules of which vibrate around their mean positions. The extent of the pressure variation has a subjective correlate in loudness. The frequency, or number of waves passing a point in a second, has a subjective correlate in pitch. Frequency is measured in cycles per second, known as hertz (Hz).

2. The particle velocities produced by a pressure variation depend on the impedance of the medium. If the impedance is high, high pressures are needed to produce a certain velocity.

3. When a sound pressure wave meets a boundary between two media of different impedance, some of the sound energy is reflected.

4. Complex sounds can be analysed by Fourier analysis, that is, by splitting the waveforms into component sine waves of different frequencies. The cochlea seems to do this too, to a certain extent.

5. In a linear system, the output to two inputs together is the sum of the outputs that would have been obtained if the two inputs had been presented separately. Moreover, in a linear system, the only Fourier frequency components that are present in the output are those that were present in the input. Neither is true for a non-linear system.

2. *The Outer and Middle Ears*

The outer ear modifies the sound wave in transferring the acoustic vibrations to the eardrum. First, the resonances of the external ear increase the sound pressure at the eardrum, particularly in the range of frequencies (in man) of 2–7 kHz. It therefore increases the efficiency of sound absorption at these frequencies. Secondly, the change in pressure depends on the direction of the sound. This is an important cue for sound localization, enabling us to distinguish above from below, and in front from behind. The middle ear apparatus then transfers the sound vibrations from the eardrum to the cochlea. It acts as an impedance transformer, coupling sound energy from the low impedance air to the higher impedance cochlear fluids, substantially reducing the transmission loss that would otherwise be expected. The factors allowing this will be described, and the extent to which the middle ear apparatus acts as an ideal impedance transformer will be discussed. Transmission through the middle ear can be modified by the middle ear muscles, and their action, and possible hypotheses for their role in hearing, will be described.

A. The Outer Ear

The outer ear consists of a partially cartilaginous flange called the pinna, which includes a resonant cavity called the concha, together with the ear canal or external auditory meatus leading to the eardrum or tympanic membrane (Fig. 2.1). The effect of the outer ear on the incoming sound has been analysed from two approaches. One is the influence of the resonances of the outer ear on the sound pressure at the tympanic membrane, the other is the extent to which the outer ear provides directionality cues for help in sound localization.

1. The Pressure Gain of the Outer Ear

The concha, meatus and tympanic membrane provide the main elements of a complex acoustic cavity, such a cavity being expected to increase

12

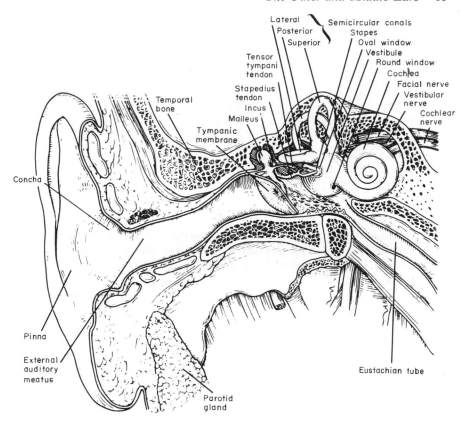

Fig. 2.1   The external, middle and inner ears in man. From *Tissues and Organs: A Text-Atlas of Scanning Electron Microscopy*, by R.G. Kessel and R.H. Kardon. W.H. Freeman and Company. Copyright © 1979.

and decrease the sound pressure at the tympanic membrane at different frequencies. The data of Wiener and Ross (1946) in man showed a broad peak of 15–20 dB at 2.5 kHz in the sound pressure gain at the tympanic membrane. A more recent synthesis of the available data by Shaw (1974) shows very much the same effect (Fig. 2.2). Shaw studied the contributions of the different elements of the external ear by adding the different components sequentially in a model. The results of such an analysis are shown in Fig. 2.3. The 2.5-kHz peak is provided by a resonance of the combination of the meatus and concha. This resonance occurs where the combination is a quarter of a wavelength long. The 5.5-kHz peak is due to a resonance in the concha alone. It appears that the main pressure gains are complementary,

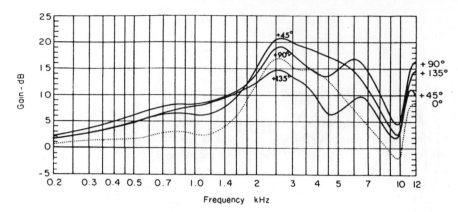

Fig. 2.2   The average pressure gain of the external ear in man. The gain in pressure at the eardrum over that in the free field is plotted as a function of frequency, for different orientations of the source in the horizontal plane ipsilateral to the ear. Zero degrees is straight ahead. From Shaw (1974, Fig. 5).

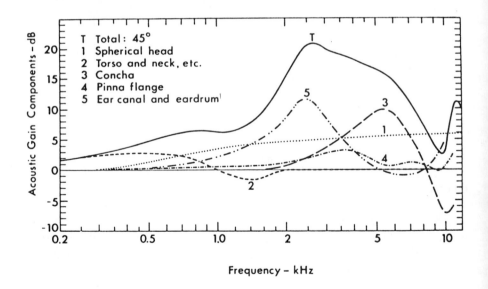

Fig. 2.3   The average pressure gain contributed by the different components of the outer ear in man. The analysis assumes that the components are added in the order shown. e.g. curve 5 is the change produced when the ear canal and eardrum are added to the other components listed. Stimulus in the horizontal plane, 45° from straight ahead. From Shaw (1974, Fig. 11).

increasing the sound pressure relatively uniformly over the range from 2 to 7 kHz. The resonances increase the efficiency of transmission around these frequencies (Shaw and Stinson, 1983; Rosowski *et al.*, 1986).

2. The Outer Ear as an Aid to Sound Localization

The most important cues for sound localization in man are the intensity and timing differences in the sound waves at the two ears. The sound wave from a source on the right will strike the right ear before the left and will be more intense in the right ear. However, this does not account for our ability to distinguish in front from behind, or above from below. The information for such localization comes from the pinna and concha.

When the wavelength is short compared with the dimensions of the pinna, the pinna can show a high degree of directional selectivity in the reception of sound. We expect the pinna to be useful in this way only in the high kHz range of frequencies. In the cat, the pinna can produce a gain of up to 28 dB in sound pressure in this frequency range, for sound sources in a narrow region directly in line with the axis of the pinna (Phillips *et al.*, 1982).

The frequency range of man is too low for the pinnae to be of much use in producing a similar degree of directional selectivity. Nevertheless, the pinnae do provide useful directional cues. When a sound is behind the ear, the wave transmitted directly interferes with the wave scattered off the edge of the pinna, reducing the response in the 3–6 kHz region (Shaw, 1974). It is in this region that there are the greatest intensity changes as a sound source is moved in the horizontal plane (Fig. 2.2). The obvious dip at 10 kHz is due to out-of-phase reflections off the back wall of the concha. The low-frequency cut-off frequency of the dip is raised when the sound source is elevated, and there is psychophysical evidence that such information is used to judge the elevation of sound sources (Gardner and Gardner, 1973).

The external ear therefore produces a spectral modulation of the incoming sound. In using such a coloration to make directional judgements, we are obviously required to make subtle judgements about the modulation of the spectra of perhaps unknown sound sources. Exploratory movements of the head can be used to provide additional information.

B. The Middle Ear

1. Introduction

The middle ear couples sound energy from the external auditory meatus to the cochlea, and by its transformer action helps to match the impedance of the auditory meatus to the much higher impedance of the cochlear fluids. In the absence of a transformer mechanism, much of the sound would be reflected. The sound is transmitted from the tympanic membrane to the cochlea by three small bones, known as the *ossicles*. They are called the malleus, the incus and the stapes (Figs 2.1 and 2.4). The first two bones are joined comparatively rigidly so that when the tip of the malleus is pushed by the tympanic membrane, the bones rotate together and transfer the force to the stapes. The stapes is attached to a flexible window in the wall of the cochlea, known as the oval window (Fig. 2.4).

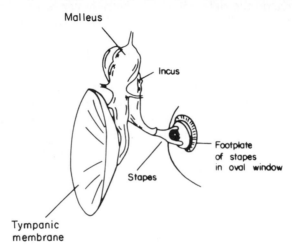

Fig. 2.4   The three ossicles, called the malleus, incus and stapes, transmit the sound vibrations from the tympanic membrane (eardrum) to the oval window of the cochlea.

A second function of the ossicles is to apply force to one window only of the cochlea. If the ossicles were missing, and the pressure of the incoming sound wave were applied to both windows equally, there would be a reduced flow of cochlear fluids. Nevertheless, in many species the other window of the cochlea, the round window, is shielded from the incoming sound wave by a bony ridge. In these cases, if the ossicles are missing, the sound pressure is still primarily applied to one window of the cochlea, and some hearing, although without the benefit of the impedance matching, is still possible.

2. The Middle Ear as an Impedance Transformer

(a) The nature of the problem

The middle ear transfers the incoming vibration from the comparatively large, low impedance, tympanic membrane to the much smaller, higher impedance, oval window. As was explained in Chapter 1 (p. 5), when a sound wave meets a higher impedance medium, normally much of the sound energy is reflected. The middle ear apparatus, by acting as an acoustic impedance transformer, reduces this attenuation substantially.

Following the tentative suggestion made by Wever and Lawrence (1954), many authors have said that the cochlear fluids would have an impedance approximately equal to that of sea-water, namely 1.5×10^6 N sec/m^3, and this led to the calculation, detailed above (p. 5), that if the sound met the oval window directly, only 0.1% of the incident energy would be transmitted. As pointed out by Schubert (1978), and indeed by Wever and Lawrence themselves, while the numerical result may be approximately correct, the physical reasoning behind it is not. Specific impedances are defined for progressive acoustic waves in an effectively infinite medium. In the range of audible frequencies, the cochlea is far smaller than a wavelength of sound in water, and so cannot develop such waves. The actual cochlear impedance is determined entirely by the fact that cochlear fluid flows from one flexible window, the oval window, to another, the round window, and the cochlear impedance depends on the way the fluids flow, and on their interaction with the distensible cochlear membranes. The input impedance of the cochlea has been determined either theoretically (e.g. Zwislocki, 1965) or experimentally (e.g. Lynch *et al.*, 1982). The direct measurements of Lynch *et al.* in the cat suggest a cochlear impedance of about 1.5×10^5 N sec/m^3 at 1 kHz,* much lower than expected from Wever and Lawrence's approximation.

(b) The mechanism of the impedance transformer

In matching the impedance of the tympanic membrane to the much higher impedance of the cochlea, the middle ear uses three principles.

1. The area of the tympanic membrane is larger than that of the stapes footplate in the cochlea. The forces collected over the tympanic membrane are therefore concentrated on a smaller area, so increasing the pressure at the oval window. The pressure is increased by the ratio of the two areas (Fig. 2.5A), This is the most important factor in achieving the impedance transformation.

---

*This is derived from their measurement of 1.2×10^6 c.g.s. acoustic ohms, by multiplying by their value for a stapes footplate area of 1.26×10^{-2} cm^2 to get a specific impedance, and converting to SI units by multiplying by 10.

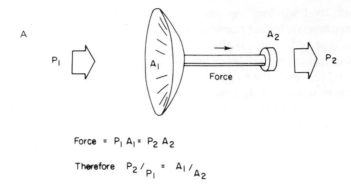

$$\text{Force} = P_1 A_1 = P_2 A_2$$

$$\text{Therefore} \quad P_2/P_1 = A_1/A_2$$

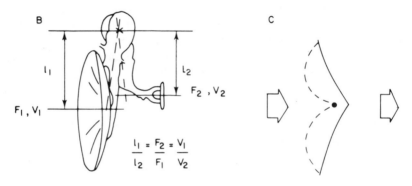

Fig. 2.5   The three mechanisms of the middle ear acoustic impedance transformer. (A) The main factor is the ratio of the areas of the tympanic membrane and oval window. The middle ear bones are here represented by a piston.(B) The lever action increases the force and decreases the velocity. (C) A buckling motion of the tympanic membrane also increases the force and decreases the velocity. A, area; F, force; L, length; P, pressure; V, velocity.

2. The second principle is the lever action of the middle ear bones. The arm of the incus is shorter than that of the malleus, and this produces a lever action that increases the force and decreases the velocity at the stapes (Fig. 2.5B). This is a comparatively small factor in the impedance match.

3. The third factor is more subtle, and depends on the conical shape of the tympanic membrane. As the membrane moves in and out it buckles, so that the arm of the malleus moves less than the surface of the membrane

(Fig. 2.5C). This again increases the force and decreases the velocity (Khanna and Tonndorf, 1972). It is also a comparatively small factor.

(c) Calculation of the transformer ratio

It might be thought that determining the transformer ratio would be a matter of comparatively simple anatomy, and would have been settled in an un-controversial way a long time ago. This is not so; the actual transformer ratio depends on the exact way the structures vibrate in response to sound. As the movements are microscopic or submicroscopic, and probably depend on the physiological state of the animal, the determination of the transformer ratio is a rather complex measurement. For instance, Khanna and Tonndorf (1972) had to use a method such as laser holography to determine the movement of the middle ear structures in the cat. It is their results which will be used here.

The most important factor is the ratio of the areas of the tympanic membrane and the oval window. In the cat, the tympanic membrane has an area of 0.42 cm^2, and the stapes footplate about 0.012 cm^2. The *pressure* on the stapes footplate is therefore increased by $0.42/0.012 = 35$ times.

The effective length of the malleus is about 1.15 times that of the incus, and so the lever action multiplies the force 1.15 times. However, the velocity is *decreased* 1.15 times. The lever action therefore increases the impedance ratio (being the pressure/velocity ratio) $1.15^2 = 1.32$ times.

The buckling factor was assessed to decrease the velocity two-fold and increase the force two-fold. The impedance ratio is therefore changed four-fold.

The final transformer ratio, calculated here as an impedance ratio, can be obtained by multiplying all these factors together. The ratio was assessed as $35 \times 1.32 \times 4 = 185$.*

Does this theoretical transformation ratio give the ideal transformation required to match the air to the cochlea? In order to answer this we need to know the input impedance of the cochlea, a measurement which has been subject to some variability. In the cat, Lynch *et al.* (1982) sealed a tube around the stapes and oval window, the other middle ear bones being detached. They applied known pressures directly to the tube, and measured the resulting displacement of the stapes by the Mössbauer technique. This technique involves the detection of γ-radiation from a vibrating source, and will be explained in more detail in Chapter 3. They found the impedance

---

*The reader may be puzzled to see very different numbers in the literature. This may be for two reasons. First, the impedances will probably be defined in acoustic ohms (see p. 6). Secondly, the transformer ratio is often quoted as the *square root* of the impedance ratio in acoustic ohms. Such a ratio is also equal to the pressure transformation ratio. The latter is calculated on p. 22.

of the cochlea at 1 kHz to be about 1.5×10^5 N s/m^3. The impedance transformer of the middle ear will make this appear to be $1.5 \times 10^5/185$ = 810 N s/m^3 at the tympanic membrane. This is rather higher than the impedance of air, which is 430 N s/m^3. The middle ear transformer ratio is not therefore quite adequate for perfect transmission.

The theoretical value of 810 N s/m^3, calculated from the input impedance of the cochlea and the transformer ratio, can be compared to direct measurements of the input impedance of the middle ear as seen at the tympanic membrane. In the cat, Møller (1965) measured the impedance of the tympanic membrane to be 1680 N s/m^3 at 1 kHz. This is about twice the expected value. It is so high because of friction in the middle ear system. Indeed, Rosowski *et al.* (1986) showed that at this frequency only about half the energy absorbed at the eardrum was transmitted to the cochlea.

Derivations of the middle ear transformer ratio have been a matter of disagreement over the years. For instance, von Békésy (1960) showed that the eardrum in man was hinged on one side, so that it flapped like a door rather than moving in and out like a piston. Obviously, a point near the hinge will contribute less to the total force transmitted than a point near the free edge, and this has led to the use of an "effective area" for the tympanic membrane which is less than the real area. Similarly, the way the tympanic membrane moves will affect the effective lever ratio of the middle ear bones, and Wever and Lawrence (1954) for this reason took a lever ratio of 2.5, rather than the 1.15 of Khanna and Tonndorf.

The transformer ratio was calculated for one species, the cat, and applies in one frequency range, around 1 kHz. It seems that at other frequencies, additional factors affect the movement. For instance, above 2 kHz the buckling motion of the tympanic membrane breaks up into separate zones, and as the frequency is raised further the effective area of the tympanic membrane becomes progressively reduced, until it becomes equal to the area of the arm of the malleus. This will reduce transmission (Khanna and Tonndorf, 1972). Transmission through the middle ear is also affected by factors such as elasticity and friction in the middle ear bones and their ligaments, particularly at low frequencies. The inertia of the middle ear bones and their imperfect coupling, in addition to acoustic resonances in the middle ear cavity, will also affect transmission. If we wish to determine the way in which the middle ear affects the transmission of sound over a range of frequencies, it is therefore necessary to measure the transmission experimentally. This can be done by measuring the *transfer function*, that is, the ratio of the output to the input, as a function of frequency.

(d) The transfer function of the middle ear

The middle ear transforms the sound pressure variations of the ear canal into a sound pressure variation in the scala vestibuli of the cochlea. The transfer function can be shown by plotting the ratio of the two pressures, at different stimulus frequencies.

Nedzelnitsky (1980) measured the pressure in the cochlear duct of the cat, just behind the oval window, for constant sound pressures at the tympanic membrane. Figure 2.6 shows the pressure gain as a function of frequency. The curve has a bandpass characteristic, greatest transmission being seen around 1 kHz. There, the sound pressures are 30 dB greater than those at the tympanic membrane. The response shows an irregularity around 4 kHz, but otherwise declines smoothly towards low and high frequencies.

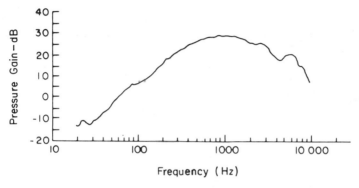

Fig. 2.6   The transfer function of the middle ear, according to Nedzelnitsky (1980). The gain of pressure in the cochlea (the scala vestibuli, basal turn) over that at the tympanic membrane, is shown as a function of frequency. From Nedzelnitsky (1980, Fig. 7).

We can attempt to identify some of the factors governing this bandpass characteristic. One factor, which attenuates the response at low frequencies, is an *elastic stiffness*. This has been ascribed to an elasticity in the tympanic membrane and the ligaments of the middle ear bones, and to a compression and expansion of the air in the middle ear cavity. For instance, as the tympanic membrane moves in and out, air in the middle ear cavity is compressed and expanded, so reducing the movement of the tympanic membrane. The importance of this factor can be shown experimentally, because when the middle ear cavity is vented to the atmosphere, transmission is increased at low frequencies but not at high frequencies (Guinan and Peake, 1967). But why should elastic stiffness be particularly important at low frequencies? This follows simply from the mathematical relation between the pressure of the sound wave and the displacement of the air, and so of the tympanic membrane. Recall from equation 3 (p. 3), that for a constant sound pressure level, the displacement of the air varies inversely with the frequency. At low frequencies, a constant sound pressure will produce a comparatively large displacement of the tympanic membrane and middle ear structures. The forces to overcome an elasticity depend on displacement, and so the forces will increase as the frequency drops. This explains why transmission is reduced at low frequencies.

The drop at high frequencies is affected by many factors, and their relative importance is not known. For instance, Khanna and Tonndorf (1972) showed that at high frequencies the vibration pattern of the tympanic membrane broke up into separate zones, reducing the effectiveness of the transmission. We would also expect the mass of the middle ear bones to have a significant effect at high frequencies. A constant sound pressure level corresponds to a constant velocity. The accelerations, and so the forces on the structures involved, therefore increase in proportion to frequency. Further, the ossicular chain begins to flex at high frequencies, also reducing transmission (Guinan and Peake, 1967).

The position is in addition complicated by acoustic resonances in the middle ear cavity, responsible for the peak and dip in the transfer function near 4 kHz. The middle ear cavities of many small animals are enlarged by a bony bulge, called the bulla, extending below the skull (incidentally, this probably serves to increase the low-frequency response of the middle ear, because it will reduce the low-frequency stiffness of the system). In many animals the bulla is divided into two by a bony wall, called the septum. The septum has a small hole, and the two cavities with a small intercommunicating hole form coupled acoustic resonators.

In the mid-frequency range, around 1 kHz, many of the factors affecting transmission at lower and higher frequencies will be small. Møller (1965) showed that in this frequency region it was the input impedance of the cochlea itself that was the main factor governing transmission. He disconnected the cochlea by disarticulating the joint between the incus and stapes, and showed that the input impedance at the tympanic membrane fell substantially. It is in this frequency region that the theoretical calculation of the transformer ratio, described above, will be most nearly accurate; the transfer here will be least affected by factors other than the input impedance of the cochlea. Therefore, we can see here whether the actual pressure gain observed by Nedzelnitsky (1980) agrees with the value expected from Khanna and Tonndorf's transformer ratio, calculated from the displacements of the middle ear structures. Khanna and Tonndorf would lead us to expect the area ratio to increase the pressure by 35 times, the lever ratio to increase it by 1.15 times, and the buckling factor to increase it by 2 times. The product is 80.5 times, or 38 dB. This is in the same range as, although rather greater than, the 30 dB increase in pressure observed in the same species, and at the same frequency, by Nedzelnitsky. The disagreement is likely to be a result of transmission losses. As pointed out above, at this frequency only about half the power absorbed at the tympanic membrane is eventually transmitted to the cochlea (Shaw and Stinson, 1983; Rosowski *et al.*, 1986).

Consideration of transmission through the middle ear is not complete without a description of the linearity of the response (see p. 10). Guinan and Peake (1967) found that the stapes movement increased in proportion to the input up to 130 dB SPL below 2 kHz, and up to 140 – 150 dB above. This suggests that the movements are linear up to these intensities. It also suggests that there are unlikely to be significant harmonics or inter-modulation products at much lower intensities. Guinan and Peake were not able to see any harmonics, although their method only allowed them to detect 10–20% of odd harmonics. In view of these measurements, it is likely that the middle ear is linear in the usual range of physiological and psychophysical measurements.

3. The Middle Ear Muscles

Transmission through the middle ear can be controlled by means of the middle ear muscles. They are two small striated muscles attached to the ossicles. The *tensor tympani* is attached to the malleus near the tympanic membrane, and is innervated by the trigeminal (fifth) cranial nerve. The other muscle, the *stapedius muscle* is attached to the stapes and is innervated by the facial (seventh) cranial nerve.

Contraction of the muscles increases the stiffness of the ossicular chain. As was explained on p. 21, below 1–2 kHz transmission through the middle ear is stiffness-controlled. The stiffness arises from the elasticity of the tympanic membrane and the ligaments of the ossicles, as well as from the compression and expansion of air in the middle ear cavity. The stiffness reduces the transmission of sounds of low frequency. When it is augmented by a stiffening of the ossicular chain, the low-frequency response is atten-uated still further. On the other hand, at high frequencies, above 1–2 kHz, where transmission is not stiffness-controlled, the response is much less affected by the middle ear muscles (Pang and Peake, 1986). Although this seems to be the main mechanism of middle ear muscle action, the real position is more complicated, because Pang and Peake showed there were still some effects in the high-frequency range. In addition, in the cat, the position of the notch in the transfer function around 4 kHz, arising from resonances in the bulla, is changed as well (Simmons, 1964).

Contraction of the middle ear muscles can be elicited by loud sound (more than 75 dB above absolute threshold), vocalization, tactile stimulation of the head, or by general bodily movement (Carmel and Starr, 1963). In some cases, the middle ear muscles can be contracted voluntarily without any other discernible movement.

Several functions have been suggested for the middle ear muscles.

1. The contraction to loud sound suggests that the reflex might be of use in protecting the inner ear from noise damage. This indeed seems to be the case. For instance, Zakrisson and Borg (1974) showed that in patients with unilateral Bell's palsy, where the stapedius muscle is paralysed, low-frequency noise produced greater temporary threshold shifts in the abnormal than in the normal ear. A temporary threshold shift, which may continue for minutes or hours after an auditory stimulus, is a sign that fatigue, and hence potential damage, has been produced in the auditory system. But the reflex is too slow to protect the ear against impulsive noises.

2. It has been suggested that the middle ear muscles may be able to keep intense low-frequency stimuli near a lower part of the intensity range. Wever and Vernon (1955) showed that the reflex kept the intensity of the input to the cochlea relatively constant when the intensity of the stimulus was varied. This near-perfect automatic gain control functioned for a range of 20 dB above the reflex threshold, and only applied to low-frequency stimuli.

3. The middle ear muscles may also have a beneficial effect on the frequency response of the middle ear. As mentioned above, the transmission characteristic shows a sharp dip near 4 kHz, due to resonances in the bulla. Simmons (1964) showed that the middle ear muscles could shift the frequency of the dip slightly. In cats which were awake and had intact middle ear muscles, the dip was not apparent, suggesting that the continually fluctuating tone in the muscles had averaged it out.

4. At high intensities, low-frequency stimuli can mask higher-frequency stimuli over a wide range of frequencies. Selective attenuation of low frequencies by the middle ear muscles can therefore be expected to affect the perception of complex stimuli with low-frequency components, such as speech, at high intensities.

C. Summary

1. The outer ear has two roles in transmitting sound to the tympanic membrane or eardrum. It aids sound localization by altering the spectrum

of the sound, depending on the direction of the source. It also, by resonances, increases the sound pressure at the tympanic membrane.

2. The middle ear apparatus couples sound energy from the tympanic membrane to the oval window of the cochlea. The sound is transmitted by three small bones, the ossicles, called the malleus, incus and stapes. The middle ear acts as an acoustic impedance transformer, coupling energy from low-impedance air to the higher-impedance cochlear fluids, so reducing the reflection of sound energy that would otherwise occur.

3. The middle ear transformer uses three principles. The area of the oval window is smaller than that of the tympanic membrane, increasing the pressure. The lever action of the ossicles increases the force, and decreases the velocity. The buckling motion of the tympanic membrane does likewise.

4. Transmission through the middle ear depends on the frequency of the stimulus. Greatest transmission is produced (in the cat) in the range around 1–2 kHz. Below that frequency, transmission is reduced by the stiffness of the middle ear structures and by compression and expansion of air in the middle ear cavity. Above that frequency, many factors, including the mass of the ossicles, and less efficient modes of vibration of the structures, reduce transmission. There are also dips in the response arising from acoustic resonances in the middle ear cavity.

5. Transmission through the middle ear is affected by the middle ear muscles, which reduce the transmission of low-frequency sounds. They may serve to protect the ear to some extent from noise damage, reduce the masking effects of low-frequency stimuli on higher-frequency stimuli, act as an automatic gain control for low-frequency stimuli over a narrow range of intensities, and reduce the perturbing effects of middle ear resonances.

D. Further Reading

The outer ear has been comprehensively discussed by Shaw (1974).

The middle ear has been described by Dallos (1973a, Chapter 3, pp. 83–126, and parts of Chapter 7, pp. 465–501), Møller (1974) and Pickles (1987).

The overall efficiency of the ear in transferring sound from the external field to the cochlea has been discussed by Rosowski *et al.* (1986).

3. *The Cochlea*

The chapter on the cochlea is a key one, and forms the foundation for much of the rest of this book. The anatomy of the cochlea will be described first. This is followed by a description of cochlear mechanics, starting with von Békésy's pioneering observations, followed by the more recent measurements. A non-mathematical description of some of the theories of cochlear mechanics is given. The electrophysiology of the cochlea will then be discussed, starting with the standing potentials and hair cell transduction, followed by hair cell responses, their relation to grossly-recordable potentials, and nerve excitation. Many of the ideas discussed in this chapter, particularly the basis for sharp mechanical tuning in the cochlea and the mechanism of transduction, are still controversial, and will be discussed in more depth in Chapter 5.

A. Anatomy

1. General Anatomy

Figure 2.1 (p. 13) shows the position of the human cochlea in relation to the other structures of the ear. It is embedded deep in the temporal bone. Overall, the cochlea stands about 1 cm wide and 5 mm from base to apex in man, and contains a coiled basilar membrane about 35 mm long. Figure 3.1A shows the turns of the cochlea in more detail, and in particular the longitudinal division into three scalae. The scalae spiral together along the length of the cochlea, keeping their corresponding spatial relations throughout the turns. The osseous spiral lamina divides the scala vestibuli from the scala tympani on the side near the modiolus (Fig. 3.1B). The scala media is separated from the scala vestibuli above by Reissner's membrane, and from the scala tympani below by the basilar membrane. The two outer scalae, the scala vestibuli and scala tympani, are joined at the apex of the cochlea by an opening known as the helicotrema (Figs 3.1A and C). The two outer scalae contain perilymph, a fluid which is similar to extracellular fluid in its ionic composition. The scala media forms an inner compartment which does not communicate directly with the other two. It contains endolymph, which is similar to intracellular fluid in that it has a high K^+ concentration, and a low Na^+ concentration. Endolymph is at a high positive potential (e.g. $+80$ mV), whereas the other scalae are at or near the potential of the surrounding bone.

The vibrations of the stapes are transmitted to the oval window, a membraneous window opening onto the scala vestibuli. Fluid in the cochlea is displaced to a second window, the round window, opening onto the scala tympani. The flow causes a wave-like displacement of the basilar membrane and the structures attached to it (Fig. 3.1C). It is this that is responsible for the stimulation of the hair cells, and the first stage of the analysis of the incoming sound is performed by the spatial distribution of the resulting displacements.

The basilar membrane undergoes an important gradation in dimensions up the cochlea; although the cochlear duct is broad near the base and narrow towards the apex, the basilar membrane tapers in the opposite direction, the difference being filled by the spiral lamina.

The organ of Corti on the basilar membrane constitutes the auditory transducer and it is here that the nerve supply ends (Figs 3.1B and D). The nerve supply and the blood vessels of the cochlea enter the organ of Corti by way of the central cavity of the cochlea, the *modiolus*, the spiral structure of the cochlea imparting a corresponding twist to the nerve and blood vessels during development.

2. The Organ of Corti

The highly specialized structure of the organ of Corti contains the hair cells, which are the receptor cells, together with their nerve endings and supporting cells (Fig. 3.1D). The hair cells consist of one row of inner hair cells on the modiolar side of the arch of Corti, and between three and, towards the apex, five rows of outer hair cells. There are about 15,000 hair cells in each ear in man (Ulehlova *et al.*, 1987), and 12,500 in the cat (Schuknecht, 1960).

The organ of Corti itself sits on the basilar membrane, a fibrous structure dividing the scala media from the scala tympani. Sometimes, though misleadingly, the whole complex is referred to as "the basilar membrane".

The organ of Corti is given rigidity by an arch of rods or pillar cells along its length, the upper ends of the rods ending in the *reticular lamina* which forms the true chemical division between the ions in the fluids of the scala media and those of the scala tympani (Fig. 3.1D). The arch is surrounded

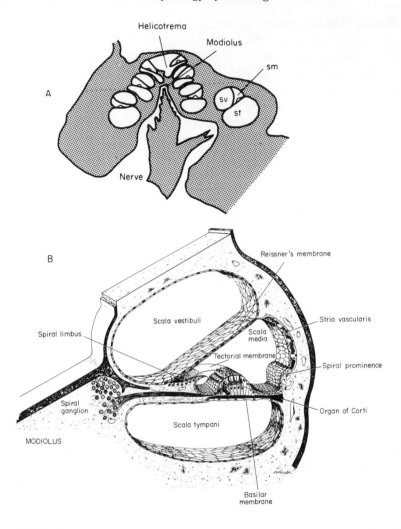

Fig. 3.1 (A) In a transverse section of the whole cochlea, the cochlear duct is cut across several times as it coils round and round. Abbreviations: sv, scala vestibuli; sm, scala media; st, scala tympani. (B) The three scalae and associated structures are shown in a magnified view of a cross-section of the cochlear duct. From Fawcett (1986, Fig. 35.11).

by phalangeal cells; that is, by cells with processes which end in a plate in the reticular lamina. The inner phalangeal cells completely surround the inner hair cells. The outer phalangeal cells, which are also known as Deiters' cells, form cups holding the basal ends of the outer hair cells. The outer

C

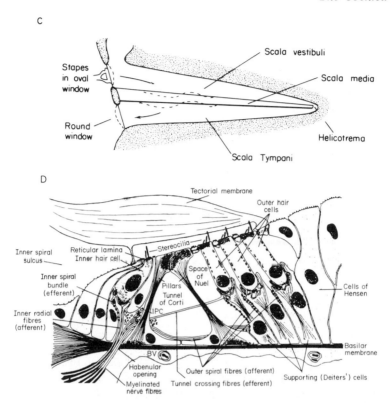

D

Fig. 3.1 (cont.) (C) The path of vibrations in the cochlea is shown in a schematic diagram in which the cochlear duct is depicted as unrolled. (D) Cross-section of the organ of Corti, as it appears in the basal turn. Deiters' cells send extensions (phalanges) up to the reticular lamina, running in the spaces around the outer hair cells, although they are not shown on this particular cross-section. The modiolus is to the left of the figure. BV, blood vessel; IPC, inner phalangeal cell. From Ryan and Dallos (1984, Fig. 22–4, slightly modified).

phalangeal cells send fine processes, or phalanges, up to the reticular lamina, leaving spaces between the outer hair cells. External to the outer hair cells there is a row of supporting cells known as Hensen's cells, and on the modiolar side of the organ of Corti there is a further row of supporting cells. The distribution of the supporting cells changes in the different turns of the cochlea (Fig. 3.2).

The organ of Corti is covered by a gelatinous and fibrous flap, the tectorial membrane. The tectorial membrane is fixed only on its inner edge, where it is attached to the limbus, although it is joined to the reticular lamina by small trabeculae or processes. Some investigators believe it to join and seal

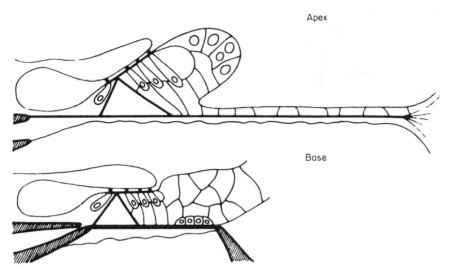

Fig. 3.2   The organ of Corti shows morphological differences along the length of the cochlea. Moreover, near the apex the basilar membrane is wide, and near the base it is narrow. From Spoendlin (1972, Fig. 1).

onto the reticular lamina along its outer edge (Kronester-Frei, 1979). The longer of the hairs on the outer hair cells are shallowly but firmly embedded in the under-surface of the tectorial membrane (Engström and Engström, 1978). The hairs of the inner hair cells are probably not embedded and fit loosely into a raised groove known as Hensen's stripe on the under-surface of the tectorial membrane.

The tectorial membrane is attached only on one side and is raised above

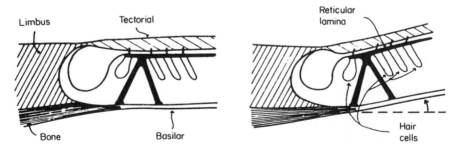

Fig. 3.3   In Davis's hypothesis for the lever action of the cochlea, the stereocilia on the hair cells are deflected as a result of vertical displacements of the basilar membrane. From Davis (1958, Fig. 8).

the basilar membrane. Therefore, when the basilar membrane moves up and down, a shear or relative movement will occur between the tectorial membrane and the organ of Corti, with the result that the hairs will be bent (Fig. 3.3). The arch of the pillar cells (the arch of Corti) would seem well suited to maintaining the rigidity of the organ of Corti during such a movement.

Figure 3.4A shows a view of the upper surface of the organ of Corti of the guinea pig once the tectorial membrane has been removed. The hairs or stereocilia of the hair cells are seen projecting through the reticular lamina. The geometric patterns on the reticular lamina between the hair cells reveal the pattern of the supporting cells making up the lamina. The pattern is formed by small microvilli, which cover the apical surfaces of the supporting cells, bunching more thickly around their edges. On each inner hair cell of the guinea-pig the hairs are arranged in three to five closely spaced rows, the rows being slightly curved (Fig. 3.4B), whereas on the outer hair cells there are three closely spaced rows, with the rows making a V or W shape (Fig. 3.4C).

Figure 3.5A shows a schematic diagram of the apical portion of a hair cell. This is the portion bearing the stereocilia, and is thought to be the end involved in the initial sensory transduction. The diagram shows the structures common to acousticolateral hair cells, including those of the vestibular system. Three rows of stereocilia are shown in cross-section. The stereocilia themselves are composed of packed actin filaments, bonded together closely in what has been called a "paracrystalline array" (DeRosier et al., 1980). This gives the stereocilia considerable stiffness, so that they behave as rigid levers in response to mechanical deflection (Flock, 1977). The stereocilia are bonded together by sideways-running links, so that all the stereocilia in a bundle tend to move together. There are also fine links emerging from the tips of the shorter stereocilia, which may well be involved in coupling the stimulus-induced movements to the actual transducer channels in the membrane (Pickles et al., 1984). In the guinea-pig, stereocilia are 2–4 μm long and 300 nm in diameter in the row of tallest stereocilia on the hair cell, tapering to some 500 nm long and 100 nm wide for the shortest stereocilia.

Some of the actin filaments of the stereocilia continue, closely-packed, into a rootlet which anchors the stereocilium into the cuticular plate. The cuticular plate is also composed of actin filaments, this time forming a dense matrix (Flock et al., 1982). There is a gap in the cuticular plate, with a *basal body* situated in the gap. The basal body is a centriole-like structure, and may be important in development. In vestibular hair cells and in embryonic cochlear hair cells, but rarely in mature ones, a cilium of different appearance known as the kinocilium emerges from the basal body. Unlike the stereocilia,

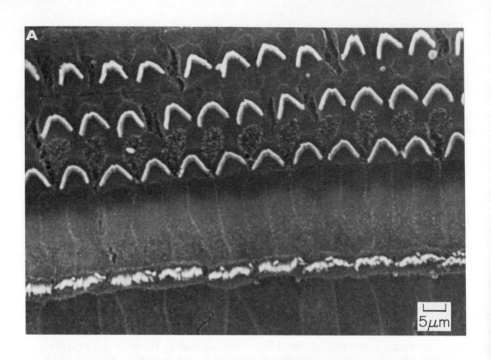

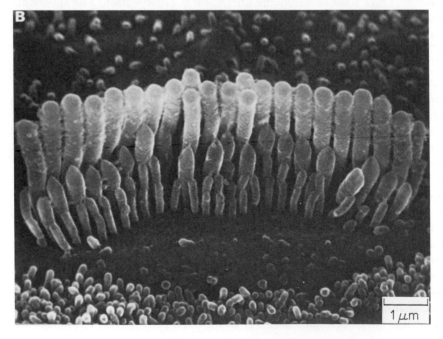

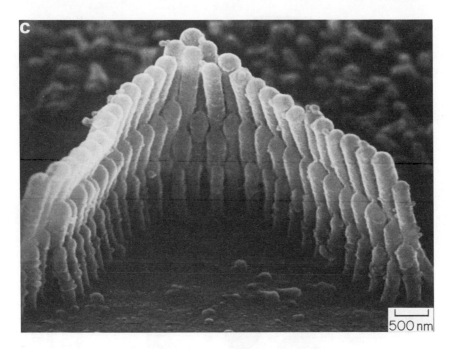

Fig. 3.4 (A) A scanning electron micrograph of the upper surface of the organ of Corti, when the tectorial membrane has been removed, shows three rows of outer hair cells (top) and one row of inner hair cells (bottom). Scale bar: 5 μm. (B) On inner hair cells, the stereocilia form nearly straight rows. Three to five rows of stereocilia are visible. The stereocilia are viewed from the side nearest the modiolus (i.e. looking upwards in Fig. 3.4.A). Scale bar: 1 μm. (C) On outer hair cells the stereocilia form V or W-shaped rows. In the guinea-pig there are three rows of stereocilia, evenly graded in height. The stereocilia are viewed from the side nearest the modiolus. Scale bar: 500 nm. Guinea-pig.

the kinocilium is a true cilium, in that it consists of tubulin-containing microtubules, nine pairs of microtubules being arranged in a ring around the outside of the cilium, with one pair in the centre. In the cell body itself we have the usual intracellular organelles, and synaptic junctions at the extreme basal end.

An inner hair cell is shown in Fig. 3.5B. It is about 35 μm in length, and about 10 μm in diameter at the widest point (chinchilla: Smith, 1968). The shape is commonly likened to that of a flask. The nucleus is central, the mitochondria are scattered, though denser above the nucleus, and the cellular organelles are most prevalent at the apex, near the cuticular plate. The nerve endings are situated near the base of the cell. These terminals are associated with the afferent fibres of the auditory nerve, conveying

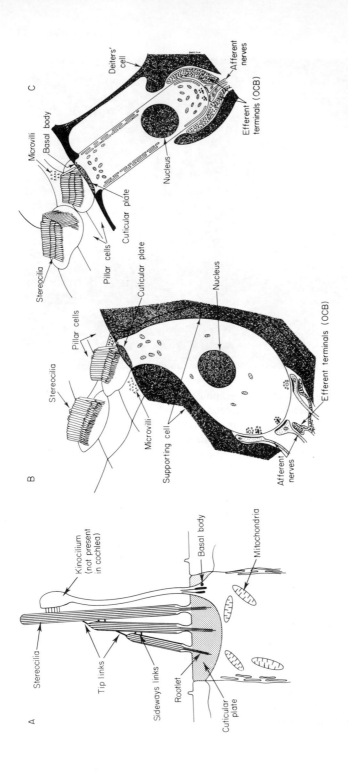

Fig. 3.5 (A) The common structures on the apical portion of acousticolateral hair cells include rows of stereocilia which are graded in height and joined by cross-links. The tip links may be involved in transduction, opening the transducer channels. The kinocilium is not present in the mature cochlea, although it is present in vestibular cells. (B) and (C) Inner hair cells are shaped like a flask (B), and outer cells are shaped like a cylinder (C). OCB, olivocochlear bundle. © J.O. Pickles 1987.

information from the cochlea to the brainstem. On the presynaptic membrane, the synapse is often marked by small dense synaptic bars or invaginations at right angles to the cell wall, or by rounded synaptic bodies, together with a few vesicles (Ades and Engström, 1974).

Outer hair cells are approximately 25 μm long in the basal turn and 45 μm long in the apical turn, and 6–7 μm in diameter (chinchilla: Smith, 1968). They have a cylindrical shape. The nucleus is located basally (Fig. 3.5C). The mitochondria are primarily situated at the base below the nucleus, and at the apex below the cuticular plate. The afferent terminals are faced with short synaptic bars, but not round synaptic bodies (Ades and Engström, 1974).

In both types of hair cell the row of tallest stereocilia and the gap in the cuticular plate are situated on the side of the hair cell furthest away from the modiolus, and the row of shortest stereocilia are situated nearest to the modiolus.

3. The Innervation of the Organ of Corti

The cochlea is innervated by about 50,000 sensory neurones in the cat, and about 30,000 in man (Schuknecht, 1960; Harrison and Howe, 1974a). There are also about 1800 "efferent" or centrifugal neurones, by means of which

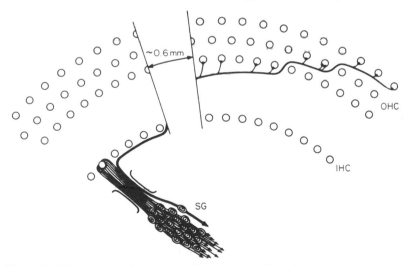

Fig. 3.6 The great majority of auditory nerve fibres (type I fibres) connect with inner hair cells. A few fibres (type II) pass to outer hair cells, after running basally for about 0.6 mm. IHC, Inner hair cells; OHC, Outer hair cells; SG, spiral ganglion. From Spoendlin (1978, Fig 8).

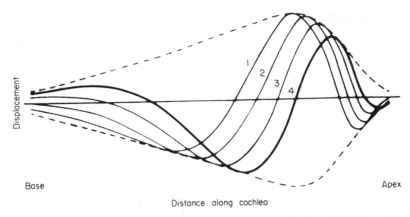

Fig. 3.7   Travelling waves in the cochlea were first shown by von Békésy. The full lines show the pattern of the deflection of the cochlear partition at successive instants, as numbered. The waves are contained within an envelope which is static (dotted lines). Stimulus frequency: 200 Hz. From von Békésy (1960, Fig. 12.17).

vibration decreased. However, the frequency of vibration at any point was, of course, the same as that of the input.

Von Békésy's plots were made in two ways. By opening a length of cochlea it was possible to see the pattern of movement distributed along the membrane, and so plot the waveforms and their envelopes for sounds of different frequencies. The vibration envelopes found by von Békésy are shown in Fig. 3.8. They show the important point that as the frequency of

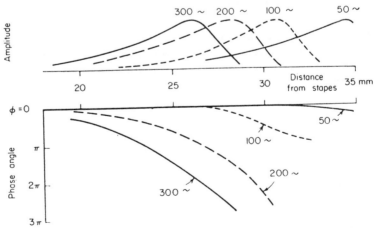

Fig. 3.8   Displacement envelopes on the cochlear partition are shown for tones of different frequency. The lower plot shows the relative phase angle of the displacement. From von Békésy (1960, Fig. 11.58).

the stimulus was increased, the position of the vibration maximum moved towards the base of the cochlea. Thus high-frequency tones produced a vibration pattern peaking at the base of the cochlea. Low-frequency tones, in contrast, produced most vibration at the apex of the cochlea, although, because of the long tail of the vibration envelope, there was some response near the base as well.

A second way in which von Békésy measured the vibration pattern is indicated in Fig. 3.9. He opened the cochlea at certain points, and measured the vibration at those points as the frequency was varied. Figure 3.9 shows his results for six points on the membrane, the peak-to-peak stapes displacement being kept constant as he varied the frequency at each point. Note that, as before, it is the most basal point that responds best to the highest frequencies. Note the shallow slope on the low-frequency side, and the much steeper slope on the high-frequency side. In going from the space axis of Fig. 3.8 to the frequency axis of Fig. 3.9 the direction of variation of the parameters marked on the curves and the abscissa have to be reversed, although the positions of the steep and shallow slopes are the same.

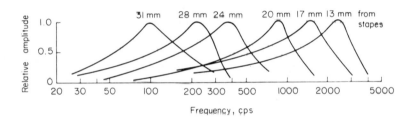

Fig. 3.9   Frequency responses are shown for six different points on the cochlear partition. The amplitude of the travelling wave envelope was measured as the stimulus frequency was varied with constant peak stapes displacements. The position of the point of observation is marked on each curve. From von Békésy (1960, Fig. 11.49).

Von Békésy's results can be summarized as follows: vibration of the stapes gives rise to a travelling wave of displacement on the basilar membrane. For a vibration of a particular frequency, the vibration on the basilar membrane grows in amplitude as the wave travels towards the apex, and then, beyond a certain point, dies out rapidly. The wave travels more and more slowly as it travels up the cochlea. Low-frequency sounds peak a long way along the membrane, near the apex, and high-frequency sounds only a short way along, near the base. The frequency selectivity shown by single points on the basilar membrane is very poor, with a very shallow slope on the low-frequency side, though a steeper slope on the high-frequency side. On the basis of his results, therefore, the basilar membrane appears to act substantially as a low-pass filter.*

2. Recent Measurements of the Travelling Wave

(a) Historical introduction

Von Békésy's observations have been questioned on two grounds. Visual observations meant that the vibration amplitude had to be at least of the order of the wavelength of light, and the high intensities (130 dB SPL) necessary make extrapolation to a more physiological range unjustified. Secondly, his measurements were performed on cadavers. It is now known that not only does the experimental animal have to be alive, but the cochlea has to be in extremely good physiological condition, to show a satisfactory mechanical response. The history of the measurements of basilar membrane vibration over the years is the story of the progress that has gradually been made in meeting these requirements. The impetus has come from studies on the auditory nerve, which showed that single auditory nerve fibres, innervating single points on the basilar membrane, could show sharply tuned bandpass characteristics, at variance with von Békésy's observations.

Since von Békésy, only Kohllöffel (1972) has opened the cochlea over a distance and measured the responses as a pattern distributed over space. Like von Békésy, he showed that the travelling wave had a very shallow slope towards the base, and a steep slope towards the apex. However, it is now recognized that this approach is too traumatic for the cochlea and more recent measurements have been confined to single points on the basilar membrane. In those cases, the response is measured as a function of stimulus frequency. Two types of plot are used. The frequency response may be described as an *amplitude response*, in which the amplitude of the vibration is shown as a function of stimulus frequency, the stimulus intensity being kept constant. Secondly, a *tuning curve* or *frequency-threshold* curve may be measured, in which the intensity of the stimulus is adjusted so as to keep the amplitude of the basilar membrane vibration at a constant value, while the stimulus frequency is varied. The necessary stimulus intensity is plotted as a function of stimulus frequency.

The earlier results of Johnstone and Boyle (1967) and Wilson and John-

---

*In Fig. 3.9 the peak-to-peak stapes displacement was kept constant as the stimulus frequency was varied. However, as described in Chapter 1, constant sound pressure level at different frequencies corresponds to constant peak *velocities*. If the data are recalculated for constant peak velocity, the low-frequency slopes in Fig. 3.9 become flat, i.e. the cochlea appears to act purely as a low-pass filter (Eldredge, 1974).

stone (1975) showed substantially low-pass responses, like those of von Békésy. However, the results of Rhode (1971, 1978) presaged the more recent measurements by showing some degree of a bandpass characteristic. He also showed that the basilar membrane vibration was nonlinear, so that the responses became relatively more sensitive, and relatively more sharply tuned, if measured at lower intensities. The more recent, and probably definitive, measurements of Sellick *et al.* (1982) have shown this indeed to be the case.

(b) Current measurements of basilar membrane responses

Sellick *et al.* (1982) used the Mössbauer technique to measure the vibration of the basilar membrane of the guinea-pig at the basal, high-frequency, end. The Mössbauer technique involves placing a small piece of stainless steel enriched with ^{57}Co on the basilar membrane. The ^{57}Co decays to ^{57}Fe in an excited state, and the ^{57}Fe decays emitting a γ-ray. If the emitted radiation has exactly the right frequency, it will be absorbed by a piece of ^{57}Fe-enriched foil nearby. If the source frequency is changed by a Doppler shift due to movement of the source, a smaller proportion of the radiation will be absorbed, and an increase in transmitted radiation can be detected. The spectral lines are so incredibly narrow that velocity disparities of 0.2 mm/sec can be detected; because it is a velocity detection, the technique does not require very rigid stabilization of the preparation, and is most sensitive at high frequencies.

Figure 3.10A shows a plot of basilar membrane responses, measured at the 18-kHz point on the basilar membrane. The lowest curve is for a stimulus of 20 dB SPL, and shows that as the stimulus frequency is varied, a sharp peak of vibration is produced around one frequency. Two points can be noted. First, the basilar membrane is very responsive at this frequency, because the relatively low-intensity stimulus (20 dB SPL) produces a relatively large (3.3 nm – but still very small!) response. Secondly, the system is very sharply tuned at this intensity, because changing the stimulus frequency by only a little produces a large change in the amplitude of vibration. The basilar membrane therefore shows a *bandpass* filter function, with a high degree of frequency selectivity.

At higher intensities of stimulation, the sharp tuning disappears (e.g. at 80 dB SPL). The response now shows only a broad peak. This occurs, because as the stimulus intensity is raised, the response grows only slowly in the region of the peak, and faster for lower frequencies.

Because of the importance of the basilar membrane data for our knowledge of the frequency selectivity of the auditory system, the results of Sellick

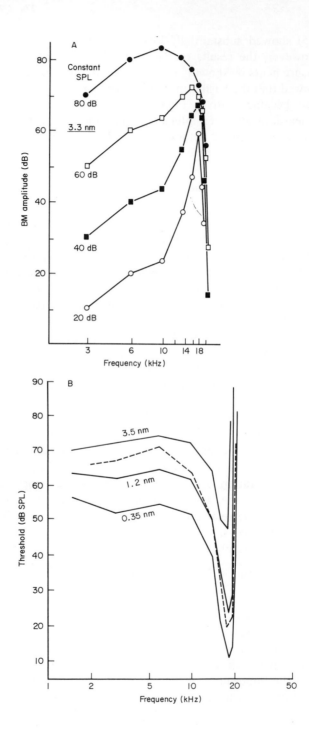

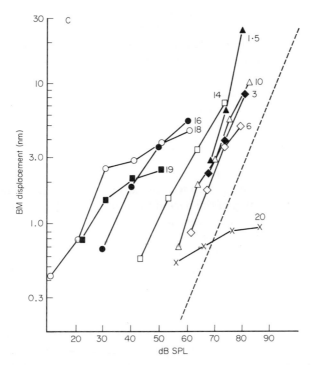

Fig. 3.10   (A) The amplitude of basilar membrane vibration, plotted as a function of stimulus frequency for different stimulus intensities, shows that the response is sharply tuned at low stimulus intensities. However, the response becomes more broadly tuned and peaks at lower frequencies as the stimulus intensity is raised. The responses were measured at the 18-kHz point on the guinea-pig basilar membrane by means of the Mössbauer technique. From Johnstone *et al.* (1986, Fig. 4A). (B) Tuning curves for the 18-kHz point on the basilar membrane (solid lines), made for vibration criteria of 0.35, 1.2 and 3.5 nm (numbers on lines). Dotted line: tuning curve of an auditory nerve fibre. Data from Sellick *et al.* (1982, Fig. 10). (C) Intensity functions for basilar membrane vibration show saturation near and above the best frequency (18 kHz), and linear responses at low frequencies. Frequency of stimulation is indicated by the numbers on the curves. Dotted line: slope of one, indicating linear response. From Sellick *et al.* (1982, Fig. 5A, modified to show amplitude, rather than velocity, of response).

et al. will be presented in the different ways that are used to describe auditory frequency selectivity.

Figure 3.10B shows a set of basilar membrane responses plotted as tuning curves, or frequency-threshold curves (FTCs). The graph shows the relations between stimulus frequency and stimulus intensity necessary to produce different, criterion, amplitudes of vibration. The lowest curve, like the lowest curve in Fig. 3.10A, shows that the system is very sensitive at one frequency, 18 kHz, and very sharply tuned. It is sensitive, because at 18 kHz a response of this amplitude is produced by a stimulus intensity of only 12 dB SPL. It is sharply tuned because, if the stimulus frequency is moved away from 18 kHz, the stimulus intensity has to be increased markedly to produce the same, criterion, response. An advantage of presenting the data in this way, is that it is possible to compare directly the tuning from different stages of the auditory system. For instance, if we go through a similar procedure and construct a tuning curve for an auditory nerve fibre innervating the same region of the cochlea, using as our criterion a certain number of evoked action potentials per second, we obtain a tuning curve which is similar in general shape to the mechanical one (dotted line, Fig. 3.10B). This shows that the tuning of the basilar membrane is reflected in the tuning of the nerve fibres which innervate it.

A third way that the data of Sellick *et al.* can be presented is shown in Fig. 3.10C. The figure shows plots of the response displacement as a function of stimulus intensity, for different stimulus frequencies. Because the scales are log–log scales, a response which grows in proportion to the stimulus intensity will have a slope parallel to the dotted line, which has a slope of one. The plot emphasizes how, for stimulus frequencies at and above the characteristic frequency of 18 kHz, the responses grow relatively slowly with stimulus intensity, over part of their range. In fact, the slopes for 18, 19 and 20 kHz go as low as 0.2, meaning that for these stimulus frequencies, the vibration amplitude increases as the stimulus amplitude is raised to the power of 0.2.

Results confirming these findings have been found by Khanna and Leonard (1982) in the cat, Robles *et al.* (1986a) in the chinchilla, and LePage (1987) in the guinea-pig.

A hint as to the possible mechanism behind the sharply-tuned tip in the basilar membrane tuning curve, and behind the nonlinearity, is given by Fig. 3.11. The condition of the cochlea deteriorated gradually during the experiment. As this occurred, the sharply-tuned tip of the tuning curve disappeared (Fig. 3.11A), the thresholds rose, and the intensity functions became linear (Fig. 3.11B). After the death of the experimental animal, the responses became yet more broadly tuned, and the thresholds were raised further, so that the sensitivity was still more reduced (Fig. 3.11A). The pattern of vibration was now similar to the insensitive, and substantially low-pass filter response, originally found by von Békésy. This suggests that when the cochlea is in good physiological condition, a sensitive and sharply-tuned component of the response is added to an insensitive and broadly tuned component. The sensitive component disappears when the cochlea is in a poor condition.

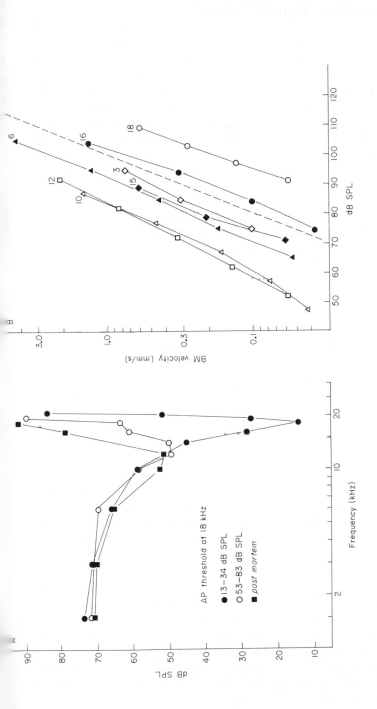

Fig. 3.11 (A) Changes in tuning as the condition of the cochlea deteriorated. ●, Cochlea in good condition; ○, cochlea in poor condition; ■, after death. Tuning curves made for 0.04 mm/s constant velocity condition. From Sellick et al. (1982, Fig. 15B). (B) Intensity functions after death show that the responses had become entirely linear. In this figure, the responses are plotted as velocities, not amplitudes. Dotted line: slope for linear growth. Numbers on curves: frequency of stimulation in kHz. From Sellick et al. (1982, Fig. 14).

The most widely accepted hypothesis for the production of the sharply-tuned tip in the tuning curve, though as yet unproven, is a rather surprising and revolutionary one. The hypothesis says that the cochlea contains an *active mechanical amplifier*. The amplifier is triggered by the acoustic stimulus, and feeds mechanical energy back into the travelling wave, so increasing the amplitude of the vibration. It also produces a sharply-tuned response. Some of the thinking behind this hypothesis will be described in the next section, when theories of cochlear mechanics are discussed. However, a more detailed consideration of the evidence will be delayed until Chapter 5. Here, we note that the presence of a mechanically-active factor in cochlear mechanics could explain the extreme physiological vulnerability of the low-threshold, sharply-tuned, component of the travelling wave.

Phase responses have also been measured (Sellick *et al.*, 1983; Robies *et al.*, 1986a). Figure 3.12 shows that stimuli of very low frequency produced

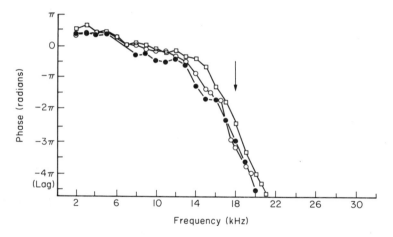

Fig. 3.12   Phase of basilar membrane vibration, as a function of frequency for the 18-kHz point on the basilar membrane of three guinea-pigs. The stimulus phase was taken as a reference. Arrow: approximate value of characteristic frequency. From Sellick *et al.* (1983, Fig. 1A).

responses at the measuring point which was in advance of the stimulus pressure by approximately $\pi/2$ radians. As the stimulus frequency was raised, a phase lag appeared, and then became greater, reaching $2.5 - 3\ \pi$ radians at the characteristic frequency, and $4.5\ \pi$ or more radians at the highest frequency at which measurements could be made. The figure shows that the waves produced by low frequency stimuli travelled rapidly from the base of the cochlea to the measuring point, while the waves produced by stimuli of higher frequency moved more slowly. Calculating back to the expected pattern of the travelling wave as distributed in space along the cochlear duct, we expect the wave produced by a tone of one frequency to move rapidly along the very basal part of the duct, and then travel more and more slowly as it approaches and passes through the point at which it peaks.

3. Theories of Cochlear Mechanics

The vibration of the cochlear partition has been investigated theoretically, initially by means of mechanical models, and more recently mathematically through analytic approaches or computer models. There seems to be a satisfactory explanation of the broadly-tuned wave which is seen in cochleae in poor physiological condition. That will be described first. The basis of the sensitive and sharply-tuned component of the wave, seen in cochleae in good physiological condition, is still controversial, and some ideas will be discussed later. The issues discussed here will be dealt with in greater detail in Chapter 5.

(a) The broadly-tuned component of the travelling wave: passive
cochlear mechanics

In spite of many different approaches that have been used to understand the broadly tuned component of the wave, a consensus seems to be emerging about its physical basis (see, e.g. Neely, 1981; de Boer 1980; Yates, 1986; Viergever, 1986). In the most general terms, the wave is analogous to a wave on the surface of water. In such a water wave, once the energy is introduced, it is carried passively along the wave by the inertia of fluid motion in the horizontal direction. Gravity provides the restoring forces in the vertical direction. The passive cochlear wave is similar, except that the restoring forces come from the stiffness of the cochlear partition (i.e. from the stiffness of the basilar membrane and all its associated structures, such as the organ of Corti and the tectorial membrane). The inertial forces include a component from the mass of the cochlear partition, as well as from the mass of the fluid. Theories are expressed in terms of shallow-water waves (also known as long waves) in which the wavelength is long compared to the depth of the cochlear duct, and deep-water waves (also known as short waves), in which the wavelength is short compared to the depth of the cochlear duct. While the rather simpler long-wave analysis is possible well basal to the peak of the travelling wave, the wavelength becomes short near the peak, and here a short-wave analysis is necessary.

The passive cochlear travelling wave differs from a water surface wave in two respects. First, it always travels from base to apex of the cochlea,

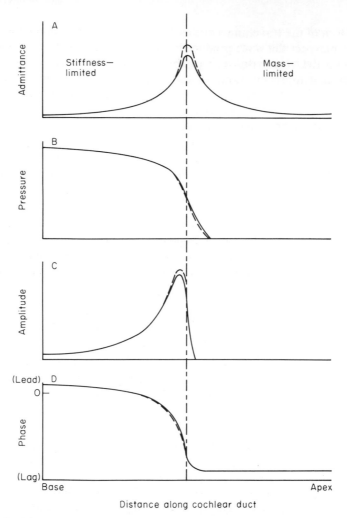

A — Admittance
Stiffness—limited   Mass—limited

B — Pressure

C — Amplitude

D — Phase
(Lead) 0
(Lag)
Base   Apex

Distance along cochlear duct

Fig. 3.13   The mechanisms giving rise to the different aspects of the broadly-tuned, passive mechanical component of the travelling wave. (A) The admittance of the cochlear partition for a stimulus of one frequency is plotted as a function of distance along the cochlea. The admittance is the membrane velocity (proportional to displacement)/driving pressure ratio. The admittance is highest at the resonant point (dash-dotted line), where the effects of stiffness-limitation and mass-limitation cancel. Dotted line: curve for decreased damping. (B) The pressure across the cochlear partition drops as the resonant point is reached. (C) The displacement of the cochlear partition, derived from the product of curves A and B for each point along the cochlea, shows a peak basal to the point of maximum partition admittance, and a sharp drop on the apical side. (D) Phase changes along the cochlear partition. Data for figure derived from de Boer and Viergever (1982) and Neely (1981).

whereas a water wave radiates away from the source in all directions. Secondly, the cochlear travelling wave has a particular dependence on the distance of travel along the cochlea, since unlike a water wave it grows in amplitude as it passes down the cochlea, comes to a maximum, and then declines sharply, the position of the peak depending on the stimulus frequency. The different behaviour in these two respects is a result of the variation in the stiffness of the cochlear partition from base to apex.

It was originally shown by von Békésy, and more recently confirmed by Gummer *et al.* (1981), that the cochlear partition is relatively stiff near the base, and relatively compliant near the apex. This affects the way that the partition vibrates in response to sound. Near the base, where the stiffness is high, stiffness is the most important factor governing the vibration of the partition. Vibrations here are known as stiffness-limited. Towards the apex, where the stiffness is lower, the masses and inertias of the system instead limit the vibration, and the vibration is known as mass-limited. In response to an applied force, a stiffness-limited system will always start to move before a mass-limited one. This means that when a pressure difference is applied across any small segment of the cochlear partition, the stiffness-limited part towards the base will move first, followed by the more mass-limited part towards the apex. The wave of deflection therefore travels *up* the cochlea, from base to apex, the direction depending only on the gradation of compliance, and not *how* the pressure difference is introduced.

The relative delay in the movement of the more apical portion induces longitudinal flow of the fluid, and the inertia of this flow carries the energy towards the apex. The travelling wave therefore primarily depends on the interaction between the stiffness of the partition in response to deflection, and the inertia of fluid moving along the duct.

Why does the travelling wave grow in amplitude as it passes up the duct, and why is there a peak in the vibration? In order to answer this, we need to know more about how the mechanical properties of the cochlear partition vary along the duct.

First, the partition becomes less stiff towards the apex, so that a certain pressure difference across the partition will induce greater amplitudes of movement towards the apex. This is shown in the plot of the admittance of the membrane Fig. 3.13A, where, for the stiffness-limited region between the base and the centre of the diagram, the admittance of the membrane increases towards the apex.

Near the apical end of the cochlea, however, the admittance becomes low again, because here the mass of the basilar membrane is large enough to limit the movement (Fig. 3.13A, mass-limited right half).

In between the stiffness-limited and the mass-limited portions, at the point of resonance, the effects of mass and stiffness become equal in magnitude while being exactly opposite in phase. Due to the phase opposition, their effects cancel, and the admittance becomes high. The partition therefore shows a relatively large amplitude of movement for a certain pressure difference across the partition.

For a wave that travels up the cochlear duct, we can trace the following sequence of events:

1. As the wave travels away from the base, the admittance of the partition increases. The amplitude of vibration of the partition is derived from the product of the admittance and the pressure difference between the scalae, and so the wave grows in amplitude.

2. When the wave approaches the resonant point, the admittance of the partition increases still further, and its movement grows further. However, towards the resonant point, the wave travels more and more slowly, and the effects of damping make the amplitude decline. The pressure across the membrane therefore also starts to drop (Fig. 3.13B). The maximum amplitude of vibration is for that reason reached rather basal to the point of maximum membrane admittance (Fig. 3.13C).

3. Beyond the resonant point, the movement of the partition becomes mass-limited rather than stiffness limited. But to produce a travelling wave, we need interaction between a stiffness (in the partition) and a mass (in the fluids). Since stiffness is no longer dominant in the partition, wave motion becomes impossible. The wave therefore dies away abruptly, with the whole apical region of the partition moving in the same phase. The phase curve therefore becomes flat (Fig. 3.13D)

4. Resonance occurs when stiffness- and mass-limitation are equal in magnitude (though opposite in phase), and this point occurs at different places along the duct, depending on the stimulus frequency. Since inertial forces are relatively greater for high-frequency stimuli, they will match the forces due to the stiffness nearer the relatively-stiff base for high-frequency stimulation, and nearer the apex for low-frequency stimulation. High-frequency waves therefore peak near the base of the cochlea, while low-frequency waves peak towards the apex.

5. Around the point of membrane resonance, the amplitude of the response is limited by damping, or friction, in the cochlear partition. If the damping is decreased, the peak in the admittance function becomes larger, and the travelling wave becomes larger and more sharply tuned (dotted lines, Fig. 3.13).

(b) Sharp tuning in the travelling wave: active cochlear
mechanics

The principles described above have been formulated in many different
mathematical and computational models, of varying degrees of complexity
and realism. It has been possible to adjust the model parameters so as to
match the older data showing a broad, low-amplitude peak in the travelling
wave (e.g. Viergever and Diependaal, 1986). However, a consensus has
been reached in that it does not seem possible to produce travelling waves
which can match the functions shown in Fig. 3.10. Although low degrees
of damping produce large and sharply-tuned peaks in the admittance
function, with the result that large and sharp peaks can be produced in the
calculated travelling wave pattern, it has not so far been possible to match
both the size of the peak and its relative width with the same set of
parameters. Figure 3.14, for instance, shows one of the most successful of

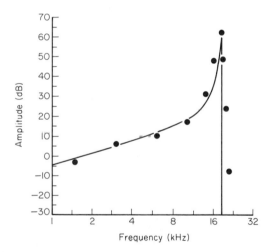

Fig. 3.14 Comparison between the predictions of one model of passive cochlear
mechanics (solid line), and the measured basilar membrane frequency response (data
points). The degree of damping in the model was adjusted so that the curves coincide
on the low-frequency slope, and at the peak of the response. The peak produced by
the model is too narrow. Basilar membrane data from Sellick *et al*, (1982). From
Viergever and Diependaal (1986, Fig. 5).

the recent attempts at matching the data of Sellick *et al.* (1982) with a
passive model. Although it has been possible to adjust the parameters so
that the amplitude of the peak of the response is the same, the bandwidth
near the tip is far too narrow. While the difficulty of making measurements

on the basilar membrane means that we must still interpret those data with caution, the conclusion holds even if we use frequency functions from single auditory nerve fibres, which can be measured with a high degree of reliability. The discrepancy reveals a fundamental inadequacy of the types of model which have been described so far, since very narrow peaks are intrinsic to the models if the peak size is made large.

It has been possible to produce functions similar to those in Fig. 3.10 only if the computations assume that there is an active source of mechanical energy in the cochlear partition, which amplifies the travelling wave as it passes through. The idea that an apparently passive mechanoreceptor might contain an active mechanical process has proved fascinating to those in the field. This has led to widespread support for the model, although the evidence in its favour cannot yet be regarded as decisive. Some indications of the possible basis of the active mechanical process will be evaluated in detail in Chapter 5.

The models which have succeeded in matching the experimental frequency functions have incorporated the active source of mechanical energy in a limited region on the basal slope of the travelling wave (Fig. 3.15). It

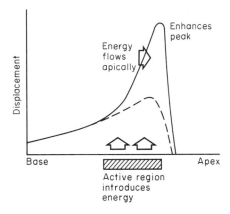

Fig. 3.15   Possible production of a sharply-tuned travelling wave by a mechanically active process on the basal slope of the wave. Dotted line: passive component of the wave.

is suggested that as the wave passes through this region, the hair cells are stimulated and, by mechanisms at present unknown, in turn feed mechanical energy of biological origin into the travelling wave (e.g. de Boer, 1983; Neely and Kim, 1983, 1986). The feedback makes the wave grow as it passes through the active region, which produces still more stimulation of the hair cells, and a still greater input of mechanical energy. The travelling wave grows more and more steeply. Soon, however, the wave passes beyond the active region, and here the amplitude drops sharply. Models incorporating these assumptions have been able to produce tuning curves similar to those seen in basilar membrane responses and in auditory nerve fibres (e.g. Neely and Kim, 1986). It leads to the idea that the observed travelling wave has two components, a relatively small-amplitude and broadly-tuned wave which depends only on passive mechanical processes, as described above, and a large-amplitude and sharply-tuned active component superimposed on that, which is only seen in cochleae in good physiological condition. The model explains many details of the data satisfactorily, such as why the peak of the wave is so dependent on the good physiological state of the cochlea, and why manipulations of the cochlea have such a large effect on its sensitivity and sharp tuning.

It is not known what is responsible for the active mechanical process. The hypothesis supported by most researchers is that the movements are produced, directly or indirectly, by the outer hair cells. This is based on the finding that if outer hair cells are damaged, the sensitive and sharply-tuned component of the travelling wave is lost, leaving the insensitive and broadly-tuned component. Active movements have been demonstrated in isolated outer hair cells (Brownell *et al.*, 1985: Zenner *et al.*, 1985, Kachar *et al.*, 1986; Ashmore, 1987), although it is not known if the movements occur *in vivo* or if they are of the correct type to feed back and amplify the travelling wave. Because the evidence so far is inadequate, the arguments are of necessity complex and indirect, and a detailed evaluation will be delayed until Chapter 5. The mechanical amplification, if it indeed exists, uses the chemical and electrical energy stored in the fluid spaces of the cochlea.

C. The Fluid Spaces of the Cochlea

The electrochemical environment of the organ of Corti is an important determinant of the normal, sharply-tuned, travelling wave, and the normal operation of the transduction process. The environment is maintained by the division of the cochlear scalae into endolymphatic and perilymphatic spaces. The high positive standing potential and high K^+ concentration of the endolymph appeal to play an essential role in cochlear function.

1. The Endolymphatic and Perilymphatic Spaces

The labyrinthine cavity is composed of two separate compartments. The larger, outer compartment, formed by the scala vestibuli and scala tympani,

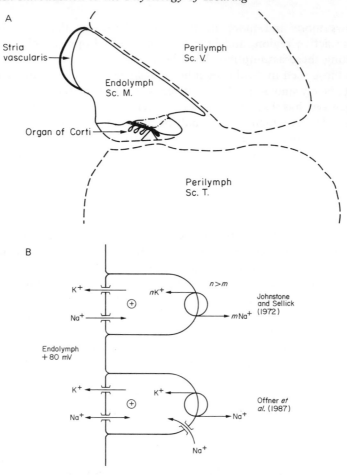

Fig. 3.16   (A) The boundaries of the scala media, formed by occluding tight junctions, are shown by solid lines. The reticular lamina is shown by a very thick line. From Smith (1978, Fig. 3). (B) Suggested ion pumping in the basolateral walls of the marginal cells of the stria vascularis, according to the models of Johnstone and Sellick (1972) and Offner *et al.* (1987). Other suggested models incorporate an electrogenic K^+ pump in the endolymphatic (lumenal) wall as well (Sellick and Johnstone, 1974).

runs the length of the system, and is filled with perilymph. There is a smaller, inner compartment also extending the entire length of the system, which is formed by the scala media and is filled with endolymph. In the cochlea itself the inner, endolymphatic space is bounded above by Reissner's membrane, laterally by the stria vascularis, and below by the reticular lamina on the upper surface of the organ of Corti. Figure 3.16A shows the generally-accepted borders of the endolymphatic space. The ionic composition of the endolymph is very different from that of the perilymph, being high in K^+ and low in Na^+. In accordance with this, the endolymphatic space is bounded on all sides by occluding tight junctions, known to inhibit the movement of ions (e.g. Smith, 1978).

2. The Endolymph

(a) The composition and electric potential of the endolymph

Smith *et al.* (1954) first measured the ionic content of the cochlear endolymph by microsampling. They found high levels of K^+ (150 mM) and low levels of Na^+ (40 mM); however, the measurements are tricky, because the endolymphatic space is small, and the samples are easily contaminated by ions from the surrounding perilymph. Their figure for Na^+ has accordingly been thought to be too high. Johnstone and Sellick (1972) suggest that 2 mM, as found with ion-selective electrodes, is more accurate. Recent values have been reviewed by Anniko and Wroblewski (1986). The endolymph has an ionic composition similar to that found intracellularly, and is unique among extracellular fluids of the body for this reason.

Von Békésy (1952) explored the fluids of the cochlea with microelectrodes. His results showed that the potential of the scala vestibuli in the guinea-pig was +5 mV with respect to the scala tympani, itself near the potential of the surrounding bone, and that the scala media had a potential of some +80 mV. The values of this endocochlear potential declined from 80 to 120 mV near the base of the cochlea to 50 to 80 mV in the higher turns. Values around +100 mV have been found in the cat (Peake *et al.*, 1969). Both the chemical and electrical properties of the endolymph therefore point to its having a special role in the cochlea, and many investigations of its functions have been undertaken. The evidence points to the endocochlear potential as being the battery that drives the transduction process and possibly the mechanically-amplifying portion of the travelling wave.

Although Fig. 3.16A shows the endolymphatic space as being bounded by the reticular lamina, the position is in fact controversial. Manley and Kronester-Frei (1980) found that the fluid in the subtectorial space (i.e. between the reticular lamina and tectorial membrane) and in the inner spiral sulcus (Fig. 3.1D) had the same potential as the perilymph (0 mV), with the endolymphatic potential being met only as the electrode entered the lower surface of the tectorial membrane. Runhaar and Manley (1987) have shown that the K^+ concentration of fluid in the inner spiral sulcus is similar to that of perilymph. Their results suggest that the border of the endolymphatic space should be the tectorial membrane rather than the reticular lamina. Studies of ion distribution with X-ray microprobes have, however, not generally supported this view, although it is difficult to ensure always that the ions do not move during the freezing process involved in the technique. Ryan *et al.* (1980) and Anniko *et al.* (1984) both found that crystals from the subtectorial space had the high K^+ and low Na^+ concentrations characteristic of endolymph, and not perilymph.

(b) The origin of the endolymph and the endocochlear potential

It is generally agreed that both the endolymph and the endolymphatic potential are produced by the stria vascularis. The cells here have the morphological and biochemical properties of secretory cells. Under the light microscope, the cells of the stria vascularis can be divided into superficially located darkly-staining cells, called the marginal cells, and more lightly-staining basal cells (e.g. Fawcett, 1986). Under the electron microscope, it can be seen that the marginal cells have long infoldings on the side furthest away from the endolymph, and contain many mitochondria. The marginal cells appear to be involved in the secretion of endolymph, and in maintaining its ionic and electrical state.

Although the endolymph has an ionic composition similar to that found inside cells, its electric potential, unlike that found intracellularly, is strongly positive. This immediately suggests that the endocochlear potential is not a K^+ diffusion potential such as is found inside nerve cells; that is, it is *not* due to K^+ ions diffusing passively down their concentration gradient taking positive charge with them and leaving a net negative charge behind. Nor does it seem to be a Na^+ diffusion potential, although the Na^+ concentration difference is in the right direction. Johnstone and Sellick (1972) increased the Na^+ concentration in scala media by perfusing with 20 mM Na^+ Ringer's solution. They found that the endocochlear potential actually increased; if it were a Na^+ diffusion potential it should have decreased, because the Na^+ concentrations in the endolymph and perilymph had become more equal.

The positive endocochlear potential is, in contrast, generally thought to be directly dependent on energy-consuming, ion-pumping processes in the stria vascularis. Anoxia, which of course inhibits the energy consuming processes rapidly, causes the endocochlear potential to decay to zero within 1–2 min (Johnstone and Sellick, 1972). However, a diffusion potential should be maintained even during anoxia as long as the ionic concentration differences do not decay. In fact, after a few minutes of anoxia when the presumed oxidative processes have disappeared, the positive endocochlear potential is replaced by a negative potential, called $-EP$, which can take several hours to disappear as the ions equilibrate. This negative potential is thought to be a diffusion potential, arising from the high K^+ concentration in the endolymph (Johnstone *et al.*, 1966). Evidence that the stria vascularis is the source of the endocochlear potential comes from several sources:

1. Tasaki and Spyropoulos (1959) destroyed Reissner's membrane so that the endolymph and perilymph could mix. The stria vascularis was the only site from which positive potentials could then be recorded. That any potentials at all could be recorded is evidence that the endocochlear potential is not a diffusion potential, because the ionic gradient will have been destroyed.

2. The stria vascularis is a site of high metabolic activity. Kuijpers and Bonting (1969) analysed Na^+–K^+ linked ATPase in several structures of the cochlea, and found that its activity was some 12 times higher in the stria vascularis than in any other cochlear structure. The ATPase was inhibited by ouabain, a specific inhibitor of Na^+–K^+ linked ATPase, and showed a concentration function for its effect on the ATPase similar to that for its effect on the endocochlear potential (Kuijpers and Bonting, 1970).

3. Ototoxic agents such as ethacrynic acid which have a preferential effect on the stria vascularis also reduce the endocochlear potential. Correspondingly, it was shown by Tasaki and Spyropoulos (1959) that the organ of Corti was not the site of production of the endocochlear potential, because the endocochlear potential is present in a strain of mice known as waltzing mice, in which the organ of Corti is congenitally absent.

4. Microelectrode penetrations in the stria vascularis show that the marginal cells have a high positive potential, more positive even than the endolymph (Melichar and Syka, 1987; Offner *et al.*, 1987). It is the only position in the cochlea from which such high positive potentials have been recorded. This would be expected if the endocochlear potential depended on ion pumps in the marginal cells, and the ion pumps were on the basolateral (i.e. blood capillary, or non-cochlear) side of the cells. The pumps would be expected to move K^+ into the cell cytoplasm, from which it would leak out into the endolymph, and possibly move Na^+ out. Mees (1983) has localized the Na^+–K^+ ATPase to the basolateral wall of the marginal cells.

The exact nature and operation of the pumps is controversial. Kuijpers and Bonting (1970) and Johnstone and Sellick (1972) suggested that the positive potential arose from a *positively-electrogenic* pump, with a Na^+–K^+ coupling ratio such that more K^+ ions were pumped into the endolymph than Na^+ out. Such pumps are, however, not known elsewhere in mammals. Offner *et al.* (1987) argued that it was possible to produce a high K^+ concentration and a positive endocochlear potential from a conventional Na^+–K^+ ATPase, if the membrane permeabilities to K^+ and Na^+ were arranged appropriately. Figure 3.16B shows possible suggested arrangements. Because the pump moves one or both types of ion up their concentration gradients, energy is required, and this is supplied by the ATP (e.g. Johnstone and Sellick, 1972; Dallos, 1973a). The observed final potential will be the net result of subtracting the negative K^+ diffusion potential, always expected where the ions can move down their concentration gradients, from the positive actively-produced potential.

In addition, the stria vascularis has a high concentration of enzymes such as carbonic anhydrase and adenyl cyclase, which are also involved in transport systems. Feldman (1981a,b) and Drescher and Kerr (1985) give reviews of the biochemistry of the stria vascularis, and its relation to the cochlear fluids. Once produced, the endolymph flows out of the cochlea by way of the *ductus reuniens*, through the endolymphatic duct, and is absorbed in the endolymphatic sac. Malabsorption or obstruction of the flow causes an increase in the pressure of the endolymph, known as endolymphatic hydrops, such as is found in Menière's disease (Kimura, 1967b).

3. The Perilymph

The ionic composition of the perilymph, situated in the outer fluid compartments of the cochlea, is similar to that of extracellular fluid or cerebrospinal fluid (Anniko and Wroblewski, 1986). Its electric potential is close to that of the surrounding plasma, being reported to be $+7$ mV in the scala tympani, and $+5$mV in the scala vestibuli (Johnstone and Sellick, 1972). It is undecided whether perilymph is produced from the cerebrospinal fluid, with which it communicates via the cochlear aqueduct, or is an ultrafiltrate of blood plasma, produced by the capillaries in the walls of the perilymphatic space, or whether there is a dual origin (Feldman, 1981b).

X-ray microanalysis of frozen tissue shows that the spaces within the organ of Corti have the same ionic content as perilymph (Anniko and Wrobleski, 1986). However, Ryan *et al.* (1980) did not manage to produce crystals with their freezing technique, and this led them to suggest that its protein content was higher than that of perilymph. The fluid in this space has been termed "cortilymph" (Engström, 1960). The fluid within the extracellular spaces of the organ of Corti has a potential of approximately 0 mV (Dallos *et al.*, 1982).

D. Hair Cell Responses

The measurement of hair cell responses was one of the important recent landmarks in the progress of cochlear physiology. Inner hair cells of the mammalian cochlea were first recorded from by Russell and Sellick (1978). Since the very great majority of afferent auditory nerve fibres make their synaptic contacts with inner hair cells, it must be presumed that it is the job of inner hair cells to signal the movements of the cochlear partition to the central nervous system. The above reports were the first to give information which permitted links to be made between the mechanical travelling wave, not known to be sharply tuned at the time, and auditory nerve fibre responses. Outer hair cells are much more difficult to record from. We have as yet only a few reports, and the role of the outer hair cells in cochlear function, possibly in amplifying the mechanical travelling wave, is still obscure.

Hair cell responses will be dealt with in terms of the resistance-modulation and battery theory of Davis (1958), as it appears in the light of recent evidence, put forward by for instance Hudspeth (1985) and Pickles (1985a). In this theory, the endocochlear potential and the negative hair cell intracellular resting potential combine to form a potential gradient across the apical membrane of the hair cell. Movement of the cochlear partition produces deflection of the stereocilia, as shown in Fig. 3.3. The deflection opens ion channels in the stereocilia. Ions are driven into the cell by the potential gradient, causing intracellular voltage fluctuations. Intracellular depolarization causes release of transmitter, activating the auditory nerve fibres (Fig. 3.17). These stages, and the evidence for them, will be dealt with in more detail in Chapter 5. Hair cell responses from non-mammalian cochleae and the vestibular system, which give basic information on the transduction process, are also dealt with in Chapter 5.

1. Inner Hair Cells

(a) Intracellular potentials

Russell and Sellick (1978) recorded intracellularly from inner hair cells of the mammalian cochlea. They found resting potentials of some −45 mV.

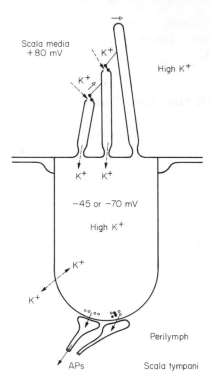

Scala media
+80 mV

K⁺

High K⁺

K⁺

K⁺

K⁺   K⁺

−45 or −70 mV

High K⁺

K⁺

K⁺

Perilymph

APs   Scala tympani

Fig. 3.17   In a modern version of the Davis battery theory, the transducer channels at the apex of the hair cells act as variable resistances. Ions flow into the cell, driven by the battery of the endolymphatic potential and the intracellular potential. Intracellular depolarization causes release of transmitter, and auditory nerve fibre activation. Increased current flow through the hair cells also makes the scala media less positive, and the scala tympani more positive. In this diagram, the transducer channels are drawn at the tips of the shorter stereocilia, although that is only a guess. In accordance with the suggestion of Pickles *et al.* (1984), the channels are shown as being opened by links between the tips of the stereocilia. Open arrows show the movements produced by deflection of the stereocilia in the excitatory (depolarizing) direction. The resting potentials marked (−45 and −70 mV) are for inner and outer hair cells respectively. Modified from Pickles (1987, Fig. 2.16).

This is rather more positive than most nerve cells, which commonly have intracellular potentials of some −70 mV. In response to a tone of low frequency, in which the individual cycles of the response waveform could be distinguished, the hair cells gave potential changes which followed a distorted version of the input stimulus (upper traces, Fig. 3.18). The evidence which will be presented later, suggests that the positive excursions are caused by ion channels opening in the stereocilia, allowing K⁺ ions to enter

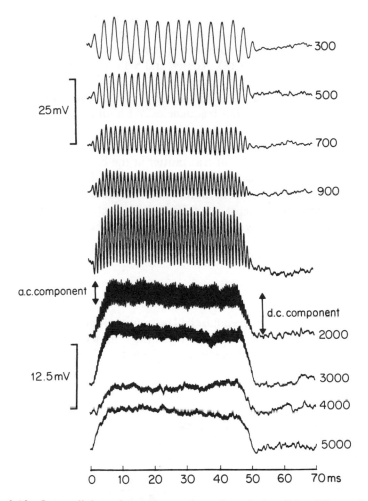

Fig. 3.18 Intracellular voltage changes in an inner hair cell for different frequencies of stimulation, show that the relative size of the a.c. component declines at higher stimulus frequencies (numbers on right of curves). Note change of scale for the lower four traces. From Palmer and Russell (1986, Fig. 9).

the cell, so producing depolarization (Fig. 3.17). The ion channels, which are open to some extent even at rest, shut completely during the opposite phase, stopping the ion flow, and produce relative hyperpolarization. The excursions in the positive direction are greater than the excursions in the negative direction, so that it is possible conceptually to divide the potential changes into an a.c. response at the stimulus frequency, and a sustained d.c. depolarization.

As the stimulus frequency is raised, the a.c. component of the voltage response declines relative to the d.c. component, so that at frequencies of a few kHz and above, the a.c. component is much smaller than the d.c. component (lower traces, Fig. 3.18). Russell and Sellick (1983) ascribed this to the capacitance of the hair cell membranes. Hair cell membranes, like all cell membranes, have a capacitance, and capacitances offer a low impedance to a.c. currents at high frequencies. At high frequencies, therefore, the a.c. current was short-circuited by the low impedance of the hair cell membranes, reducing the a.c. voltage response in the cell.

Depolarization leads to release of transmitter at the base of the hair cells, and so to the production of action potentials in the auditory nerve fibres, which make their synaptic contacts there. We expect release of transmitter, to a first approximation at least, to follow the waveforms of Fig. 3.18. Post-synaptic potentials, as well as action potentials, have been recorded intracellularly from afferent terminals or dendrites in the space below the inner hair cells (Siegel and Dallos, 1986)

(b) Relation to basilar membrane responses

The close correspondence between hair cell responses and basilar membrane responses can be shown in several ways. First, the tuning curve for an inner hair cell is shown in Fig. 3.19. The curve has a low-threshold, sharply-tuned

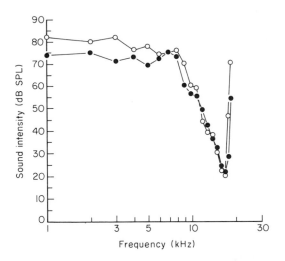

Fig. 3.19   Inner (○) and outer (●) hair cells have very similar tuning curves. The curves are also very similar to those for the mechanical response of the basilar membrane (Fig. 3.10B). From Cody and Russell (1987, Fig. 7).

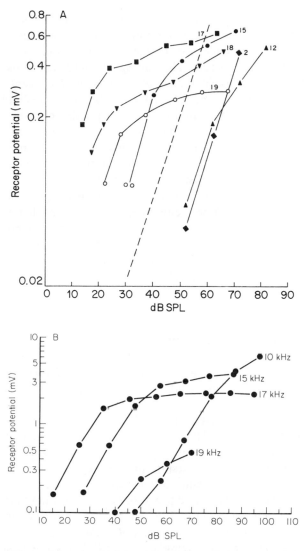

Fig. 3.20 (A) a.c. intensity functions for an inner hair cell show an approximately linear increase in potential at the lowest intensities at each frequency, followed by a saturation. The parameter marked on each curve shows the frequency of stimulation in kHz; this cell was most sensitive at 17 kHz. Dotted line: slope for linear growth. From Russell and Sellick (1978, Fig. 3).(B) Intensity functions for an outer hair cell are similar to those of inner hair cells, although the response saturates more abruptly at high intensities. The a.c. response is plotted. In part B, but not part A, the voltages have been compensated for the attenuating effects of the current flowing through the capacitance of the hair cell walls. From Cody and Russell (1987, Fig. 6).

tip, at the best or "characteristic" frequency (CF). There is also a high threshold and broadly-tuned tail, stretching to low frequencies. The shape corresponds to the tuning curve for the mechanical response of the basilar membrane shown in Fig. 3.10B. The tuning of inner hair cells therefore appears to be derived from the tuning of the basilar membrane, although it is possible that there are some differences in the high threshold tail.

Correspondingly, intensity functions for inner hair cells are similar to those of the basilar membrane mechanical response (Fig. 3.20A; compare with Fig. 3.10C). Around the CF (17 kHz for this particular hair cell), the response grew linearly at first, at the rate of a 10-fold voltage change for a 20-dB increase in stimulus intensity, parallel to the dotted line. When the intensity was raised further, the response at and near CF grew nonlinearly, i.e. with a more shallow slope (remember these are log–log scales). At frequencies above CF the response saturated to a low maximum output voltage, and well below the CF (e.g. 2 kHz) the response grew entirely linearly. Again, both of these indicate that inner hair cell responses closely follow basilar membrane mechanical responses.

(c) Input–output functions

The relations between the acoustic stimulus and the a.c. and d.c. components of the responses can be described by the *input–output* function. Here, the instantaneous value of the input stimulus pressure is plotted against the instantaneous value of the intracellular potential (Fig. 3.21A). It is possible to construct these functions only for low-frequency stimuli, where the a.c. component of the intracellular response has not been attenuated by the capacitance of the cell walls. The function is asymmetric in various ways. First, the maximum excursion in the positive, depolarizing direction is greater than the maximum excursion in the negative, hyperpolarizing direction. Secondly, the maximum depolarization is reached much more gradually and at much greater sound pressure excursions, than the maximum hyperpolarization. The function in Fig. 3.21A describes how, in response to sinusoidal stimulation, the depolarizing potential changes in the hair cell are greater than the hyperpolarizing ones, and shows the basis for the production of the d.c. component superimposed on the a.c. component of the response. It illustrates the way that the d.c. component appears as a distortion component of the a.c. response.

Inner hair cell responses have also been reported by, among others, Nuttall *et al.* (1981), Goodman *et al.* (1982), Brown *et al.* (1983a,b), Russell (1983), Russell and Cowley (1983), Russell and Sellick (1983), Dallos (1985, 1986) and Russell *et al.* (1986a,b). Mechanisms of transduction will be dealt with in detail in Chapter 5.

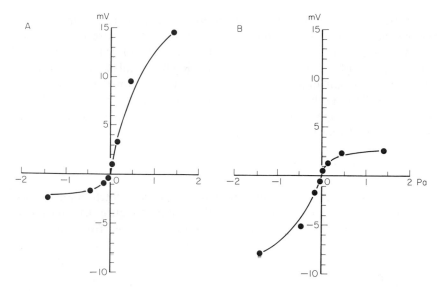

Fig. 3.21   (A) Input–output functions are made by plotting the instantaneous value of the intracellular voltage change (vertical axis) against the instantaneous value of the stimulus pressure during sinusoidal stimulation (horizontal axis). The function for an inner hair cell shows voltage changes in the depolarizing direction (upwards) that are greater than those in the hyperpolarizing direction (downwards). Stimulus frequency: 600 Hz. From Cody and Russell (1987, Fig. 2.B). (B) Input–output for an outer hair cell in the base of the guinea-pig cochlea shows voltage changes in the hyperpolarizing direction that are larger than those in the depolarizing direction. This curve was made for a 600-Hz stimulus. However, the input–output function must change form for higher frequencies of stimulation, since basal turn OHCs do not show d.c. changes in response to moderate levels of stimulation with high-frequency sinusoids. From Cody and Russell (1987, Fig. 2D).

2. Outer hair cells

(a) Intracellular potentials

Outer hair cells have proved to be particularly difficult to record from. One reason may be that outer hair cells are suspended by their apical and basal ends within the space of Nuel, so that an advancing electrode tends to push them aside rather than penetrate. In contrast, inner hair cells are closely surrounded on all sides by supporting cells. The position of outer hair cells halfway across the cochlear duct, rather than near the edge, may also mean that they can move more easily away from the electrode. It is also likely that stimulus-induced vibrations are greater at the outer hair cell position, so that the electrode is thrown out more easily when acoustic stimulation is applied. We have only a few reports of outer hair cell recordings. Dallos *et al.* (1982) and Cody and Russell (1987) have recorded from outer hair cells in the guinea-pig cochlea. They reported resting potentials to be considerably more negative than inner hair cells (-70 mV as against -45 mV). Like inner hair cells, outer hair cells can show both a.c. and d.c. voltage responses. However, the voltage responses in outer hair cells are only one-half to one-third the size of those of inner hair cells.

(b) Relation to basilar membrane responses

Like inner hair cells, outer hair cells have responses which closely follow the mechanical response of the basilar membrane. Figure 3.19 shows the tuning curve for an outer hair cell a.c. response, recorded from the base of the cochlea. The frequency selectivity is very similar to that of the inner hair cell which was recorded immediately adjacent to it (Cody and Russell, 1987). Figure 3.20B shows intensity functions for the response of an outer hair cell. Again, they are generally similar to the functions for an inner hair cell (Fig. 3.20A) and the basilar membrane (Fig. 3.10C), although they may saturate rather more abruptly.

(c) a.c. and d.c. components

While there is agreement about the sharpness of tuning and the shape of the intensity functions found by the different groups of workers, there seem to be some differences in the relative sizes of the a.c. and d.c. components of the response. Dallos and his colleagues recorded in the apical (low-frequency) end of the guinea-pig cochlea (Dallos *et al.*, 1982; Dallos, 1985, 1986). With stimulation at moderate intensities at the CF, they showed that there was a depolarizing d.c. response superimposed on the a.c. component. Unlike the position in inner hair cells, however, the d.c. component could become hyperpolarizing at frequencies just below CF. Figure 3.22 shows how the d.c. component changed polarity with stimulus frequency. If the stimulus intensity was raised, the frequency range over which depolarization could be obtained spread, so that the response became consistently depolarizing.

Russell and his colleagues, recording in the basal (high-frequency) end of the guinea-pig cochlea, have obtained slightly different results (Russell *et al.*, 1986a; Cody and Russell, 1987). Stimuli at the CF of the cell produced only a.c. responses, without any sustained d.c. components at all unless the stimulus intensity became rather high (above 90 dB SPL). However, at low frequencies, a long way below CF in this case, they found that the response

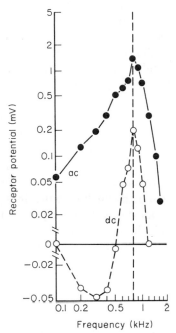

Fig. 3.22.   a.c. and d.c. responses of an outer hair cell at the apical (low-frequency) end of the guinea-pig cochlea. The hair cell shows a d.c. depolarization for stimuli near the best frequency (800 Hz), and a d.c. hyperpolarization for stimuli below 500 Hz. The stimulus intensity was kept constant at 30 dB SPL. From Dallos (1986, Fig. 5).

became hyperpolarizing at medium intensities. When the intensity was raised, the response became depolarizing at all frequencies, although this happened at some 90 dB SPL, a higher intensity than in the experiments of Dallos (Dallos, 1985).

The hyperpolarizing d.c. component seen with low-frequency, medium-intensity, stimuli in outer hair cells can be explained by an input–output function in which the excursions in the hyperpolarizing direction are larger than those in the depolarizing direction. Cody and Russell (1987) showed just such an input–output function in outer hair cells driven by low-frequency sinusoids (Fig. 3.21B). Possible reasons for the difference from inner hair cell responses will be discussed in Chapter 5.

Russell's *high-intensity* depolarizing component was rather different from he d.c. components described hitherto. Unlike the other d.c. components, vhich appeared and disappeared instantaneously with the a.c. response, he high-intensity d.c. component in basal turn outer hair cells took several cycles to develop, and several cycles to disappear after the end of the stimulus. This suggests that it cannot simply be thought of as a distortion component of the a.c. response, and that a description in terms of an input–output function (Fig. 3.21) is not appropriate. Here, the d.c. response might, for instance, be metabolic, or due to factors such as changes in the ionic environment of the hair cells, produced as a result of the acoustic stimulation.

The lack of a d.c. component in basal turn outer hair cells, at moderate and low intensities of stimulation by high-frequency sinusoids, has an interesting implication. Since the a.c. currents will be shunted through the capacitance of the cell walls, the intracellular a.c. voltage change can be expected to be small at high frequencies. This is confirmed by experimental recording (Russell *et al.*, 1986a). Basal turn outer hair cells will therefore show neither significant a.c. nor d.c. voltage responses intracellularly. This suggests that if outer hair cells have any function in hearing, it is the a.c. *currents* through the outer hair cells that are most likely to be important.

E. The Gross Evoked Potentials

When electrodes are placed in or near the cochlea, it is possible to record gross stimulus-evoked potentials, which are derived from the massed activity of large numbers of the individual receptor and nerve cells. The gross evoked potentials can be divided into three groups. First, the cochlear microphonic (CM) is an a.c. response that approximately follows the acoustic stimulating waveform (Fig. 3.23). The cochlear microphonic is derived mainly from the currents flowing through the outer hair cells. Secondly, there is a d.c. shift in the record, known as the summating potential (SP), which depends on the d.c. components generated by the hair cells. Thirdly, there are a series of deflections at the beginning, and sometimes also at the end, of the stimulus, called the N_1 and N_2 neural (or action) potentials. The neural potentials are produced by the summed activity of auditory nerve fibres, producing synchronized action potentials at the onset of the stimulus.

1. The Cochlear Microphonic

Wever and Bray (1930) placed a wire electrode in the auditory nerve, connected it to an amplifier in a room 16 m away, and from there to a loudspeaker. As they reported,

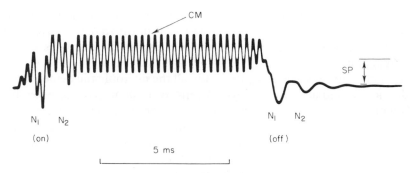

Fig. 3.23 Diagram of the response to a tone burst, recorded with gross electrodes, shows the cochlear microphonic (CM), N_1 and N_2 phases of the gross action potential to both the beginning and the end of the stimulus, and, in the d.c. shift of the microphonic from the baseline, the summating potential (SP).

> "the action currents, after amplification, were audible in the receiver as sounds which, so far as the observer could determine, were identical with the original stimulus. Speech was transmitted with great fidelity. Simple sounds, commands, and the like were easily received. Indeed, under good conditions the system was employed as a means of communication between operating and sound-proof rooms."

They thought that they were recording the massed action potentials of the auditory nerve, but Adrian (1931) suggested that they were in fact recording potentials evoked earlier in the transducer chain, and called them the cochlear microphonics.

(a) Generation

Tasaki *et al.* (1954) presented clear evidence as to the site of production of the cochlear microphonic. They advanced a microelectrode from the scala tympani, through the basilar membrane and the organ of Corti, into the endolymphatic space and the scala media. They recorded the cochlear microphonic as they advanced the electrode, and showed that the microphonic reversed polarity at the same time as the endocochlear potential appeared. This fixes the site of generation of the cochlear microphonic as the border of the endolymphatic space, namely the reticular lamina. The reticular lamina is the surface carrying the transducing structures of the hair cells, the stereocilia. It is now believed that the cochlear microphonic is generated by the hair cells. When transducer channels open, so that current flows into the hair cells, making them more positive, current is drained from the scala media, making it less positive (Fig. 3.17). When the channels shut, the current flow is reduced, and the scala media moves more positive. Potential changes are therefore produced, which in the scala media are in the opposite phase to the potential changes in the hair cells and the scala tympani.

The outer hair cells generate practically all the microphonic demonstrable in the normal animal. Dallos (1973b) selectively damaged the outer hair cells with the ototoxic drug kanamycin. This drug in the right doses can destroy the outer hair cells while leaving the inner hair cells intact, at least as judged under the light microscope. Dallos found that the cochlear microphonic was reduced to 1/30th of its normal size, suggesting that, if the inner hair cells were indeed unaffected, it is the outer hair cells that contribute practically all the microphonics recordable in the normal animal.*

(b) Spatial localization

As might be expected from the theory of its production, the cochlear microphonic is spatially localized in the same way as the mechanical travelling wave. Low frequencies give the greatest response near the apex, and high frequencies near the base. The position of the peak of the response at low intensity compares well with the position of the peak of the travelling wave envelope (Eldredge, 1974). In addition, when the intensity is raised, the peak of the response moves towards the base of the cochlea. Part of the shift will be due to the basalward shift in the travelling wave envelope at high stimulus intensities (Fig, 3.10A). However, there is a further reason for the shift, which depends on the way that an extracellular electrode sums the potentials produced by hair cells situated at different points on the travelling wave.

In the basal part of the travelling wave, the basilar membrane deflections all occur in nearly the same phase (Figs 3.7 and 3.12). The sign of the potential changes produced in, say, the scala media, will therefore be the same for all the hair cells in the basal part of the wave, and an electrode in the scala media will strongly sum their responses. However, in the apical part of the travelling wave, the phase changes at a greater and greater rate with distance. Some hair cells will tend to make the scala media go positive at the same time as adjacent ones tend to make it go negative. The effects

---

*While that interpretation was unexceptionable at the time, care is now required, since it appears that normal outer hair cells are necessary for the production of the low-threshold, sharply-tuned, component of the mechanical travelling wave. In other words, the loss of the outer hair cells could have reduced the magnitude of the travelling wave and so the stimulus to the remaining hair cells. However, the results of Dallos (1973b) held up to very high intensities (e.g. 120 dB SPL), at which we would expect any outer hair cell contribution to the mechanical travelling wave to have become negligible.

on a remote electrode in the scala media will therefore tend to cancel. In their contribution to the net recorded cochlear microphonic, hair cells around the peak and the apical slope of the travelling wave therefore tend to be under-represented (Fig. 3.24). This leads to the dominance of the basal part of the travelling wave in the final recorded microphonic. The effect is particularly large at high intensities, when the peak of the travelling wave is in any case reduced relative to the basal slope (Fig. 3.10A). The peak microphonic therefore tends to be situated rather basal to the peak of the travelling wave.

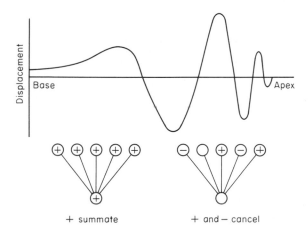

Fig. 3.24   The influence of spread of current on the magnitude of recorded microphonics. Because electrodes integrate over a length of the cochlea, the recorded microphonic is attenuated near the peak of the travelling wave, where the phase changes relatively rapidly with distance.

(c) Intensity functions

Figure 3.25 shows intensity functions for the cochlear microphonic, recorded between one electrode in the scala vestibuli and one in the scala tympani, in response to a 1020-Hz tone. The response in the electrode pair in the third turn, near the peak of the travelling wave, grows nearly linearly only for the lowest intensities, and then starts to saturate. This behaviour is similar to that seen in the growth of basilar membrane vibration (Fig. 3.10C). The response in the basal turn electrodes, recording from the basal part of the travelling wave, is smaller at low intensities. However, the response continues growing linearly up to higher intensities, so that at the highest intensity the response is largest more basally. This phenomenon again can be seen in the mechanical travelling wave (Figs 3.10A and C).

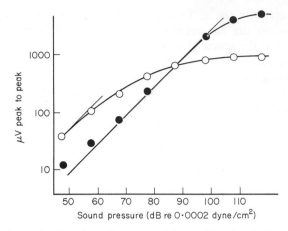

Fig. 3.25   Intensity functions for the cochlear microphonic in response to a 1020-Hz stimulus, recorded differentially between the scala vestibuli and scala tympani in the third turn (○) and basal turn (●). The thin straight lines are drawn for linear growth. From Tasaki *et al.* (1952, Fig. 6).

2. The Summating Potential

The d.c. change produced in the cochlea in response to a sound is known as the summating potential (SP), and is visible as a base-line shift in the recorded signal of Fig. 3.23. Depending on the circumstances, it is recorded as a sustained positive or negative deviation in the scala media during acoustic stimulation. The summating potential has correlates in the d.c. stimulus-evoked potentials in hair cells, and probably is derived directly from them. When the input–output functions of hair cells are such as to produce intracellular d.c. depolarization of the hair cells, the d.c. component of the current into the cells is greater than normal. The scala media will tend to move more negative, because the endolymphatic potential is partially being short-circuited, and the scala tympani will tend to move more positive. On the other hand, during intracellular d.c. hyperpolarization, the net current into the hair cells is smaller than normal, and the sign of the summating potential will be reversed. Figure 3.26 shows how the summating potential recorded with gross electrodes depends on the stimulus frequency, and how it relates to the cochlear microphonic. The similarity to the intracellular d.c. and a.c. changes in outer hair cells (Fig. 3.22) is striking. Note however that the microphonic shows a slight relative leftward shift, consistent with the dominance of the basal part of the travelling-wave envelope in the recorded gross microphonic.

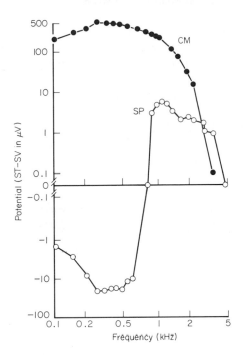

Fig. 3.26   The frequency response of the cochlear microphonic (CM), and of the summating potential (SP). The potentials were recorded with gross electrodes in the scalae, and are here plotted as the potential in the scala tympani minus the potential in the scala vestibuli. Modified from Dallos *et al.* (1972, Fig 21).

As the stimulus intensity is raised, the intracellular depolarization of apical turn outer hair cells can be produced over a greater and greater frequency range (Dallos, 1985; see above). We would expect this to lead to increased negativity above the reticular lamina. Correspondingly, as the intensity is raised, the negative component of the summating potential (negative in the scalae media and vestibuli, that is) spreads over a wider and wider range of stimulus frequencies, until at the highest intensities only negative summating potentials are produced.

As expected from the theory of its origin, the summating potential reverses sign around the reticular lamina. Konishi and Yasuno (1963) advanced an electrode through the organ of Corti and found just such a reversal of the sign of the summating potential, at the same time as the endocochlear potential appeared, and the microphonic reversed phase.

Like the d.c. depolarization in outer hair cells at high intensities, the summating potential sometimes reaches its maximum value only slowly over time (Honrubia and Ward, 1969; Cody and Russell, 1987). This again suggests that the two phenomena are directly dependent upon each other. However, we cannot rule out other contributions to the slow component of the summating potential, from for instance metabolic changes in the supporting cells and the stria vascularis.

Inner hair cells also seem to contribute a component to the summating potential. As we have seen, kanamycin selectively affects outer hair cells. Dallos and Wang (1974) showed that after kanamycin, whereas the negative component was reduced by 26 dB, the positive component was completely abolished and replaced by a negativity. This suggests the possibility that the inner hair cells make a contribution to the remaining negative component (negative in scala media). This would agree with the direction of stimulus-evoked d.c. changes in inner hair cells, which are always in the depolarizing direction, producing negativity above the reticular lamina.

3. The Gross Neural Potentials

Electrodes that are too large to record from single neural elements can nevertheless record summed neural activity produced by the simultaneous activation of many neurones. An electrode in or near to the cochlea will record a gross neural response of the cochlea to a tone onset, as in Fig. 3.23.

The two dominant waves of the gross neural response are known as N_1 and N_2, the first coming about 1 ms after the start of the microphonic, and the second about 1 ms after that. As the intensity is raised from low levels, N_1 is the first to appear, followed at higher intensities by N_2. At still higher intensities the response becomes more complex, with the appearance of other components of different latency (e.g. Antoli-Candela and Kiang, 1978). The gross potential is interpreted as the summed effect of massed action potentials travelling in a volley down the auditory nerve. Detailed experiments have supported this conclusion. An analysis of the mechanisms behind this potential is useful because it is the gross potential, rather than the activity of single fibres of the auditory nerve, that can be recorded clinically.

Antoli-Candela and Kiang (1978) sampled the activity of a large number of auditory nerve fibres in response to a click. They showed that fibres had to have best (or characteristic) frequencies above 4 kHz if they were to fire in synchrony with the N_1 and N_2 phases of the gross neural action potential. We can understand how this result arises, because a gross summed potential can be produced only by neural action potentials that are substantially in synchrony. The travelling wave travels most rapidly over the basal part of the cochlea, and more and more slowly thereafter. Therefore, it is only the fibres from the high-frequency, basal region that will be activated in synchrony. We therefore expect fibres from this region to dominate the grossly-recorded N_1 and N_2 potentials.

F. Summary

1. The cochlea is a coiled tube, divided lengthways into three scalae. The three divisions are known as the scala vestibuli, the scala media and the scala tympani. The two outer scalae, the scala vestibuli and scala tympani, contain perilymph which is like normal extracellular fluid in composition and is at or near ground potential. The scala media contains endolymph which is more like intracellular fluid, and has a positive potential. The positive potential arises partly at least from Na^+/K^+-linked ATPase ion pump in the stria vascularis.

2. The organ of Corti is the auditory transducer. It sits on the basilar membrane dividing the scala media from the scala tympani. The transducing cells are called *hair cells*. Hair cells are of two types, known as *inner* and *outer* hair cells. They have many hairs, or stereocilia, projecting from their apical surface. Deflection of the hairs initiates transduction.

3. Deflection of the hairs is produced by deflection of the basilar membrane. The latter occurs as a result of a sound-induced displacement of the cochlear fluids, which interacts with the stiffness of the basilar membrane, to produce a progressive travelling wave on the membrane. The wave passes up the cochlea from base to apex.

4. Travelling waves produced by sounds of high-frequency come to a peak near the base of the cochlea. High-frequency sounds are therefore transduced near the base of the cochlea. The travelling wave produced by low-frequency sounds comes to a peak further up the cochlea, and low-frequency sounds are transduced near the apex.

5. The travelling wave has a sharply-tuned peak. The sharp peak underlies the sensitivity and frequency selectivity shown by the rest of the auditory system, and indeed the sensitivity and frequency selectivity shown by the whole organism. The mechanism behind the sharply-tuned peak is at the moment unknown. However, many investigators suggest that it is produced by outer hair cells generating movements in response to auditory stimulation.

6. Deflection of the stereocilia by the travelling wave opens and closes ion channels in the stereocilia. This modulates the current being driven into the hair cells by the combined effects of the positive endocochlear potential and the negative intracellular potential. The current produces potential changes that can be measured both in the hair cells with fine microelectrodes, and grossly in the cochlea with larger electrodes.

7. Inner hair cells have resting potentials of about -45 mV. They produce both an a.c. voltage and a steady d.c. depolarization in response to sound. The potentials have sharp tuning curves and amplitude functions similar to those shown by the basilar membrane vibrations. Outer hair cells have resting potentials of about -70 mV. They show a.c. potential changes in response to sound and, depending on the circumtances, either no, or a depolarizing or hyperpolarizing, d.c. response. Like inner hair cells, they show sharp tuning curves and amplitude functions similar to those shown by basilar membrane vibrations.

8. The potential changes in inner hair cells serve to govern the release of transmitter, to produce action potentials in the auditory nerve fibres. The function of outer hair cells is not certain, but it has been suggested that outer hair cells are responsible for mechanical amplification of the travelling wave, triggered by voltage changes or current flows within the cells.

9. The cochlear microphonic is the extracellular correlate of the a.c. current flowing through hair cells. It is generated predominantly by outer hair cells. The summating potential is the extracellular correlate of the d.c. component of the current flowing through hair cells. Depending on circumstances, it can be recorded as either a positive or negative shift in scala media. The summating potential probably receives contributions from both inner and outer hair cells, and possibly from other sources.

10. The massed activity of auditory nerve fibres, the N_1 and N_2 potentials, can be recorded with gross electrodes in response to stimulus onsets. Fibres with best frequencies of 4 kHz and above seem to make most contribution.

G. Further Reading

Aspects of cochlear anatomy have been reviewed by Ades and Engström (1974), Lim (1980, 1986) and Pickles (1985a). Cochlear biochemistry has been reviewed by Feldman (1981a,b) and Drescher and Kerr (1985).

The older measurements of cochlear mechanics were reviewed by Dallos (1973a, Chapter 4, pp. 127–217) and Rhode (1980). The recent measurements were reported by Sellick *et al.* (1982), and have been reviewed by Pickles (1985a) and Johnstone *et al.* (1986).

Cochlear potentials and their relation to earlier results on cochlear mechanics were reviewed by Dallos (1973a, Chapter 5, pp. 218–390). More recent work has been reviewed by Dallos (1985, 1986), Mountain (1986), Nuttall (1986) and Russell *et al.* (1986a).

Mechanisms of transduction are further dealt with in Chapter 5 of the present work. Reviews are given by Hudspeth (1985) and Pickles (1985a).

4. *The Auditory Nerve*

We now have a comprehensive description of the responses of auditory nerve fibres to a variety of stimuli in normal, albeit anaesthetized, animals. We are also beginning to understand some of the changes that occur in auditory nerve activity during cochlear pathology. The responses of auditory nerve fibres underlie the responses of the later stages of the auditory system, and closely relate to the psychophysical capabilities of the intact organism. For these reasons, a knowledge of the material presented in this chapter is essential for the understanding of the later chapters on the central auditory system (Chapters 6–8), psychophysical correlates of auditory physiology (Chapter 9) and sensorineural hearing loss (Chapter 10). Those whose later interest is primarily in Chapter 10 need here only read up to and including Section B.2 (ending on p. 93).

A. Anatomy

Auditory nerve fibres, with their cell bodies in the spiral ganglion, provide a direct synaptic connection between the hair cells of the cochlea and the cochlear nucleus. Each ear has about 50,000 fibres in the cat and 30,000 in man (Harrison and Howe, 1974a). It was once thought that a substantial proportion of auditory nerve fibres were directed to the outer hair cells, which are of course the more numerous of the hair cells (e.g. Fernandez, 1951). Now, however, it is recognized that only about 5–10% of the spiral ganglion cells are connected to the outer hair cells, the majority connecting directly and exclusively to the inner hair cells (Spoendlin, 1972; Morrison *et al.*, 1975). A diagram summarizing Spoendlin's scheme is shown in Fig. 3.6. The reader is reminded that the fibres innervating the inner hair cells innervate the hair cells nearest their point of entry into the cochlea, whereas those innervating the outer hair cells run basally for about 0.6 mm before terminating. About 20 fibres innervate each inner hair cell, whereas about six fibres innervate each outer hair cell. Each fibre to the inner hair cells connects with one and only one hair cell, whereas those to the outer hair cells branch and innervate about 10 hair cells. There is evidence that the differentiation in the targets is associated with a morphological differentiation in the cell bodies and axons. A total of 95% of cells (in the cat) connect with the inner hair cells, have bipolar cell bodies in the spiral ganglion, and have myelinated cell bodies and axons (Spoendlin, 1978). They are called Type I cells. Type II cells connect with outer hair cells, are monopolar, and are not myelinated. Retrograde transport of horseradish peroxidase, from the cochlear nucleus to the cochlea, confirms that both types send axons to the cochlear nucleus (Ruggero *et al.*, 1982).

It is generally thought that the great majority of auditory nerve responses have been measured from Type I cells. Type I cells are typically driven strongly by tones, and can show high levels of spontaneous activity. We have only one confirmed instance, by tracing a horseradish peroxidase injection, of a record from a Type II cell (Robertson, 1984). Here, the cell was silent. We do not therefore know the function of the Type II system in audition, although some ideas will be discussed in Chapter 5.

B. Physiology

We must assume, provisionally at least, that all, or substantially all, the auditory nerve fibres or cells that have been recorded from innervate the inner hair cells. Although our knowledge of inner hair cell physiology is relatively sketchy, we do have a fairly complete knowledge of the responses of the auditory nerve. Our knowledge has accumulated over the last 25 years or so, and has depended on a careful control of stimulus and physiological parameters, together with the surveying of large populations of fibres, sometimes as many as 418 fibres in one animal (Kim *et al.*, 1980). This has been achieved in spite of the inaccessibility of the nerve deep in the bone, and the lack of mechanical stability of the adjacent brainstem, which means that nerve fibres can easily be lost in recording. In fact, it was not until 1954 that the first responses of auditory nerve fibres were published (Tasaki, 1954), and it is now recognized that the records indicate that the cochlea must have been in poor physiological condition. Tasaki's approach was to drill through the temporal bone until the auditory nerve was encountered in the internal auditory meatus. The approach commonly used nowadays is to open the occipital bone at the back of the skull and, by inserting a retractor around the edge of the cerebellum, to retract the cerebellum and brainstem medially away from the wall of the skull until the stub of the nerve running between the internal auditory meatus and the cochlear nucleus becomes visible. Microelectrodes can then be inserted under direct vision. Alternative approaches are to record from cells of the spiral ganglion directly through holes in the cochlear wall, or to record from within the internal auditory meatus by means of microelectrodes inserted stereotaxically through the brainstem. When appropriate measures are taken to stabilize the preparation, fibres or cells can now be recorded for many tens of minutes, compared with the 10 s or so managed by Tasaki.

With microelectrodes with a tip size 0.3 μm or less, single fibres can be recorded from, and give waveforms corresponding to those in Fig. 4.1. Many fibres show random spontaneous activity. There tends to be a bimodal

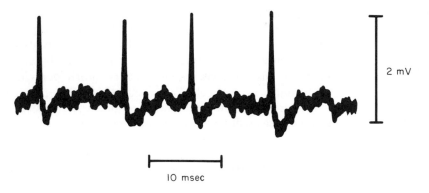

Fig. 4.1 Action potentials recorded from a single auditory nerve fibre. The waves are initially positive and nearly monophasic. Photograph by courtesy of G. Leng,

distribution of spontaneous discharge rates. About a quarter of the fibres discharge at below 20/s, and most of these discharge at 0.5/s or less. The other group has a mean of 60–80 discharges/s, with a maximum of 120/s (Liberman and Kiang, 1978; Evans, 1972).

1. Response to Tones

(a) Frequency selectivity

Fibres are responsive to single tones, and in the absence of other stimuli the tones are always excitatory, never inhibitory. The responses can be demonstrated by means of a post- (or peri-) stimulus time histogram (PSTH). In making such a histogram, a stimulus is presented many times, and the occurrence of each action potential is plotted on the histogram by incrementing the count on the column, or bin, corresponding to the time after the beginning of the stimulus. Tone bursts produce a sharp onset response, which drops rapidly over the first 10–20 ms (Fig. 4.2), and then

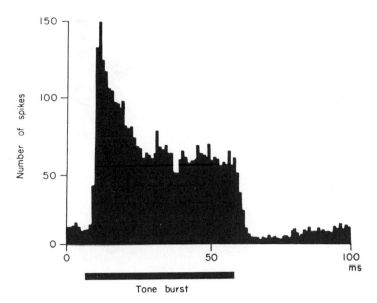

Tone burst

Fig. 4.2 Single fibres of the auditory nerve show an initial burst of activity at the beginning of a tone pip, a gradual decline, and a transient off-suppression of the spontaneous activity at the end of the stimulus. Here, a post-stimulus–time histogram was made by presenting tone pips many times and incrementing the count at the corresponding point on the histogram whenever an action potential occurred. Reprinted from *Discharge Patterns of Single Fibers in the Cat's Auditory Nerve* by N. Y.-S. Kiang *et al.*, by permission of The MIT Press, Cambridge, Massachusetts. © The MIT Press, 1965.

more and more slowly over the next several minutes. The fibres can be characterized by their threshold as a function of frequency of the tone. The intensity of a tone burst is adjusted until an increment in firing is just detectable. This increment is commonly between 5 and 30 spikes/s, depending on the spontaneous firing rate of the fibre and the method used to detect the increment. The procedure is repeated for different frequencies of stimulation. Examples of the resulting tuning curves relating threshold to frequency are shown in Fig. 4.3. Each fibre has a low threshold at one frequency, the "characteristic" or "best" frequency, and the threshold rises rapidly as the stimulating frequency is changed. Figure 4.3 shows the typical change in shape of tuning curves across frequencies, if the frequency scale is logarithmic. At low frequencies, below 1 kHz, tuning curves are symmetric. At higher frequencies the curves become increasingly asymmetric, with steep high-frequency slopes and less steep low-frequency slopes. A distinction between two parts of the tuning curve also becomes obvious in high-frequency units. There is a very sensitive, frequency-selective "tip"

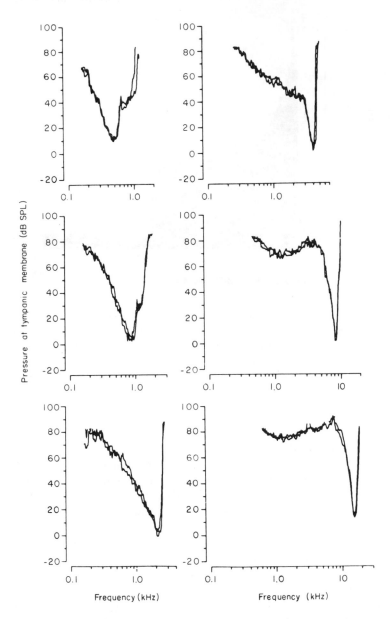

Fig. 4.3   Representative tuning curves (frequency threshold curves) of cat auditory nerve fibres are shown for six different frequency regions. In each panel, two fibres from the same animal, of similar characteristic frequency and threshold are shown, indicating the constancy of tuning under such circumstances. From Liberman and Kiang (1978, Fig. 1).

of the tuning curve, and a long, broadly-tuned "tail", stretching to low frequencies. The tail has a broad dip around 1 kHz. This is probably derived from the boost given to the input by the middle ear characteristics, since it disappears if the stimulus intensity is plotted with respect to constant stapes velocity (Kiang *et al.*, 1967). Single auditory nerve fibres therefore appear to behave as bandpass filters, with an asymmetric filter shape. The frequency selectivity is similar to that of the basilar membrane and the hair cells, from which their frequency selectivity is almost certainly derived (See Chapter 3; Sellick *et al.*, 1982; Russell and Sellick, 1978).

All mammals investigated show tuning curves broadly similar to those of Fig. 4.3, although details such as the degree of frequency selectivity and depth of the tip may vary from species to species.

Our ideas as to the distribution of the fibres' thresholds at the characteristic frequency have had a chequered history. The early report of Katsuki *et al.* (1962) suggested that there was a wide distribution of fibre thresholds. He thought that this could be associated with the different thresholds of inner and outer hair cells, on the then current ideas of cochlear innervation. It was thought that the fibres innervating the outer hair cells had low thresholds, and those innervating the inner hair cells had high thresholds. Later, Kiang (1968) showed that when care was taken to calibrate the sound system properly, and when sufficient fibres were recorded from in each animal rather than pooled across animals, the range in any one animal was 20 dB or less. The "high-threshold" units of Katsuki *et al.* (1962) were probably the high-threshold, broadly-tuned tails of the tuning curves of fibres of high characteristic frequency. It became dogma that all fibres had similar, low thresholds. Liberman and Kiang (1978) and Liberman (1978) have more recently shown that there is a 60 – 80 dB spread of fibre thresholds at any one characteristic frequency and in any one animal if precautions are taken to include units with very low spontaneous rates and high thresholds. Nevertheless, the majority, perhaps 70%, of fibres have thresholds within the bottom 10 dB of the range, and 80% within the bottom 20 dB. The remainder, which have particularly low spontaneous rates, are spread over the rest of the range (Fig. 4.4). The position is discussed further by Geisler *et al.* (1985).

The degree of frequency selectivity has been expressed in two ways. One is by the slopes of the tuning curve above and below the characteristic frequency. The slopes are a function of the characteristic frequency of the fibres concerned, with, in many species, the fibres in the 10-kHz region showing the steepest slopes. Here the high-frequency slopes measured between 5 and 25 dB above the best threshold range from 100 to 600 dB/octave, and the low-frequency slope from 80 to 250 dB/octave (Evans, 1975a). Further up the slope of the tuning curves, the low-frequency slopes

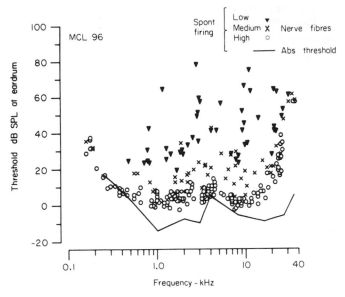

Fig 4.4   Distribution of best thresholds of auditory nerve fibres in one cat. Fibres with high spontaneous firing rates (O, $\geqslant 18/s$) have low thresholds, and those with low spontaneous firing rates (▼, $<0.5/s$) have high thresholds. Fibres with intermediate spontaneous firing rates ($\times$) have thresholds in between. The behavioural absolute threshold of the cat, expressed in terms of the intensity at the eardrum, lies just below the lowest thresholds of the auditory nerve fibres. Neural data from Liberman and Kiang (1978, Fig. 2). Behavioural data from Elliott *et al.* (1960).

become shallower as the 'tail' is approached, but the high-frequency slopes become even steeper, sometimes increasing to as much as 1000 dB/octave (Evans, 1972).

A second way that resolution can be expressed is by measuring the bandwidth of the tuning curve at some fixed intensity above the best threshold. By analogy with the practice in the measurement of the bandwidths of electrical filters, it might be thought appropriate to measure the half-power bandwidth, that is, the bandwidth 3 dB above the best threshold. However, because it is difficult to measure thresholds sufficiently accurately, bandwidths 10 dB above the best threshold have been used instead. The 10-dB bandwidths plotted as a function of characteristic frequency show a restricted spread (Fig. 4.5A). A related way in which the resolution can be expressed is by analogy with the electrical "quality" or "Q" factor of a filter, defined as the centre frequency divided by the bandwidth, the bandwidth here being defined at 10 dB above the best threshold. The quality factor so defined is called Q_{10}, and a high-quality factor, or high Q_{10},

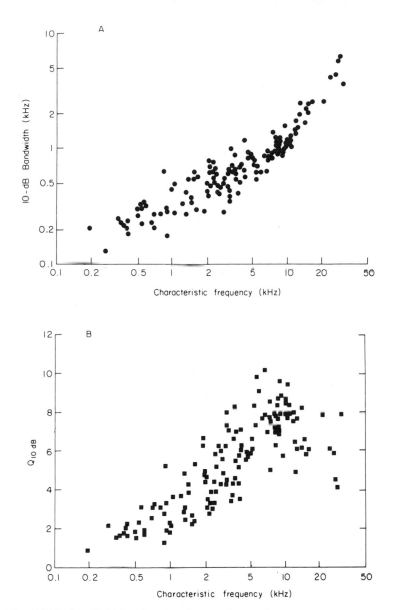

Fig. 4.5 (A) The bandwidths of tuning curves of cat auditory nerve fibres are plotted as a function of the fibres' characteristic frequency. Here, the bandwidths were measured 10 dB above the best threshold. Data calculated from Q_{10}s of Evans (1975a). (B) Q_{10}'s of auditory nerve fibres are shown as a function of characteristic frequency. (Q_{10} = characteristic frequency/bandwidth measured 10 dB above best threshold.) From Evans (1975a, Fig. 10).

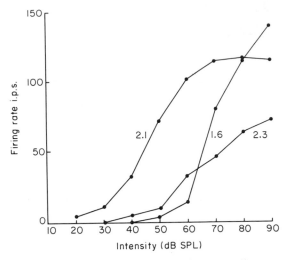

Fig. 4.6   Rate-intensity functions are shown for one auditory nerve fibre for different frequencies of stimulation. At the characteristic frequency (2.1 kHz) the fibre goes from threshold to saturation in about 40 dB. The maximum firing rate to the 1.6-kHz stimulus is greater, and that to the 2.3-kHz stimulus is less, than to the tone at the characteristic frequency. Parameter on curves: frequency of stimulation in kHz. From Pickles (1986a, Fig. 2.4B).

corresponds to a narrow bandwidth. Figure 4.5B shows that for the cat the minimum relative bandwidth occurs around 10 kHz, where it averages about one-eighth of the characteristic frequency. Comparable values have been measured for basilar membrane and hair cell responses (Sellick *et al.*, 1982; Robles *et al.*, 1986a; Russell and Sellick, 1978).

    The tuning curves of Fig. 4.3 show the intensity necessary to raise the firing rate above the spontaneous rate by a certain criterion amount, plotted as a function of frequency. We can also measure the firing rate as a function of intensity for different frequencies, giving rate-intensity functions (Fig. 4.6). The functions show a sigmoidal shape, saturating at each frequency at an intensity some 20–50 dB above the threshold at that frequency. Thus the dynamic range at any one frequency is limited to 20–50 dB. In some fibres the dynamic range is a little greater for frequencies above the characteristic frequency (Nomoto *et al.*, 1964; Evans, 1975a). The maximum, or saturated, firing rate depends on the stimulus frequency. The maximum rate for frequencies below the characteristic frequency commonly occurs at higher rates, and the maximum rate at frequencies above the characteristic frequency commonly occurs at lower rates, than at the characteristic frequency itself (Fig. 4.6). The mechanical responses of the basilar membrane and the electrical responses of hair cells show a similar dependence upon stimulus frequency (Figs 3.10C and 3.20).

The suprathreshold response can be plotted in three ways. We can, as in Fig. 4.6, plot the firing rate at a constant frequency for different intensities of stimulation (rate-intensity functions). We can continue the analogy of the frequency threshold curve to higher firing rates by plotting the combinations of intensities and frequencies necessary to evoke a constant increment in firing rate, giving iso-response or iso-rate contours (Fig. 4.7A). Or the firing rate can be plotted as the frequency is varied, the curves being called iso-intensity plots (Figs 4.7B–D). Each of the ways is best for showing a different property. The iso-rate, iso-response or tuning curves are best at showing the degree of frequency selectivity, at least for intensities below saturation. The iso-rate or iso-response curves show that the frequency selectivity generally improves a little as a higher rate criterion is used (Fig. 4.7A), although it later deteriorates as the fibre saturates. The iso-intensity functions show that the frequency evoking the highest firing rate can shift as the intensity is raised, moving upward for fibres with characteristic frequencies below 1 kHz, and downwards for fibres with characteristic frequencies above 1 kHz (Figs 4.7B and C). The curves in Fig. 4.7A are similar to the iso-displacement curves for the basilar membrane in Fig. 3.10B, and those in Figs 4.7C and D are similar to the iso-intensity curves for the membrane in Fig. 3.10A.

(b) Temporal relations

At high frequencies, above 5 kHz, the nerve fibres fire with equal probability in every part of the cycle. At lower frequencies, however, it is apparent that the spike discharges are locked to one phase of the stimulating waveform. That is not to say that each fibre fires once every cycle: the fibres fire randomly, perhaps as little as once every 100 cycles on average. But when they do fire, they do so in only one phase of the stimulus. The phase-locking can be most easily demonstrated by means of a period histogram. In making a period histogram, the occurrence of each spike is plotted in time, but the time axis is reset in every cycle at a constant point on the stimulus waveform, perhaps at the positive zero crossings (Fig. 4.8). The period histogram appears to follow a half-wave rectified version of the stimulating waveform. It is reasonable to suppose that this corresponds to deflection of the cochlear partition in the effective direction. Deflection in the opposite direction reduces the spontaneous activity of the fibre. It will be shown later that this is the result of actual suppression of the activity by the stimulating waveform, rather than the effect of refractoriness from previous activity. One explanation for the loss of phase-locking at 5 kHz and above is that there is some

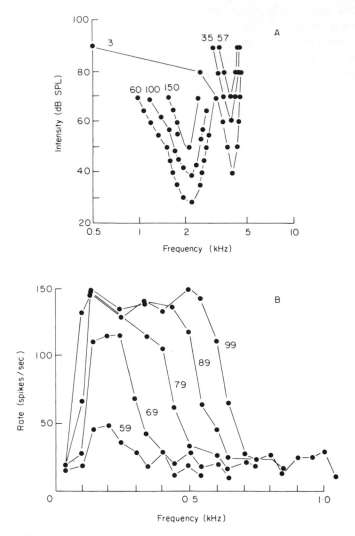

Fig. 4.7   (A) Tuning curves constructed at different firing rate criteria (rate shown by numbers on curves) become a little sharper as higher rate criteria are used. From Evans (1975a, Fig. 13).

jitter in the time of initiation of action potentials. While that may be true to some extent, it has also been suggested by Russell and Sellick (1978), on the basis of their hair cell records, that the phase-locking disappears when the cell's a.c. response becomes small in comparison with the d.c. response, owing to attenuation of the intracellular a.c. component by the capacitance

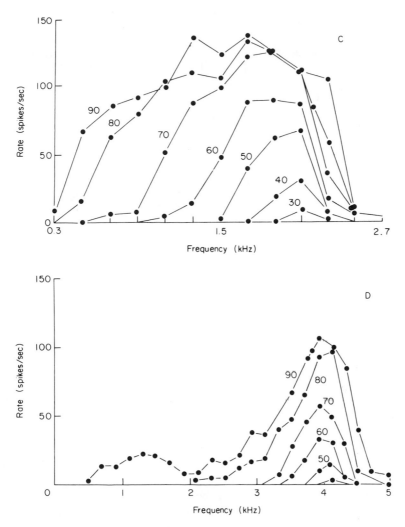

Fig. 4.7 (cont.)   (B–D) Iso-intensity functions for auditory nerve fibres show that at the lowest intensity the greatest response is produced by tones near the CF, but that at higher intensities the most effective frequency moves towards 1 kHz. In C the firing saturates at a lower rate at the CF than at lower frequencies. Numbers on curves: intensity in dB SPL. From Rose *et al.* (1971, Figs 1 and 2).

of the cell walls. In this case all the spikes would become initiated by the continuous d.c. depolarization of the cell, rather than the depolarizing half cycles of the a.c. response. Indeed, phase-locking in auditory nerve fibres

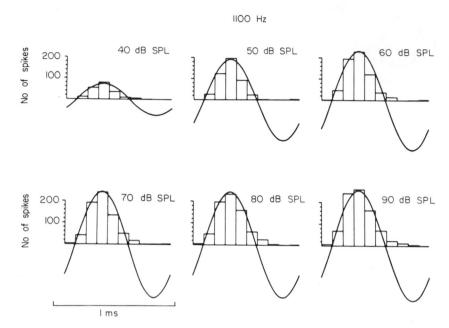

Fig. 4.8   Period histograms of a fibre activated by a low-frequency tone indicate that spikes are evoked in only one-half of the cycle. The histograms have been fitted with a sinusoid of the best fitting amplitude but fixed phase. Note that although the number of spikes increases little above 70 dB SPL, meaning that the firing is saturated, the histogram still follows the sinusoid without any tendency to square. From Rose *et al.* (1971, Fig. 10).

declines directly in proportion to the decline in inner hair cells' a.c./d.c. ratio (Palmer and Russell, 1986). Phase-locking and mean firing rates therefore can be thought of as being related to different aspects of inner hair cell function, the phase-locking being related to the a.c. component, and the mean firing rate to the non-linearity in the a.c. component, or in other words, to the d.c. response.

   Phase-locking is a sensitive indicator of the activation of a fibre by a low-frequency tone. At low stimulus intensities, a tone can produce significant phase-locking even though the mean firing rate is not increased. Tuning curves based on a criterion of a certain degree of phase-locking are similar to those based on an increase in firing rate, although for the above reason they may be more sensitive by 20 dB or so (Evans, 1975a). As the intensity is raised, phase-locking is preserved (Fig. 4.8). Note that, although the total number of spikes evoked does not increase above 70 dB SPL, meaning that the firing rate is saturated, the period histogram still follows the waveform of the stimulus, and does not show any sign of squaring. This occurs because the hair cell's a.c. response is still approximately sinusoidal at high intensities (Dallos, 1985).

2. Response to Clicks

A click, which lasts a short time, but which spreads spectral energy over a wide frequency range, can be thought of as the spectral complement of a tone, which lasts a long time but which has only a narrow frequency spread. Figure 4.9 show the post-stimulus–time histograms of the auditory nerve

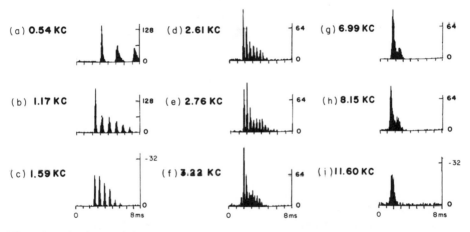

Fig. 4.9   The form of the post-stimulus–time histograms to clicks depends on the CF of the fibre. Low-frequency fibres show ringing (a–f), high-frequency fibres do not (g–i). High-frequency fibres also show a later phase of activation (f–h). Reprinted from *Discharge Patterns of Single Fibers of the Cat's Auditory Nerve* by N. Y.-S. Kiang *et al.* (Fig. 4.7), by permission of The MIT Press, Cambridge, Massachusetts. © The MIT Press, 1965.

fibres to clicks. The histograms of low-frequency fibres show several decaying peaks. It looks as though they would be produced by a decaying oscillation, i.e. as though the basilar membrane rings in response to the stimulus. The frequency of the ringing is equal to the characteristic frequency of the cell (Kiang *et al.*, 1965). This ringing at the characteristic frequency is exactly that expected if the tuning of the auditory nerve fibres were produced by an approximately linear filter. We would also expect the rate

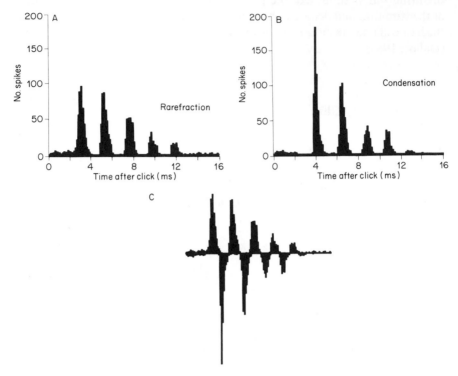

Fig. 4.10   (A) Post-stimulus–time histograms to (A) rarefaction and (B) conden-sation clicks show that the peaks and troughs occur in complementary places for the two stimuli. Fibre CF: 450 Hz. Reprinted from *Discharge Patterns of Single Fibers of the Cat's Auditory Nerve* by N. Y.-S. Kiang *et al.*, by permission of The MIT Press, Cambridge, Massachusetts. © The MIT Press, 1965. (C) A compound histogram is formed by inverting the histogram to condensation clicks under that to rarefaction clicks.

of decay of the ringing to be inversely proportional to the bandwidth of the tuning curve, so that a sharply tuned fibre would ring for a long time. Although this seems to be roughly true, there are some practical difficulties in making an exact comparison, because the number of spikes in the early peaks tends to limit, just as the response to tones saturates at high intensities.

As with the response to tones, it appears as though only one phase of the basilar membrane movement is effective. The histogram corresponds to half cycles of the decaying oscillation produced on the basilar membrane. It appears as though an upwards motion of the membrane is responsible for excitation, since at high intensities a rarefaction click produces the earliest response (Fig. 4.10A). A rarefaction click will move the oval window outwards, and so the basilar membrane upwards. A condensation rather than a rarefaction click reverses the positions of the peaks and troughs of the histogram, with the basilar membrane being driven in the opposite direction (Fig. 4.10B). An approximate picture of the excitatory oscillation can be produced by inverting the histogram for a condensation click under that for a rarefaction click, to produce what has been called a compound histogram (Fig. 4.10C). Histograms to clicks can also show that the suppression of activity during the less effective half cycle of the stimulating waveform is not due to refractoriness from previous activity, because the first sign of influence on a fibre can sometimes be a suppression of spontaneous activity produced by the less effective half cycle.

As the intensity is raised, earlier, previously subthreshold cycles of activation become effective, and so the histograms for low-frequency fibres shift to shorter latencies (Fig. 4.11A). Other complexities are also visible which cannot be fitted into the above scheme. Some fibres, which are of too high a frequency to show phase-locking, can show an earlier phase of activation at high intensities (Figs 4.11C and D). This must be of a different origin, since these units do not show phase-locking to the effective half cycles of the stimulus, and the early phase is more than $1/CF$ (CF = characteristic frequency) before the normal one. Some fibres also show a late phase in the post-stimulus–time histogram (Figs 4.11E, 4.9E–H), again far too late to be a result of a cycle of a decaying oscillation at the characteristic frequency of the fibre. The reasons for these changes in the shape of the post-stimulus–time histogram are not certain. Auditory nerve fibres of high characteristic frequency have a broad dip in the low-frequency tails of their tuning curves. A high-intensity click with significant low-frequency energy may therefore be able to superimpose a *low*-frequency oscillation on the oscillation at the characteristic frequency. This may cause the later or earlier phases of activation in high-frequency fibres.

3. Frequency Resolution as a Function of Intensity and Type of Stimulation

The tuning curves of Fig. 4.3 clearly show the degree of frequency resolving power of auditory nerve fibres; that is, they show the extent to which the fibres will respond to one tone rather than another on the basis of frequency. It is reasonable to suppose that this ability is fundamental to the frequency resolution shown by the auditory system as a whole. It is therefore important to consider the ways in which the fibres' frequency resolving power varies for different intensities and different types of stimulation.

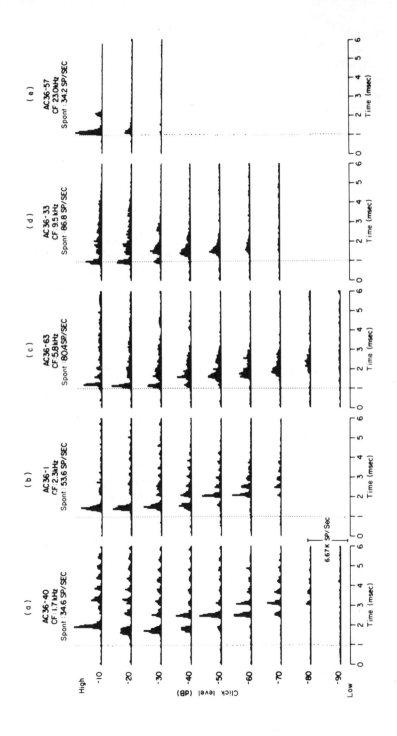

Fig. 4.11 Post-stimulus–time histograms to clicks as a function of click level. In (a) earlier cycles of the oscillation become effective. In (c–d) an early wave appears about 0.5 ms before the original peak. In (e) a late phase appears. From Antoli-Candela and Kiang (1978, Fig. 10).

(a) Frequency resolution with broadband stimuli

The tuning curves are produced in response to individual tones; yet we know that a considerable frequency selectivity must be shown in the response to clicks, which have broadband spectra, because of the ringing shown by fibres. In principle, we can calculate the tuning curve of the filter behind the ringing response by taking a Fourier transform of the compound histogram of Fig. 4.10C. If this is done, the frequency selectivity of the tuning curve is found to be approximately comparable to that determined with pure tones (Goblick and Pfeiffer, 1969; Pfeiffer and Kim, 1973). This shows that the frequency selectivity of the auditory nerve is, roughly at least, the same to a broadband stimulus as to a narrowband stimulus, and rules out theories that the sharp tuning is due to lateral inhibition. Lateral inhibition is widespread in sensory systems, and increases the contrast of the peaks and troughs of intensity in a sensory pattern; but a narrow click, which has a broad spectrum, has no spectral peaks and troughs, at least in the frequency range of interest.

A second way of showing sharp tuning with broadband stimuli involves the correlation of the firing pattern with the input stimulus, which is broadband noise. Broadband noise will of course stimulate the fibre, which will fire with an irregular pattern of discharge. We can imagine the noise as being made of a random collection of waves of different durations, phases and frequencies (Fig. 4.12). If there is a particular wavelet of just the right frequency to stimulate the nerve fibre in question, and if it is of sufficient amplitude, the fibre will be activated and an action potential will be recorded. The action potential will be phase-locked to the stimulating wavelet if the fibre's characteristic frequency is below 4–5 kHz. Of course, because the noise is random, other frequency components will be present, all of which will be in random phase and amplitude relations to the action potential. Therefore, if we add together all the samples of the original broadband noise occurring just before the recorded action potentials, all the component wavelets will cancel, *except* those which were in the right frequency and phase relations to fire the fibre. Mathematically, the resulting waveform turns out to be the same as the impulse response (i.e. the response to a click) of the nerve fibre, only reversed in time. This technique is known as reverse correlation and was first used by de Boer (de Boer and de Jongh, 1978). Once we have obtained the impulse response, we can perform a Fourier transformation on the impulse response to obtain the frequency response, or tuning curve. Figure 4.13 shows for two fibres the Fourier transforms calculated in this way, and compares them with the tuning curve obtained with pure tones. The transformed responses at all intensities in the 1-kHz fibre, and for the lowest noise intensities in the 2-kHz fibre,

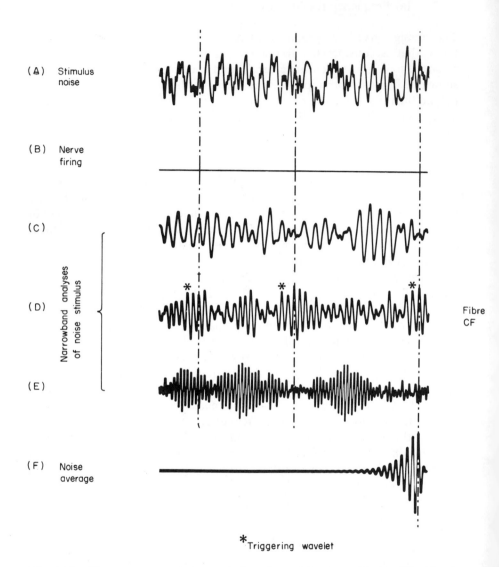

(A) Stimulus noise

(B) Nerve firing

(C)

Narrowband analyses of noise stimulus

(D)   Fibre CF

(E)

(F) Noise average

*Triggering wavelet

Fig. 4.12   The reverse correlation technique is explained graphically. The fibre is stimulated with noise (top trace) and gives action potentials (second trace). In the lower traces the results of passing the noise through narrowband filters of different centre frequencies are shown. One of the filters has the same centre frequency as the fibre. The bottom trace shows the result of adding together all the samples of the original noise waveform that triggered an action potential.

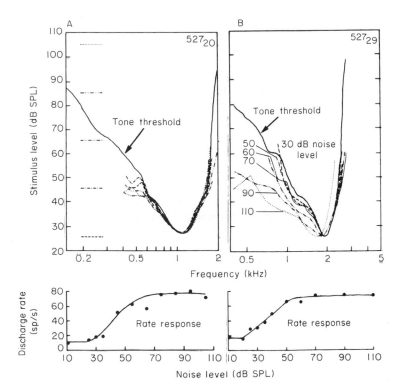

Fig. 4.13 Two fibres studied by the reverse correlation technique. (A) The Fourier transformation of the impulse response (broken lines), as recovered by the reverse correlation technique, shows good agreement with the tuning curve obtained with tones at threshold (solid line), over the bottom 15 dB of the tuning curve, even though the mean firing rate is 40 dB into saturation (rate response: bottom). (B) The 2-kHz fibre shows some deterioration of tuning as the intensity is raised, together with a downward shift in best frequency. Some tuning is still preserved, even though the intensity is 60 dB above that producing saturation of the firing. From Evans (1977, Figs 4 and 5).

showed excellent agreement with the pure-tone tuning curve, at least over the bottom 15–20 dB of the tuning curve, the range of intensities over which the method is reliable. Again, the result shows that the tuning to a broadband stimulus is approximately the same as that to a narrowband stimulus.

(b) Frequency resolution as a function of intensity

We are now in a position to assess how the frequency resolution of auditory nerve fibres changes with intensity. The width of the tuning curve is smallest at low intensities, and this has sometimes been taken to mean, erroneously, that the fibres' frequency resolving power is necessarily greatest near threshold, and deteriorates as the intensity is raised and the tuning curve becomes wider. This is not so: if the fibre is not driven into saturation the *relative* importance of stimuli near and away from the characteristic frequency is unchanged. Tuning curves constructed at higher firing rate criteria (iso-rate or iso-response curves) show the same or greater frequency selectivity as the intensity is raised (Fig. 4.7A). It is only when the fibre is driven into saturation that the frequency resolution as shown by the mean firing rate deteriorates. The fibre is now firing as fast as it can, and the firing rate does not change over a wide frequency range. This is shown by the flat tops of the iso-intensity plots of Figs 4.7B and C.

Does this mean, however, that the mechanism behind the fibre's filter function has become inoperative? It is possible, using the reverse correlation technique, to calculate the fibre's frequency resolving power in spite of a saturation of the firing rate. Such a calculation is possible because the reverse correlation technique depends only on the *accuracy* of the timing of the action potentials and not on the mean rate. As was shown in the period histograms of Fig. 4.8, the temporal relations of the firings are preserved unchanged at high intensities in spite of saturation of the mean firing rate. The results of the reverse correlation technique in Fig. 4.13 show that although the firing rates were saturated at noise intensities of 60 dB or so, the impulse responses of the fibres' filter functions were substantially unchanged for 40 dB above that. Transformation of the computed impulse responses showed that for the 1-kHz fibre the shape of the tip of the tuning curve was unchanged as the intensity was raised. At low intensities the computed frequency filter function of the 2-kHz fibre of Fig. 4.13B was comparable to that of the tuning curve. At 70 dB, however, when the fibre was 15 dB into saturation, the frequency resolution started to deteriorate, and the best frequency started to move to lower frequencies. Nevertheless, the fibre still showed substantial frequency resolution well into the range of saturation. The different behaviour of these two units appears to be the reflection of a general phenomenon. Møller (1977), using a comparable technique involving the correlation of the firing pattern with the noise stimulus, has shown that fibres with characteristic frequencies near 1 kHz did not change their frequency resolving power or best frequency as the noise level was raised 40 dB into saturation, whereas fibres of higher characteristic frequency deteriorated by up to 60% in their frequency resolving power and shifted to lower frequencies. The downwards shift in best frequency in high-frequency fibres, and the broadening of tuning, have correlates in the basilar membrane responses of Fig. 3.10A.

These studies are important for auditory physiologists because they show that the mechanism behind the frequency resolution of the auditory nerve is preserved at high intensities. We are therefore in a position to use the data to make theories about, for instance, the relation between basilar membrane motion and transducer resolution over a wide range of intensities. Whether the phenomena shown by these studies is as important for the animal itself is open to question. We do not yet know the extent to which the auditory system is able to retrieve the information in the temporal pattern of discharges, or whether it is able to response only to the *mean* rate of firing. Possible ways in which we are able to make detailed discriminations at these intensities will be discussed in Chapter 9.

4. Response to Complex Stimuli

(a) Two-tone suppression

It was stated above that single tones produce excitation in auditory nerve fibres, and never sustained inhibition. However, the presence of one stimulus can affect the *responsiveness* of nerve fibres to other stimuli, and if the relative frequencies and intensities of two tones are arranged correctly, the second tone can inhibit, or suppress, the response to the first. This can occur, even though the second tone produces no inhibition of spontaneous activity when presented alone. Figure 4.14 shows the post-stimulus–time histogram produced by a suppressing tone superimposed on a continuous excitatory tone. The pattern of response to the suppressing tone looks like the inverse of the pattern to an excitatory one. The suppressing tone produces an initial maximum of suppression when turned on, and produces a prominent rebound of activity when turned off. The dip in activity at the beginning of the suppressing tone looks like the transient suppression seen at the end of an excitatory stimulus, and the activity at the end looks like the onset burst seen at the beginning of an excitatory stimulus (cf. Fig. 4.2). This suggests that the suppressing tone simply turns the effect of the excitatory tone off. The fact that only stimulus-evoked, and not spontaneous, activity can be suppressed makes the same point. Arthur *et al.* (1971) made detailed measurements of the relative latencies of excitation and suppression. Although the latencies in individual fibres could differ either way by as much as 2.5 ms, on average excitation and suppression only differed in latency by 0.1 ms. These latencies suggest very strongly that suppression is not the result of inhibitory synapses in the cochlea, even if possible synapses had been demonstrated anatomically, because there is no time for synaptic delay (about 1 ms). The latency argument also means that the suppression cannot be a result of the activity of the olivocochlear bundle, the "feedback" pathway from the brainstem nuclei to the hair cells (see p. 236). Neverthe-less, Kiang *et al.* (1965) tested the possibility directly, and showed that

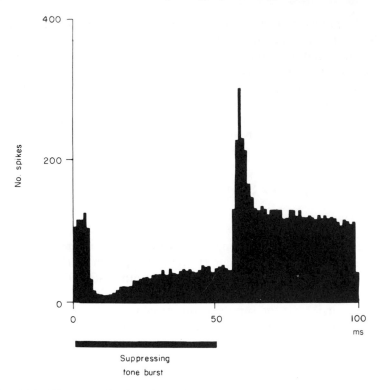

Fig. 4.14   The post-stimulus–time histogram of a suppressing tone burst superimposed on a continuous excitatory tone. Reprinted from *Discharge Patterns of Single Fibers of the Cat's Auditory Nerve* by N. Y.-S. Kiang *et al.*, by permission of The MIT Press, Cambridge, Massachusetts. © The MIT Press, 1965.

suppression survived the sectioning of the olivocochlear bundle. Because it is believed that two-tone suppression is not the result of inhibitory synapses, the more neutral term "suppression" rather than "inhibition" is often used. Of course, "inhibition" is still used for the process mediated by inhibitory synapses which is seen in the later stages of the auditory system, in for instance the cochlear nucleus. "Suppression" is only used for the process occurring in the cochlea. The interpretation of the mechanism is that the presence of one sound will affect the cochlea's responsiveness to another. It is now believed that the interference occurs right at the mechanical stage, and depends on the nonlinearity of the mechanics. Under the right circumstances, and with the correct analysis, a reduction of up to 60 dB in the effective driving intensity has been demonstrated (Javel, 1981).

   In support of the notion that two-tone suppression is a cochlear phenomenon, two-tone suppression can be demonstrated in the basilar membrane mechanics (Patuzzi *et al.*, 1984b; Robles *et al.*, 1986b) and in inner hair cells (Sellick and Russell, 1979). Recordings from hair cells have the advantage over those in the auditory nerve that the effects of the exciting and suppressing tones can be assessed separately. In the hair cell shown in Fig. 4.15, the excitatory tuning curve was first assessed from the d.c. response to an excitatory tone. The cell was then stimulated with a continuous tone (the probe), having the intensity and frequency indicated by the triangle in Fig. 4.15. A suppressive tone was then swept across the response area, and

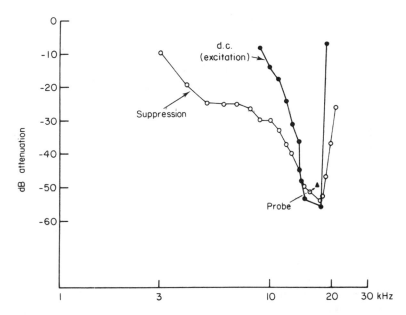

Fig 4.15 Excitatory and suppressive tuning contours for an inner hair cell. The d.c. contour is the d.c. depolarization to a single tone and shows the excitatory tuning curve. The triangle shows the frequency and intensity of the probe tone, and the open circles the contour for 20% suppression of the a.c. response to the probe. All stimuli within the excitatory contour excite, and all within the suppressive contour suppress. Stimuli within both contours both suppress and excite. From Sellick and Russell (1979, Fig. 3).

the contours for 20% suppression of the a.c. response at the *probe's* frequency, determined (Sellick and Russell, 1979). The results show that the suppressing tone can reduce the response to the exciting tone when presented over a wide range of frequencies. The suppressive area is more broadly tuned than the excitatory response area, and overlaps it at the tip. In other words, a stimulus can suppress even though it does not excite, and can still suppress the response to another stimulus, even though it excites when presented alone. The overlap of excitatory and suppressive areas that has been demonstrated in inner hair cells can also be shown in auditory nerve fibres, for tones of low frequencies, by taking advantage of the fact that the firing will follow the waveform of an exciting stimulus. If two tones are presented, the firing will follow the waveform of the sum of the two in an appropriate combination of amplitude and phase. By looking at the degree of modulation of the firing pattern at the frequency of one tone in a complex, it is possible to calculate the degree to which the fibre is activated by that frequency component and to measure the extent to which the response to that component is suppressed by the other stimulus. In this way, Javel *et al.* (1983) showed that the narrow excitatory response area was overlaid by a broader suppressive area.

If, of course, the second tone is in the suppressive area but outside the excitatory area, it will be easy to measure the suppression by measuring the total firing rate to the stimulus complex. The second tone produces only suppression, and does not contribute any excitation. The overall mean firing rate will then be a measure of the activation produced by the excitatory tone and the extent to which it is suppressed. Plots of the combinations of intensity and frequency necessary to reduce the mean firing rate in response to a constant excitatory tone by a certain criterion amount (20% has been commonly taken) show the suppressive areas where they flank the excitatory area (Sachs and Kiang, 1968; Arthur *et al.*, 1971; Fig. 4.16). When the suppressing tone reaches the boundary of the excitatory area it will begin to activate the fibre on its own account, and so the total number of action potentials will increase. Suppression areas plotted in this way therefore stop at, or near, the boundary of the excitatory area.

The detailed mechanism of two-tone suppression is a matter of debate. Travelling waves on the basilar membrane behave nonlinearly in their peak region, with a saturating nonlinearity (Fig. 3.10C). This means that the response to one stimulus is reduced in the presence of another. Such a nonlinearity explains two-tone suppression in the case where both tones are in the excitatory part of the tuning curve, such that their travelling waves overlap. Similarly, we can suggest how higher-frequency tones might be able to suppress, even though they do not excite (e.g. in the upper shaded two-tone suppression area of Fig. 4.16). A high-frequency suppressor will produce a travelling wave on the basilar membrane, with a peak basal to that produced by the exciter tone. It is possible that the travelling wave produced by the lower-frequency tone is reduced in amplitude as it passes through the suppressor tone's peak region. Further up the cochlea, the travelling wave to the high-frequency suppressor will be filtered out by the mechanics, leaving the wave to the lower tone, but at a reduced amplitude. Note, however, that there are some problems with this explanation, which

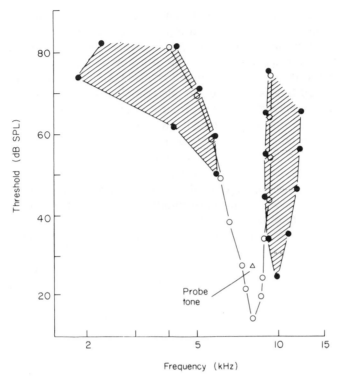

Fig. 4.16 The suppression areas of an auditory nerve fibre (shaded) flank the excitatory tuning curve (open circles). A stimulus in the suppression areas was able to reduce the mean firing rate found with the probe by 20% or more. From Arthur *et al.* (1971, Fig. 2).

will be pointed out in Chapter 5. In addition, this explanation of how a stimulus can suppress without causing excitation does not apply to the *low-frequency* suppression area (Fig. 4.16). The detailed mechanism of two-tone suppression is at the moment controversial, and will be discussed further in Chapter 5.

Figure 4.16 indicates that a stimulus is able to reduce the driven response of fibres tuned to neighbouring frequencies. Two-tone suppression is therefore able to increase the contrast in a complex sensory pattern, so that, for instance, the peaks of activation produced by dominating frequencies will tend to stand out in stronger contrast against the background. The possible role of two-tone suppression in discrimination will be discussed in Chapter 9.

(b) Masking

Masking denotes the general phenomenon which one stimulus obscures or reduces the response to another. Two-tone suppression could therefore provide one mechanism of masking. There is in addition another, and probably more important, mechanism of masking operative as well.

An important mechanism of masking is known as the "line busy" effect. If one stimulus has pre-empted the firing of a fibre, superimposed stimuli will not be able to provoke an increment in firing. In one statement of the hypothesis, if the firing is saturated to one stimulus, superimposed stimuli will not be able to increase the rate further (Smith, 1979). This mechanism will also be operative to some extent below saturation. If one signal has a greater effective intensity than the other, the less intense one will add negligible activity of its own. Such an effect will be greater than might appear at first sight, because the summation of effective intensities will occur on a linear scale rather than the logarithmic scale of decibels. For instance, a signal added 10 dB below another will produce an increase in net stimulus intensity of only 0.4 dB.

Figure 4.17 shows an example of masking by the line busy effect as well as masking by suppression. The figure shows rate-intensity functions for a tone both with and without wideband masking noise. The noise alone produced a firing rate of 160/s. Tones less intense than 50 dB SPL did not produce a greater firing rate than this, and so did not increase the response. In this intensity range, the tone was masked by the line busy effect.

Figure 4.17 also shows masking by suppression. Noise reduced the

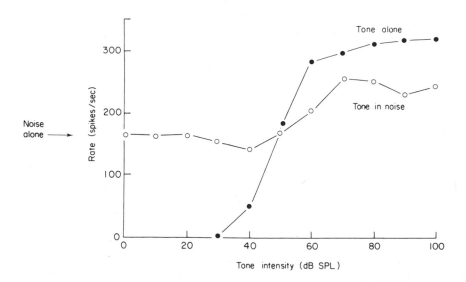

Fig. 4.17   Rate-intensity functions to a tone with and without masking noise. The tone was presented at 2.9 kHz, the CF. Noise band: 2.5–4 kHz. Adapted from Rhode *et al.* (1978, Fig. 3E).

maximum firing rate to the tone, even though the tone was able to produce a clear increase in firing (at 60 dB SPL and above). This reduction, which is another case of masking, had the characteristics of two-tone supression. For instance, at moderate intensities two-tone suppression is strongest for suppressors above rather than below the characteristic frequency, and in this experiment noise bands above the fibre's characteristic frequency were most effective at reducing the firing. In general, we can expect that where maskers fall on the two-tone suppression areas, suppression will play a part.

We are now in a position to understand some of the complex interactions between the components of multicomponent stimuli. If two stimuli of comparable levels are presented, both well inside the excitatory area, each will suppress the other strongly, but both will excite the fibre even more strongly. Below saturation, the firing rate in response to both will be greater than that to either one alone, and they will therefore appear to summate in their effects. For low-frequency units, the firing will follow a waveform which can be composed of the waveforms of the component stimuli added together with suitable amplitudes and phases (Rose *et al.*, 1971). The relative amplitudes and phases giving the best fit are not necessarily those presented in the acoustic stimulus. The frequency selectivity of the fibre, as well as the mutual suppression of the components, will alter their relative amplitudes. In the case of the phases, the effects of suppression are complex and not entirely understood (e.g. Patuzzi *et al.*, 1984a; see also Chapter 5).

In a more trivial case, where one stimulus has much less influence over the fibre than the other, the most effective stimulus will dominate both the firing rate and the temporal pattern of the action potentials.

If one stimulus is moved away from the characteristic frequency, both its excitatory and suppressive effects will decline, but the excitatory effects will decline the more rapidly and the balance will be tipped in favour of suppression. A stimulus away from the characteristic frequency will suppress the response to a stimulus near the characteristic frequency, but will add only a little activation of its own. The overall firing rate will therefore be smaller than that to the most excitatory stimulus alone. Now the firing predominantly follows the waveform of the suppressor (Rose *et al.*, 1971). In the terminology of Rose *et al.*, the suppressing tone now dominates the response.

Some of the different effects of suppression and summation can be seen in the mean firing rates to bands of noise of different widths (Fig. 4.18). As a narrow band of noise of constant spectral density is widened around the characteristic frequency of a fibre, the first effect is for the firing rate to increase, because the greater number of noise components in the excitatory area summate and drive the fibre more intensely (Gilbert and Pickles, 1980). As the bandwidth becomes still broader, the extra noise components added

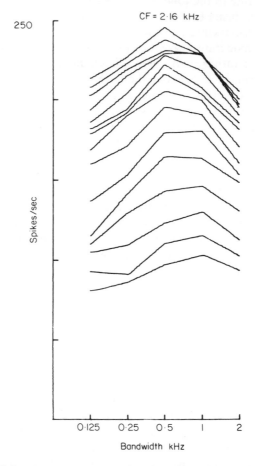

Fig. 4.18   The firing in response to a band of noise shows the effects of both summation and suppression. The noise was of variable bandwidth kept centred on the CF. The contours are at 3-dB intervals of noise spectral density, the lowest being at −25 dB SPL/Hz. From Gilbert and Pickles (1980, Fig. 2).

come to fall on the parts of the suppressive area outside the excitatory area, and now not only fail to contribute excitation, but contribute a net suppression. The firing rate therefore comes to a maxium, and then declines.

(c) Combination tones

If the ear is stimulated with two tones at the same time, combination tones may be heard which are not physically present in the stimulus. The presence of combination tones was first demonstrated psychophysically rather than physiologically. It is thought that they occur as a result of a nonlinearity in the cochlea, almost certainly in the basilar membrane mechanics. One combination tone is the difference tone, which is at a frequency $f_2 - f_1$, where f_2 and f_1 are the frequencies of the two tones, or primaries, presented. The level of the difference tone at high signal levels is almost completely independent of the frequency separation of the primaries. It was once thought that it originated as an overloading type of distortion in the middle ear (e.g. Helmholtz, 1863). Now the direct measurements of Guinan and Peake (1967) have shown that the middle ear has insufficient nonlinearity, and an intracochlear origin is suspected.

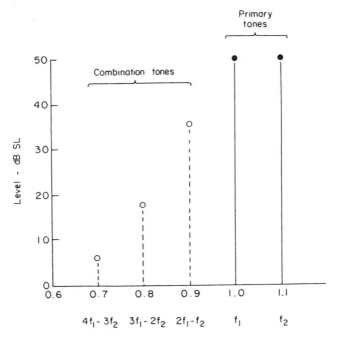

Fig. 4.19 The subjective cubic and related combination tones form a series below the primaries, of frequency spacing equal to the separation of the primaries. The levels indicated were those found by the cancellation method for primaries of 1 and 1.1 kHz, according to Goldstein (1967); 0 dB SL $\approx$ 0 dB SPL.

A second set of combination tones has certain fascinating properties and can be heard even at low sound levels. The best known representative of this group is the tone known as the cubic distortion tone or $2f_1 - f_2$, from the frequency at which it is heard; f_2 must be above f_1. Figure 4.19 shows the frequency and level of the cubic distortion tone in relation to the primaries. In this experiment, the level of each distortion tone was measured by introducing a third, cancellation tone into the stimulus, and altering its level and phase until it just cancelled the sensation of the distortion tone.

The frequency at which $2f_1 - f_2$ appears might be explained by supposing that the auditory system undergoes a distortion such that the output contains a component that is the cube of the input. If two tones of frequency f_1 are f_2 are inserted, the output distortion component $= (\cos 2\pi f_1 + \cos 2\pi f_2)^3$. This can be decomposed into a series of cosines containing frequencies such as f_1, $3f_1$, f_2, $3f_2$, $f_1 + f_2$, $2f_2 + f_1$, $2f_1 - f_2$, and $2f_2 - f_1$. Of these, only $2f_1 - f_2$ is at a frequency below the primaries and we can provisionally suppose that the others are masked or filtered out by some hypothetical high reject filter later in the auditory system. Models behind the generation of $2f_1 - f_2$ will be discussed in Chapter 5. At the moment we can note that the name "cubic distortion tone" arises because the tone can hypothetically be produced by a cubic distortion. However, it can also be produced by nonlinearities such as those shown for the basilar membrane amplitude response in Fig. 3.10C. If the basilar membrane saturating input–output function is expressed as a power series:

$$\text{output} = a_1.\text{input} + a_2.\,\text{input}^2 + a_3.\,\text{input}^3 + a_4.\,\text{input}^4 + a_5.\,\text{input}^5$$

where the a_i are constants, the cubic term will give rise to the combination tones described above, and the other odd-exponent terms will in addition give rise to terms such as $3f_1 - 2f_2$ and $4f_1 - 3f_2$, for the distortion components below f_1 and f_2 in frequency. Such additional distortion components are detectable psychophysically, although at a lower amplitude than $2f_1 - f_2$ (see Fig. 4.19).

Three important psychophysical results about the cubic distortion tone concern the physiologist. One is that the amplitude of the cubic distortion tone, in contrast to that of the high-level difference tone, is strongly dependent on the frequency separation of the primaries, declining at a rate of some 100 dB/octave as the frequency ratio of the primaries increases (Goldstein, 1967). This suggests that the site of distortion is preceded by some stage of frequency analysis. Secondly, the cubic distortion tone is heard for stimuli near threshold. This suggests that it is not merely an "overloading" type of distortion, but must be regarded as part of the normal operation of the auditory system. Thirdly, as measured by the cancellation method, the relative amplitude of the distortion tone with respect to the primaries is almost independent of the amplitude of the primaries if they are both varied in level together.

The search for a correlate of the cubic distortion tone in the responses of fibres of the auditory nerve, has been reported by Goldstein and Kiang (1968) and Kim *et al.* (1980). Goldstein and Kiang presented two tones f_1

and f_2, such that both lay outside the response area of the fibre, neither provoking a response when they were presented singly. If however the calculated frequency $2f_1 - f_2$ lay at the characteristic frequency of the fibre, it was possible for the fibre to be excited by both stimuli together. In other words, the fibre could be excited by the combination tone, in the absence of a response to the primaries. The same point was made by studies of the phase-locking to the stimuli for low-frequency fibres; it was possible to show significant phase-locking to $2f_1 - f_2$ without any phase-locking to f_1 or f_2. Moreover, Kim *et al.* (1980) showed that the fibres' frequency selectivity to $2f_1 - f_2$ was the same as that to introduced real tones of the same frequency. This shows that as far as the auditory nerve was concerned, it was as though a real tone at the frequency $2f_1 - f_2$ was present in the stimulus. In support of this, it was possible to cancel the response to the combination tone by the addition of a third tone, suitably adjusted in amplitude and phase. Studies of the amplitude behaviour of the combination tone showed a correlate with the psychophysical result: the amplitude of the combination tone calculated from the firing rate was constant relative to the amplitude of the primaries, in one case over a range of over 60 dB (Goldstein and Kiang, 1968). In addition, the response to the combination tone could be found at intensities just above the threshold of the fibre. Neither of these results is consistent with the idea that the response reflects the operation of a high-intensity, overloading type of distortion in the inner ear. In its frequency relations, too, the physiological response paralleled the psychophysical results, since the response to the combination tone dropped rapidly as the frequency separation of the primaries increased. It was only in the phase relations that a clear difference emerged between the psychophysical and physiological results. The psychophysical cancellation tone shifted strongly in phase with changes in stimulus level, but the phase of the physiological distortion tone was almost invariant (Goldstein and Kiang, 1968). The reason for the difference is not clear.

It is now thought that the cubic distortion tone is produced by the mechanical nonlinearity of the basilar membrane, as it appears in Fig. 3.10C, and which is present at low levels. It is also thought that the cubic distortion tone has its own travelling wave on the basilar membrane. Studies of the cubic distortion tone and related phenomena have great significance for our understanding of cochlear physiology. These studies will be discussed further in Chapter 5.

C. Summary

1. The very great majority of the fibres present in the auditory nerve innervate inner hair cells.

2. Single fibres of the auditory nerve are always excited by auditory stimuli, and never show sustained inhibition to single stimuli.

3. The fibres have lower thresholds to tones of some frequencies than of others. The relation between threshold and stimulus frequency is known as the "tuning curve". Turning curves show one threshold minimum, at what is known as the "characteristic frequency". The threshold rises sharply for frequencies above and below the characteristic frequency. The tuning curve therefore shows a sharp dip in this frequency region. Tuning curves of auditory nerve fibres are similar to those of hair cells and the basilar membrane mechanics.

4. The great majority (80%) of auditory nerve fibres have minimum thresholds in a 20-dB range near the animal's absolute threshold. The others have thresholds spread over a 60-dB range above that. The low threshold fibres have particularly high rates of spontaneous activity in the absence of sound.

5. Fibres show a sigmoidal relation between firing rate and stimulus intensity, in many cases going from threshold to maximum rate (saturation) in 20–50 dB at any one frequency.

6. The frequency resolving power of auditory nerve fibres has been measured by a "quality" factor, by analogy with a quality factor for filters. The quality factor is the characteristic frequency, divided by the bandwidth of the fibre to tones at an intensity 10 dB above the best threshold. This is called "Q_{10}". Therefore, fibres with a high Q_{10} have good frequency selectivity. At any one frequency, different fibres have Q_{10}'s at in a restricted range. In any one animal, the range of Q_{10}'s at one frequency is two-fold or less. The greatest Q_{10}'s in the cat are reached at around 10 kHz, where they have an average value of eight.

7. During tonal stimulation, auditory nerve fibres fire preferentially during one part of the cycle of the stimulating waveform if the stimulus is below 4 – 5 kHz. The fibres are excited by deflection of the basilar membrane in only one direction.

8. For fibres with characteristic frequencies below 4 – 5 kHz, clicks preferentially evoke responses at certain intervals after the stimulus. A histogram of action potentials made with respect to time after the stimulus suggests that the fibres are activated by the half cycles of a decaying oscillation of the mechanical resonance on the basilar mem-

brane. The frequency of the oscillation is equal to the characteristic frequency of the fibre.

9. One tone can reduce, or suppress, the response to another, even though single tones are only excitatory. This is called two-tone suppression. The suppression arises from the nonlinear properties of the basilar membrane mechanics. Two-tone suppression can also be seen in the basilar membrane mechanics and the responses of inner hair cells. Stimuli other than tones cause suppression too.

10. One stimulus can mask the response to another. Masking mainly occurs because the masking stimulus produces a greater firing rate than the masked stimulus. The other mechanism of masking is the suppression of the response to one stimulus by another.

11. When two-tone stimuli are used, auditory nerve fibres can respond to distortion products as a result of nonlinear interactions in the cochlea. One distortion tone, known as the cubic distortion tone, is at a frequency $2f_1 - f_2$, where f_1 is the lower of the tones presented, and f_2 the higher.

D. Further Reading

The auditory nerve has been reviewed by Evans (1975a). More recent information on the auditory nerve has been reviewed by Manley (1983), Javel (1986), Irvine (1986) and Pickles (1986a).

5. *Mechanisms of Transduction and Excitation in the Cochlea*

The study of cochlear processing is currently the most exciting area of auditory physiology. New and revolutionary ideas are rapidly replacing established views. It is hoped that this chapter will be able to convey some of the flavour of an area that is occupying the attention of a large proportion of auditory physiologists today. We have new and detailed information on hair cell structure and electrophysiology, giving us the hope that full evaluation of the transduction process will be possible within the next few years. The recent measurements of sharply-tuned mechanical resonances on the basilar membrane have produced new basic data for those trying to explain the basis of cochlear frequency resolution; the only theories that have been successful in modelling sharp tuning have involved an active mechanical process, and evidence on the existence of active mechanical processes, and their possible origin, will be discussed. Finally, some complexities in the relation between basilar membrane vibration and nerve activation will be evaluated, as will the possible basis of two-tone suppression in the cochlea. The chapter is written at a more advanced level than the rest of the book, and may be omitted without affecting the comprehensibility of the other chapters.

A. Introduction

Recent electrophysiological measurements suggest that deflection of the stereocilia opens ion channels near the tips of the stereocilia. Ions are driven into the hair cells by the combined effects of the positive endocochlear potential, and the negative intracellular potential. Intracellular depolarization causes the release of transmitter at the synapse, evoking action potentials in the afferent auditory nerve fibres.

112

The cytoskeletal structures in the apical portions of the hair cells are important in determining the electrophysiological responses of the hair cells to mechanical stimulation: the way in which the stereocilia deflect in response to mechanical stimulation, and the way in which the movement is transferred to the hypothesized sites of the transducer channels, are dependent on the mechanical properties of the stereocilia and the way in which the stereocilia are coupled together. The stereocilia appear to be extremely rigid, composed of closely-packed actin filaments, and bending as stiff levers at their points of insertion into the cuticular plate. A shear is developed between the different rows of stereocilia on the hair cell, and is coupled to fine links, of macromolecular dimensions, which are likely to transmit the movement to the transducer channels themselves.

There is in addition the possibility that hair cells are motile, and that they actively generate movements when stimulated. The basis of this motility is not known: it may possibly depend on the cytoskeleton in the apical portions of the hair cell, or on specialized areas of the basal membranes of the hair cell.

B. The Cytoskeleton of the Transducer Region

1. Stereocilia and Cuticular Plate

Flock and Cheung (1977) originally showed that stereocilia were composed of packed actin filaments. This can be shown by decorating the filaments with the heavy heads of myosin molecules, known as the S1 fragment. Actin filaments can form an association with the S1 fragments of myosin molecules, in the same way as actin and myosin associate in muscle. S1 decoration shows a characteristic arrowhead pattern, with the arrowheads pointing downwards into the cell body (Fig. 5.1). The arrowheads also point into the cell body after S1 decoration of intestinal microvilli, which are also composed of actin filaments. The presence of actin can also be shown by immuno-fluorescent labelling under the light microscope, or by gold-coupled immunolabelling under the electron microscope (e.g. Zenner, 1981; Slepecky and Chamberlain, 1985a).

The actin filaments within the stereocilia are packed into what is known as a paracrystalline array (Tilney *et al.*, 1980). In the chick cochlea, the filaments are closely packed so that in cross-sections of the stereocilia the filaments appear in a hexagonal pattern. Adjacent actin filaments are bonded together, and longitudinal sections of stereocilia show transverse striations corresponding to the planes in which the bonding occurs. The patterns change according to the direction in which the sections are cut.

Fig. 5.1 Actin filaments are demonstrated in a stereocilium of the lizard cochlea, by means of decoration with the S1 fragment of myosin. Some filaments end near the stereociliar membrane where it tapers just above the cuticular plate, while others continue down into the rootlet. The cuticular plate is towards the lower part of the figure. Reproduced from Tilney *et al.* (1980, Fig. 13) by copyright permission of the Rockefeller University Press.

Figure 5.2A shows how in one view the sideways striations appear evenly and closely spaced, while in Fig. 5.2B, cut in a different longitudinal plane, the striations appear in pairs, separated by gaps.

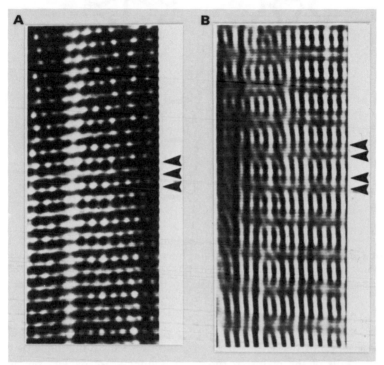

Fig. 5.2   (A) A stereocilium was cut longitudinally, and the pattern of transverse striations emphasized by digital filtering. The striations are evenly spaced at 12.5-nm intervals. (B) In a stereocilium cut longitudinally at another angle to the actin paracrystal, the striations after enhancement appear in an alternating pattern, with spaces at 12.5- and 25-nm intervals. Reproduced from Tilney *et al.* (1983, Fig. 8) by copyright permission of the Rockefeller University Press.

The pattern of striations can easily be explained by supposing that actin filaments bond wherever subunits from adjacent filaments are aligned appropriately. Since actin filaments are twisted, subunits in the six nearest filaments will be aligned three times per half-twist of the filament, if the filaments are in register in the longitudinal direction (Fig. 5.3). One half-twist of an actin filament takes 37.5 nm, and therefore bonding can occur at 12.5-nm intervals. This is the separation of the horizontal striations in Fig. 5.2A. If the direction of view is changed, so that one set of bonds is seen end-on, those links will not produce a cross-striation. The pattern changes to that shown in Fig. 5.2B, with two striations separated by 12.5

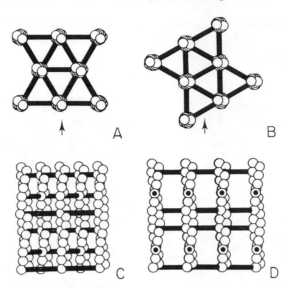

Fig. 5.3   Model of actin filaments in a hexagonally-packed stereocilium. (C) In the 1,1 view (in direction of arrow in A) the inter-filament bonds appear as striations which are approximately evenly spaced. (D) In the 1,0 view (in direction of arrow in B) the bonds appear in pairs, separated by a gap. From Tilney *et al.* (1983, Fig.12).

nm and then a gap of 25 nm. Figure 5.3 shows how the pattern depends on the orientation of the actin paracrystal with respect to the angle of view.

The details of the way that the actin filaments are packed seems to be species-dependent. In the cochlea of the alligator lizard the actin filaments are less closely and regularly spaced, so that fewer bonds are possible. This presumably will lead to a bundle with a lower rigidity. The packing in mammalian stereocilia has not yet been worked out in detail, although Itoh (1982) showed that there was close hexagonal packing in the central core and in the rootlet region of guinea-pig stereocilia.

A high degree of bonding of adjacent actin filaments can be expected to give substantial rigidity to the stereocilia, and indeed micromanipulation experiments have shown that stereocilia act as stiff levers, bending only at the point of insertion into the cuticular plate, and fracturing as though brittle when pushed too far (Flock, 1977; Flock *et al.*, 1977). After acoustic overstimulation, the paracrystal is seen to be disordered, with an irregular spacing of the filaments in longitudinal view, and loss of the horizontal striations (Tilney *et al.*, 1982). The regions of disorder are associated with a bending or kinking of the stereocilia. This suggests that the mechanical rigidity of the stereocilia is indeed associated with the integrity of the paracrystal.

Other cytoskeletal proteins have been described in stereocilia by means of immunological techniques. Thus, Flock *et al.* (1982) demonstrated the presence of fimbrin, a microfilament bundling protein, in stereocilia of the guinea-pig cochlear and vestibular systems. The fimbrin may therefore be responsible for cross-linking the actin filaments. Macartney *et al.* (1980) showed the presence of myosin, which with actin gives motility in muscle cells. However, this was not confirmed by Drenckhahn *et al.* (1982) and Flock *et al.* (1982).

The stereocilium tapers at its lower end, and continues into the cuticular plate as a dense rootlet. About 1% (counted in the lizard basilar papilla; Tilney *et al.*, 1980) of the actin filaments in the stereocilium enter the rootlet in this way, the rest ending in association with the membrane of the stereocilium where it tapers (Fig. 5.1). The narrowing of the stereocilium just before it enters the cuticular plate would allow it to flex at this point.

The cuticular plate is composed of a dense network of actin filaments. The filaments of the rootlet are bonded to the cuticular plate by means of fine 3-nm filaments (Hirokawa and Tilney, 1982). In sections cut parallel to the surface of the plate the 3-nm filaments are see to run radially outwards from the rootlet like the spokes of a wheel (Tilney *et al.*, 1980). According to Flock (1983) and Slepecky and Chamberlain (1985a), tropomyosin is found in this region, while α-actinin (a component of the muscle Z-line), myosin and Ca^{2+}-binding protein are found generally in the cuticular plate (Flock *et al.*, 1982; Drenckhahn *et al.*, 1982; Rabie *et al.*, 1983; Macartney *et al.*, 1980). It has been suggested that spectrin, originally known as a component of the erythrocyte membrane, bonds the matrix of the cuticular plate to the overlying membrane (Drenckhahn *et al.*, 1985). The dense matrix of interlinked actin filaments in the cuticular plate would be expected to give the plate considerable rigidity, although the presence of proteins normally associated with muscle motility suggests that the mechanical properties might be modifiable.

In contrast to the apparently haphazard matrix of filaments in the cuticular plate, organized actin filaments can be found running in a ring-like arrangement just inside the zonula adherens at the apical end of the hair cell (Flock *et al.*, 1981; Hirokawa and Tilney, 1982). The rings contain actin of opposite polarities. Some of the filaments are arranged in hexagonal assemblies around a dense core, an arrangement reminiscent of that of actin and myosin in muscle. Therefore, Flock *et al.* (1981) suggested that the structures contained actin and myosin, arranged so as to give constriction of the apical end of the hair cell.

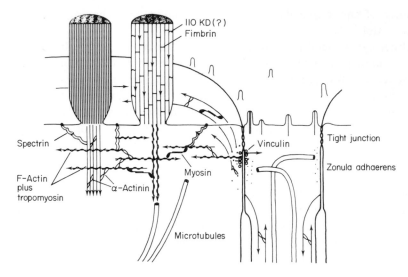

Fig. 5.4   The arrangement of actin filaments in the apical regions of hair cells. Arrows show the direction of the arrowhead complexes formed after decoration with S1-myosin. The links between the actin filaments are also shown. Dots show actin filaments end-on. From Drenckhahn *et al.* (1985, Fig. 18.40, slightly modified). Courtesy of Charles C. Thomas, Publisher, Springfield, Illinos.

The suggested arrangements of actin filaments and associated cytoskeletal structures in the apical region of the hair cells is shown in Fig. 5.4.

2. The Cross-linking of Stereocilia

The stereocilia in a hair bundle are heavily cross-linked in a variety of ways. First, stereocilia are bonded together sideways by links that run predominantly parallel to the cuticular plate. The links run between the stereocilia of the different rows on the hair cell as well as between the stereocilia of the same row (e.g. Spoendlin, 1968; Pickles *et al.*, 1984). These links probably serve to couple the stereocilia mechanically, with the result that in micromanipulation experiments all stereocilia in a bundle tend to move together when some are pushed (Flock and Strelioff, 1984). Figure 5.5 shows an example of the sideways links running between stereocilia of the same row on an inner hair cell, and Fig. 5.6 shows links joining stereocilia of different rows. The latter links are concentrated in a band just below the tips of the shorter stereocilia, and hold the tips of the shorter stereocilia in near the adjacent taller stereocilia. The rows of stereocilia therefore make triangles when seen in sideways view.

Fig. 5.5 Inner hair cell of the guinea-pig cochlea, showing links (arrows) joining stereocilia of the tallest row on the hair cell. Note also that the surface membranes of the stereocilia appear rough, particularly at the level of the links. Scale bar: 500 nm.

The links of a second set are rather different, and have generated interest because they may couple the stimulus-induced movements to the transducer areas of the stereocilia (Pickles *et al.*, 1984). A single vertically-pointing link emerges from the tip of each shorter stereocilium on a hair cell, and runs up to join the adjacent taller stereocilium of the next row. Figure 5.6 shows these tip links by transmission electron microscopy, and Fig. 5.7 by scanning electron microscopy. Each tip link consists of a fine 6-nm strand, surrounded by a variable coat (Osborne *et al.*, 1988; Pickles *et al.*, 1988). The central strand has approximately the same diameter as a single actin filament. It inserts into specialized densities in the stereocilia. The surrounding coat is probably a continuation of the glycoconjugate material that surrounds the surfaces of the stereocilia (Slepecky and Chamberlain, 1985b; Santi and Anderson, 1987). Although evidence in favour of the idea that the tip links are involved in transduction will be evaluated in a later section, at the moment it can be noted that if the tallest stereocilium in Fig. 5.6 were deflected away from the shorter stereocilia, the tip links would tend to be stretched.

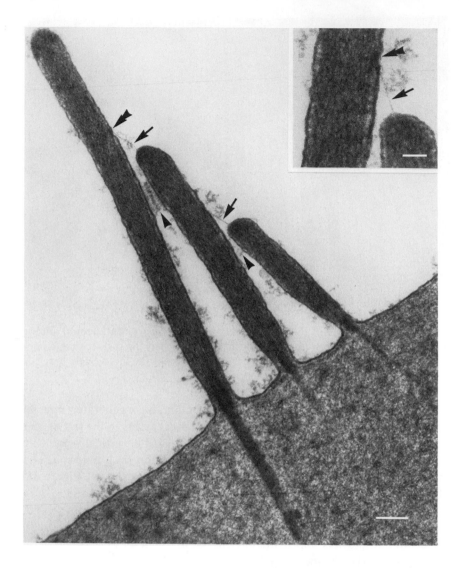

Fig. 5.6   The three rows of stereocilia on an outer hair cell are shown in cross-section. The stereocilia of the different rows are joined by horizontal (between-row) links just below their tips (arrowheads). The insert shows a higher magnification of the lower tip link. Tip links (arrows) have a fine central core, surrounded by amorphous material (see also insert). Double arrowhead: upper density. Guinea-pig. Scale bar on main figure: 200 nm, on insert: 100 nm. From Osborne *et al.* (1988, Figs 3 and 4).

Fig. 5.7   Tip links (arrows) in the apex of the "V" of stereocilia on an outer hair cell of the guinea-pig cochlea. The stereocilia were photographed nearly parallel to the axis of bilateral symmetry, showing that the tip links run parallel to that axis, i.e. parallel to the excitatory–inhibitory axis, and approximately radially across the cochlear duct. Scale bar: 200 nm.

C. Mechanisms of Transduction

It is difficult to record intracellularly from single cochlear hair cells while manipulating the stereocilia, and so the best information on transduction has been obtained from hair cells of the vestibular system. However, where data from cochlear hair cells are available, they have supported the data from vestibular cells.

Before considering the process of transduction, it is useful to review a few of the basic facts about cell membrane potentials.

1. Cell Membrane Potentials

Nerve cell membranes are more permeable to some ions than to others – for instance, in the resting state they are many more times permeable to

K^+ than to Na^+ ions. Because there is a high K^+ concentration inside the cell, and because the cell membrane is permeable to K^+, K^+ tends to diffuse out passively down its concentration gradient, taking positive charge with it, and leaving the inside of the cell at a net negative potential. This potential is known as a diffusion potential. Diffusion continues until the negative potential inside the cell is sufficient to stop the further movement of ions, at which point in an ideal system K^+ would be held in equilibrium. The value of potential at which this occurs is known as the equilibrium potential.

It is possible to think of various schemes for the modulation of ion flows in hair cells. If, for instance, the apical surface of the hair cell were faced by an endolymph which, as well as having a low electrical potential, had a low K^+ concentration, increasing the permeability to K^+ alone would generate a negative diffusion potential across the apical membrane of the hair cell. K^+ would tend to diffuse out, tending to take the membrane towards the K^+ equilibrium potential. The inside of the cell would then become hyperpolarized, that is, more negative. Increasing membrane permeability alone does not therefore necessarily produce a reduction in the membrane potential; the important point is that the membrane potential tends to move in a direction towards the equilibrium potential of the diffusible ion, or to a weighted mean of the various equilibrium potentials if more than one ion is involved.

In mammals the K^+ concentration in endolymph is approximately the same as that inside the cell, and the Na^+ concentration is very low. No substantial diffusion potentials will be produced by ion flows across the apical membrane of the hair cell, and decreasing the resistance will simply pull the intracellular potential towards the endolymphatic potential, producing a depolarization. As a bonus, in mammals the endolymphatic potential is 80 mV or so positive. While this may have evolved secondarily from the mechanism for maintaining a high K^+ concentration in the endolymph, it also serves to increase the driving potential across the apical membrane from 45 mV or so in the case of the inner hair cells to 125 mV.

It is presumably advantageous for the current to be carried by K^+ rather than say Na^+. Because K^+ is in equilibrium across the basal membrane of the hair cell, any K^+ entering the cell will diffuse out automatically, and will not accumulate inside the cell. In this case, the energy driving the current flow is ultimately derived from ion pumps in the stria vascularis. It has the advantage of allowing the main blood supply to be removed well away from the organ of Corti, with a consequent reduction in possible vascular noise.

Although there are considerable advantages in the transducer current being carried by K^+, we must not forget that other schemes are possible. For instance, if the apical surface is faced with a normal extracellular fluid high in Na^+ and low in K^+, and if the channels are non-specific for Na^+ and K^+ (which is the case – see below), then lowering the membrane resistance will tend to bring the potential to a weighted mean of the Na^+ and K^+ equilibrium potentials, i.e. to around 0 mV if there is no endocochlear potential. The cell will therefore tend to depolarize. This is in fact the position in many *in vitro* investigations of hair cell transduction. Neither must we forget the possibility that other ions, such as Ca^{2+} and Cl^-, may also make a contribution.

2. The Nature of the Transducer Channels

(a) Transduction depends on deflection of the stereocilia

The most straightforward information on transduction comes from direct manipulation of the stereocilia on single hair cells during intracellular recording. This was done for instance in hair cells of the bullfrog sacculus (Hudspeth and Corey, 1977). They excised patches of the saccular macula, removed the covering otoliths and membranes, and directly manipulated the stereocilia with a glass rod while recording intracellularly with fine microelectrodes. They showed that deflection of the stereocilia in the direction of the kinocilium or tallest stereocilia caused intracellular depolarization, and so was associated with neural excitation, while deflection in the opposite direction caused hyperpolarization (Figs 5.8A and B). They therefore confirmed the conclusion made less directly by Lowenstein and Wersäll (1959), that deflection of the stereocilia in the direction of the kinocilium was excitatory, while deflection in the opposite direction was inhibitory. More recently, Russell *et al.* (1986a,b) made the same conclusion in hair cells of the mammalian cochlea, by manipulating the stereocilia of hair cells from the mouse cochlea grown *in vitro*. In this preparation stereocilia can be manipulated directly because the tectorial membrane does not develop.

In cells which have a kinocilium as well as stereocilia, it can be shown that transduction depends only on the movement of the stereocilia. If the kinocilium is teased away from the bundle of stereocilia, transducer currents are evoked only by manipulation of the stereocilia, and not the kinocilium alone (Hudspeth and Jacobs, 1979). In cochlear hair cells, of course, there was never any question that transduction might depend on the kinocilium, since cochlear hair cells do not have a kinocilium when mature.

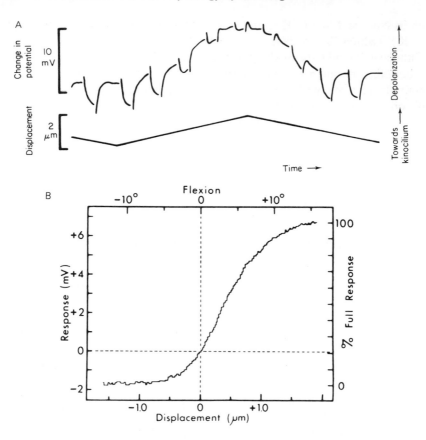

Fig. 5.8   (A) Deflection of the stereocilia towards the kinocilium produced depolarization, and deflection away hyperpolarization, in a hair cell of the bullfrog sacculus.
The small pulses on the upper trace show the voltage response to current pulses in
the recording electrode. Large voltage changes show that the membrane resistance
was high, and small changes that it was low. The membrane resistance was smaller
when the cell was depolarized. (B) The relation between hair deflection and voltage
change is asymmetric and saturating. From Hudspeth and Corey (1977, Fig. 3).

(b) Transduction opens channels in the apical membrane of the hair cell

Depolarization of the cell is associated with a decrease in membrane
resistance, while hyperpolarization is associated with an increase. Hudspeth
and Corey (1977) introduced current pulses down their recording electrode,
and found that the resulting intracellular voltage changes were small when
the stereocilia were deflected towards the kinocilium, i.e. when the cell was

depolarized, and large when the stereocilia were deflected in the other direction, when the cell was hyperpolarized (Fig. 5.8A). A small voltage response to the introduced current suggests that the introduced current had been able to leak out of the cell relatively easily. This shows that the resistance was low when the cells were depolarized, and suggests that the depolarization had been produced by ion channels opening in the cell membrane.

By separate perfusion of the apical and basal membranes of the hair cell, Corey and Hudspeth (1979a) showed that the permeability of the apical, i.e. stereocilia-bearing, membrane had been changed by the mechanical stimulus.

(c) The number of transducer channels: only a few per stereocilium

Knowing the unit conductance of a single channel, and knowing the total conductance change during maximal stimulation, one is able to calculate the total number of transducer channels on a cell. The conductances can be measured with patch-clamp electrodes which have wide (about 1 μm) tips, giving low access resistances and allowing the recording of fast electrical events inside the cell.

Ohmori (1985) used patch-clamp electrodes in a voltage-clamp circuit, which kept the membrane voltage at a certain set value, while recording the changes in transcellular current. He found step changes in transcellular current when the stereocilia were deflected (Fig. 5.9). In other systems it is known that such records can be produced by the individual membrane channels flicking between their open and closed states, and this suggested that the fluctuations represented the spontaneous opening and closing of individual hair cell transducer channels. The results fit in with the hypothesis that the channels spontaneously oscillate between their open and closed states, the proportion of time being spent in any one state being a function of the mechanical stimulus to the cell.

That may well be the case here: however, it is never possible to rule out the possibility that channels are opening and closing together in groups. Because the calculated conductance changes at 50pS* are rather larger than obtained in other experiments, it has since been suggested that the records might show the simultaneous openings and closings of more than one channel (Holton and Hudspeth, 1986).

A second, more indirect, way of characterizing the channels is to record

---

*Conductance is measured in siemens (S), which is the inverse of ohms: the conductance in siemens = current divided by driving voltage.

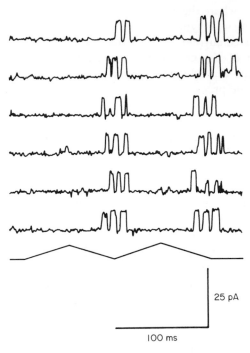

Fig. 5.9   Step changes in transducer current may represent the opening and closing of single transducer channels or groups of channels in response to deflection of the stereocilia. In this experiment the stereocilia were held with a bias towards the excitatory position, so that channels were open most of the time. The channels closed transiently (upwards deflections in top traces), when the stereocilia were displaced in the inhibitory direction (downwards movement in bottom trace). Hair cells of chick vestibular macula. From Ohmori (1985, Fig. 5).

the summed currents resulting from the activation of *all* the channels available on the cell, as the stereocilia are slowly taken through their total range of deflection (Holton and Hudspeth, 1986). The instant-to-instant variation in the net transducer current will be the summed effect of the fluctuations in the openings of all the individual channels on the cell. If the stereocilia are held in the extreme excitatory position, it can be supposed that the channels spend most of their time in the open state. Because the rate of transitions to the closed state and back is low, the variance in the net current will be low. Similarly, if the stereocilia are held in the extreme inhibitory position, we would expect the channels to spend most of their time in the closed position, also resulting in a low variance in net current. The maximum variance will be produced when the individual channels spend 50% of their time in the open position. Figures 5.10A and B show, for hair cells of the bullfrog sacculus, that the variance in the record was indeed at a maximum when the net transducer current was at half its maximum value. The theoretical relation between variance and transducer current is a parabola: fitting this curve to the data permits the estimation of the number of channels and the unit conductance of a single channel (Fig. 5.10C). In this way, the unit conductance (at 10°C) was estimated to be 12.7 pS, and the maximum number of transducer channels per cell to be 280. This corresponds to about four transducer channels per stereocilium.

How does this number of channels agree with the values found in other hair cells? Assuming that the transducer channels behave like other aqueous-pore channels, we would expect a conductance of 12.7 pS at 10°C to become 25 pS at 37°C (Holton and Hudspeth, 1986). In inner hair cells of the guinea-pig cochlea, Russell *et al.* (1986a) found the maximum conductance change during acoustic stimulation to be 2.5 nS. At a unit conductance of 25 pS, we expect there to be about 100 channels per hair cell. Outer hair cells produced maximum conductance changes of 3.1 nS, giving about 120 channels per hair cell. These numbers can be compared with the total number of stereocilia per hair cell, which for the basal turn of the guinea-pig are about 60 per inner hair cell, and 110 per outer hair cell. Thus, in these hair cells there is roughly one channel per stereocilium.

(d) The ionic selectivity of the channel

Corey and Hudspeth (1979a) varied the ionic content of the solution bathing the apical surfaces of their hair cells, and showed that a range of ions could carry the transducer current. The alkali cations, including Na^+ and K^+ could carry the transducer current to an approximately equal extent. Ca^{2+} could also carry the transducer current, though to a lesser extent. Ohmori (1985) produced generally similar results in chick vestibular hair cells, although here Ca^{2+} and other divalent ions such as Sr^{2+} and Ba^{2+} were more permeant than the alkali cations. The channel therefore appears to be relatively nonspecific, although the greater permeance of divalent cations suggests that the channel is charge selective (Ohmori, 1985). Tetramethylammonium (TMA) could also carry the current, although with a lower permeance (Corey and Hudspeth, 1979a). Since the unhydrated TMA ion has a diameter of 0.54 nm, the channel must have a diameter at least as large as this. The lower permeance of still larger ions suggests that the pore diameter is approximately 0.7 nm (Hudspeth, 1985).

Although Ca^{2+} can carry the transducer current, experiments have suggested that it has a further role, in that Ca^{2+} seems to be an essential cofactor for the operation of the channel (Sand, 1975). If the Ca^{2+} concentr-

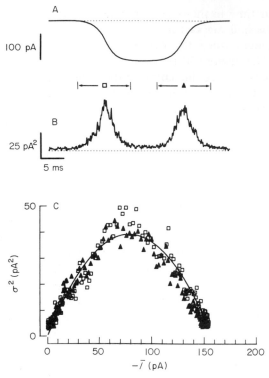

Fig. 5.10   The stereocilia on hair cells of the bullfrog sacculus were moved through their excitatory–inhibitory range many times, and the transducer current averaged. (A) The change in transmembrane current, averaged over all cycles of deflection, is plotted as a function of position in the cycle (horizontal axis). (B) The instantaneous variance of the current trace for each point of the cycle in A was calculated from the repeat variability of the observations made at that point in the cycle. The variance is at a maximum during the falling and rising phases of the transducer current. (C) The plot of variance (vertical) against transducer current (horizontal). ■, points derived from falling phase. ▲, points derived from rising phase. The points lie on a parabola (solid line). From Holton and Hudspeth (1986, Fig. 6).

ation is reduced below about 10 μm, mechanotransduction ceases (Corey and Hudspeth, 1979a). This is only a little less than the concentration of Ca^{2+} in mammalian endolymph, which for the rat is 30μm (Bosher and Warren, 1978).

By far the most abundant cation in the endolymph is K^+, and since K^+ has a high permeance through the transducer channel, it is reasonable to suppose that K^+ carries most of the transducer current in the mammalian cochlea. This is supported by experimental evidence, since K^+ accumulates in the space around the outer hair cells during acoustic stimulation (Syka *et al.*, 1987). Moreover, the reversal potential of the transducer current in inner hair cells of the guinea-pig cochlea is approximately equal to the endocochlear potential (Russell, 1983). This would be expected if the stereocilia were faced by an endolymph with the high positive potential, and if current were carried by an ion with the same concentration in the endolymph as intracellularly.

(e) Kinetic evidence on the nature of transduction

The transducer channels seem to open and close with a very short delay. Corey and Hudspeth (1979b) showed that the transducer current, in response to pulse stimuli, followed the deflection with a delay of approximately 40 μs at 22°C. The short latency agrees with the latencies that would be required to subserve hearing at the upper frequency limits of mammalian perception, for instance 120 kHz for dolphins and 130 kHz for seals (Brown and Pye, 1975). The latency has a temperature dependence, decreasing by a factor of 2.5 for every 10°C increase between 1 and 38°C.

These results suggest that the transduction process has a fairly simple mechanism, with its short latency and its low temperature dependence arguing against any great complexity (Corey and Hudspeth, 1983). By contrast, one could imagine rather more complex models, in which, for instance, a protein kinase was activated, or a second messenger released, opening the channels by a biochemical reaction. A further continuously active enzyme would then return the channels to their non-conducting state. The likelihood of these models is limited by the short latency of channel opening, and the speed at which the channels can close at the end of a stimulus, which mean that the biochemical reactions would have to occur at unrealistically high rates (Corey and Hudspeth, 1983).

Rather more plausible are models in which the displacement opens the channel by a direct mechanical effect. Put in terms of a kinetic analysis, the deflection alters the energy difference between the open and closed states of the channel. Thermal energy would continuously change the channel between its opened and closed states, the proportion of the time spent in one state depending on the energy difference and so on the displacement. Such a model explains some otherwise curious results, such as the way that the time-constants of the electrical response decrease for large stimulus steps (Corey and Hudspeth, 1983). In addition, it suggests that the channel does not have any absolute threshold for opening, since the opening is probabilistic. Such processes are required by models which account for the very low threshold of auditory sensitivity.

Since it is supposed that the channel oscillates between its open and closed states under the influence of thermal energy, the relative probabilities of the two states will be given by the Boltzmann distribution (Corey and Hudspeth, 1983; Holton and Hudspeth, 1986). If it is assumed that the stimulus-induced displacements are coupled to the channel through a mechanical structure which can deform elastically, it is possible to relate the displacement of the stereocilia to the tension in the coupling structure, and so to the energy difference as the channel oscillates between its two states. Where the model has only two distinguishable kinetic states (the open and closed states), the resulting calculated input–output function is a symmetrical sigmoid, similar to that shown experimentally in Fig. 5.11A. Apart from the position of the curve along the horizontal axis, the only free parameter that can be adjusted in fitting the theoretical function is the rate at which energy is coupled into the channel, which in the case illustrated was set at 6.2 kcal/mol for every micron of displacement.

Other experimental paradigms have given other shaped functions: when the total transepithelial current of the sacculus is measured, the function can be asymmetrical (Fig. 5.11B). A function such as this can be fitted by a three-state kinetic model, in which there are *two* kinetically-distinguishable closed states, and one open state (Corey and Hudspeth, 1983). The factors leading to the differences between the results are not known.

3. The Site of the Transducer Channels

(a) Electrophysiological experiments

Hudspeth (1982) measured the extracellular potentials around a bundle of stereocilia, while deflecting the stereocilia with a probe. He showed that the greatest potential changes were produced near the top of the bundle, rather than near the bottom. In fact, the greatest potential changes of all were recorded over the tapering part of the bundle, just above the tips of the multiple ranks of stereocilia (Fig. 5.12). This suggests that the transducer current was flowing into the cell at or near the tips of the individual stereocilia.

(b) Hypothesis on the site of the transducer channels

There have been no convincing anatomical demonstrations of the actual transducer channels themselves – for instance, membrane specializations corresponding to the channels known in other systems, appearing as membrane densities in transmission electron micrographs or as membrane particles in freeze-fracture. One difficulty is that if there are only a very few

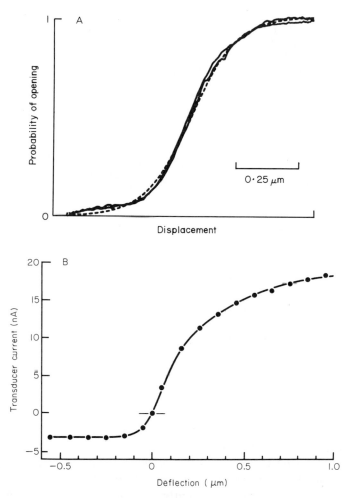

Fig. 5.11 (A) Symmetrical change in transducer current, as recorded with gigohm-seal electrodes from single cells of the bullfrog sacculus. The stereocilia were taken through a full cycle of deflection. The curves for the transducer current (solid lines) have been fitted with a curve calculated from the two-state kinetic model (dotted line). The cell was held at a constant intracellular voltage in a voltage-clamp circuit. From Holton and Hudspeth (1986, Fig. 12). (B) Net transducer current through a patch of saccular macula (data points), fitted with a curve calculated from the three-state kinetic model (solid line). From Corey and Hudspeth (1983, Fig. 2).

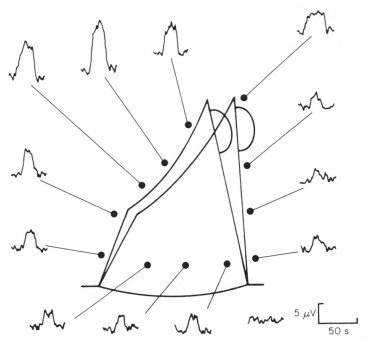

Fig. 5.12   Changes in current (small traces) recorded from an extracellular electrode in different positions around a bundle of stereocilia, as the stereocilia were deflected. The changes in current were largest when the electrode was over the curved part of the bundle, just above the tips of the stereocilia. The semicircle on one side of the bundle represents the bulb at the top of the kinocilium (see Fig. 3.5A). From Hudspeth (1982, Fig. 3).

transducer channels per stereocilium, then it would be very difficult indeed to find them without a theoretical guide as to where to look.

A key to analysing the possible site of the transducer channels comes from the relation between the functional and morphological polarization of hair cells – in other words, the apparently consistent finding that deflection of the stereocilia in the direction of the kinocilium or the tallest stereocilia opens the transducer channels, while deflection in the opposite direction closes them (Lowenstein and Wersäll, 1959; Hudspeth and Corey, 1977; Russell *et al.*, 1986b).

Early hypotheses related transduction either to the kinocilium, or to the point of entry of the stereocilia into the cuticular plate (Hillman, 1969; Dallos, 1973a). However, these hypotheses either do not satisfactorily account for the relation between the morphological and functional polarization of hair cells, or do not fit with evidence from micromanipulation of the cilia. For instance, Hudspeth and Jacobs (1979) separately manipulated the stereocilia and the kinocilium, and showed that transducer currents were produced by manipulation of the stereocilia rather than the kinocilium.

A hypothesis which relates the functional polarization to the graded heights of the stereocilia, and which does not suffer from these shortcomings, was proposed by Pickles *et al.* (1984). They suggested that stretch of the tip links (i.e. the fine links which emerge from the tips of the shorter stereocilia in the bundle) opened membrane channels in the stereocilia at the points of insertion of the links.

(c) The suggested site of transduction

In the scheme proposed, the bundle is held together by the sideways links which run between the different rows of stereocilia, just below their tips. The result will be that, during deflection of the bundle, a vertical shear or sliding movement will be developed between the rows. The fine links running from the tips of the shorter stereocilia will be in a position to detect the shear. If a pull on the links opens the channels, and relaxation of the links closes them, then the functional polarization of hair cells is accounted for in a very simple manner (Fig. 5.13). The transducer channels would be at one or both points of insertion of the tip links.

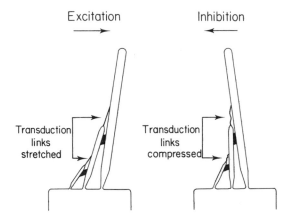

Fig. 5.13 A hypothesis for transduction in hair cells. It is suggested that stretch of the tip ("transducer") links opens the transducer channels, and that relaxation closes them. From Pickles *et al.* (1984, Fig. 9).

The hypothesis is entirely consistent with the electrophysiological results from micromanipulation experiments on hair cells. For instance, it explains the finding that the transducer current flows into the stereocilia near their tips (Hudspeth, 1982), since it is here that the tip links are attached. The gradation in heights of the different rows of stereocilia, seen universally in acousticolateral hair cells, also neatly fits with the hypothesis. The electrophysiological evidence also suggests that channel opening is not produced by a stretch of those links which run parallel to the cuticular plate, and which join the stereocilia in the horizontal direction. In many of Hudspeth's experiments the shorter stereocilia were *pushed* in the direction of the tallest, i.e. in the normally excitatory direction. Such a stimulus would have tended to compress the sideways links. However, the response was still excitatory.

The hypothesis has received further support, in that it was later shown that the tip links run with the horizontal component in their orientation which is always parallel to the excitatory–inhibitory axis of the cell (Comis *et al.*, 1985). This would explain why deflections of the stereocilia along that axis are effective, and why deflections at right angles are ineffective. The outer hair cell shown in Fig. 5.7 was photographed almost exactly parallel to the axis of bilateral symmetry, i.e. parallel to the excitatory–inhibitory axis of the cell, and so looking radially across the cochlear duct. The tip links run parallel to the line of view, rather than across it (see also Fig. 3.4C). Figure 5.14 shows the direction of the links plotted from another outer hair cell, seen in top view. This confirms the point made by the photograph, that tip links run parallel to the axis of bilateral symmetry. An analogous result holds true over hair cells of many different geometrical configurations. For instance, stereocilia on hair cells of lizard and bird basilar papillae are packed tightly in a hexagonal array. Here again, the tip links are oriented parallel to the cell's axis of bilateral symmetry, i.e. parallel to the excitatory–inhibitory axis (Pickles *et al.*, 1988).

Given that the tip links might be involved in transduction, does the fine structure tell us anything more about the possible sites of the transducer channels? Each tip link has a fine central filament, about 6 nm in diameter, which is surrounded by looser material, the latter apparently coextensive with the cell coat of the stereocilia (Fig. 5.6; Pickles *et al.*, 1988; Osborne *et al.*, 1988). The central presumably proteinaceous filament would be ideal for transmitting the forces involved in transduction to a very small area of cell membrane, containing the one to four possible transducer channels on each stereocilium. At its upper end, the filament runs straight to a density in the wall of the taller stereocilium, the density bridging the gap between the external membrane and actin paracrystal in the centre (double arrowhead, Fig. 5.6). At the lower end, the link runs to a conical extension of the stereociliar membrane. Beneath this, there is a dense cap over the ends of the actin filaments of the paracrystal, often attached by fine filaments to the conical extension of the membrane (Pickles *et al.*, 1988). The most likely

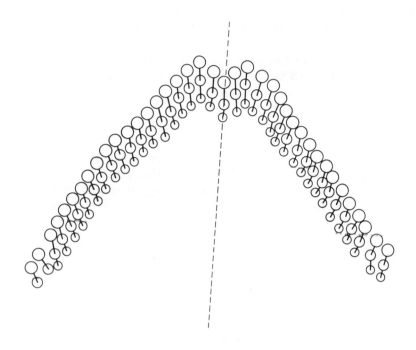

Fig. 5.14   Direction of tip links in guinea-pig outer hair cell from top view. The dotted line shows the axis of bilateral symmetry, i.e. the excitatory–inhibitory axis. From Comis *et al.* (1985, Fig. 7).

sites of the transducer channels are the upper density, and the conical extension of the membrane at the tip of the shorter stereocilium.

(d) The displacement of the transducer at threshold

It is possible to calculate the expected displacement of the transducer at threshold, to see if it is of a size that could reasonably be expected to affect membrane channels. One of the early problems of auditory physiology was the calculation, on the basis of von Békésy's measurements, that at 0 dB SPL the basilar membrane must vibrate by as little as 2×10^{-4} nm at threshold, less than 1/500th of the diameter of the hydrogen atom. The amplitude of the vibration coupled to the transducer site must have been even less than this, and it is difficult to see how such small displacements could lead to the opening of membrane channels. It is of course now recognized that the basilar membrane at threshold moves much more than von Békésy believed – for instance, the data of Sellick *et al.* (1982), shown in Fig. 3.10B, indicate that at 12 dB SPL the basilar membrane vibrates by 0.35 nm.

At the frequency of measurement, 18 kHz, the behavioural threshold of the guinea-pig is 10 dB SPL at the tympanic membrane (Evans, 1972). The data of Sellick *et al.* can be extrapolated to give a 0.3-nm displacement of the basilar membrane at this intensity. It can be shown by simple geometry that such a displacement of the basilar membrane would be expected to deflect the stereocilia by 10^{-2} degrees, and that this would be expected to extend the tip links by 0.4 Å (Pickles, 1985a). Such a value, while still subatomic, is within the bounds of reasonable possibility for threshold influences on a membrane channel. Since it is hypothesized that the channels are continually opening and closing under the influence of thermal energy, and that deflection merely causes a redistribution of the open and closed states, the actual definition of threshold becomes a statistical one, and dependent on the amount of averaging employed. Since there are some 100 channels per hair cell, and about 45 inner hair cells activated at threshold, a considerable amount of averaging is possible (Pickles, 1985a).

D. The Origin of Sharp Tuning in the Cochlea

As described in Chapter 3, current passive mechanical models of the cochlea have difficulty in reproducing the low-threshold, sharply-tuned component of the travelling wave. The only theories to be successful in matching mechanical, hair cell, and auditory nerve tuning suppose that the cochlea contains an active mechanical amplifier. The amplifier, probably involving the outer hair cells, detects the movement of the basilar membrane and feeds mechanical energy back into the travelling wave. The position will now be examined in more detail. The discussion will be based initially on models of the type discussed in Chapter 3, that is, with a cochlear partition having a gradation of stiffness, mass and damping, and which separates two fluid-filled compartments with a pressure difference between them.

1. Is an Active Process Necessary Theoretically?

The argument turns on whether tuning curves of the type observed experimentally can be produced in models containing purely passive mechanics. As described in Chapter 3, some passive models can produce large and sharply-tuned peaks, but of the wrong shape (e.g. Fig. 3.14). How reliable is this as evidence that purely passive processes are inadequate?

Diependaal *et al.* (1987) undertook a novel approach to the problem, by starting with observed basilar membrane tuning curves, and calculating back to the expected power flux along the cochlea. If the power flux stayed constant or declined from base to apex, passive models could apply, but if the power flux *increased* along the cochlear duct, then clearly extra energy must have been introduced. Their results show that the older data showing broad tuning are associated with a constant or declining power flux (Fig. 5.15A). However, the newer data, showing sharp tuning, are associated with a power flux that increases towards the peak of the travelling wave (Fig. 5.15B). The rate at which power is introduced is related to the *slope* of the power flux function, which is maximal just basal to the peak of the travelling wave. This analysis therefore suggests that power is introduced just basal to the peak of the travelling wave.

Unfortunately, the position is not as clear as that description suggests, because the calculated power flux is highly dependent on the detailed shape of the basilar membrane tuning curve. At the moment it is difficult to obtain accurate tuning curves or more than a few data points. If the experimental data points are interpolated with a process that allows more degrees of freedom (i.e. by fitting with a high-order polynomial), then the calculated increase in the power flux, though still present, is much reduced (Diependaal *et al.*, 1987).

Further points can be raised to show that purely passive models are unrealistic. (1) If large resonant peaks are to be produced in purely passive systems, then the amount of damping has to be very low indeed. This is unrealistic in view of the levels of viscous damping to be expected in the cochlea. (2) As described in Chapter 3, if the amplitude of the travelling wave is matched with the physiological data, the tuning becomes far too sharp. (3) Other attempts have been made to mimic the observed basilar membrane responses by varying the parameters of the cochlear model, and in particular to broaden the tip of the tuning curve while keeping the resonance peak large. This might for instance be done by supposing that there is appreciable longitudinal coupling along the cochlear partition. However, calculations show that the effect is to make the apical slope of the travelling wave too shallow, while still keeping an unrealistic sharply-tuned tip on the travelling wave (Viergever and Diependaal, 1986).

2. Models Incorporating an Active Mechanical Process

Models that simulate sharp tuning by means of an active process suppose that the outer hair cells, when stimulated, feed energy into the travelling

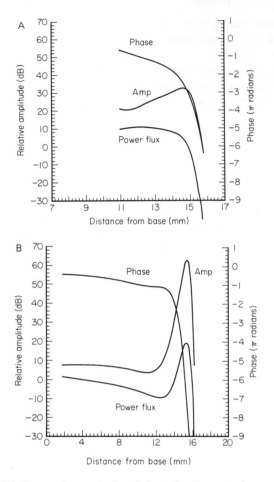

Fig. 5.15   (A) Power flux, calculated from basilar membrane velocity curves of Johnstone and Yates (1974). The power flux is at first constant and then declines along the cochlea, consistent with entirely passive cochlear mechanics. Note the broad travelling wave. (B) Power flux, calculated from the data of Robles *et al.* (1986b). The flux at first declines, but then increases sharply (between 13 and 15 mm along the cochlea), just basal to the peak of the travelling wave. From Diependaal *et al.* (1987, Figs 2 and 3).

wave. By appropriate choice of frequency and spatial dependence of the active process, it is possible to produce tuning curves of realistic shape. Moreover, the models are in agreement with the considerable body of data suggesting that normal functioning of the outer hair cells is necessary for the development of sharp tuning in the wave. The evidence on this point will be discussed below (Section D.3).

Two types of model have been presented, based on different hypotheses for the mode of vibration of the tectorial membrane. Both hypotheses suggest that the stereocilia of outer hair cells are deflected by the shear developed between the reticular lamina and the tectorial membrane, as in Davis's diagram (Fig. 3.3). In one type of model, the tectorial membrane hinges on its inner, spiral limbus edge, and does not undergo any other distortion. In these models the shear, and therefore the amount of feedback, depends only on the displacement of the basilar membrane. Models incorporating this type of active process have been made, for example, by Zwicker (1986a), who formulated an electronic model in hardware, and by Lumer (1987a,b), who formulated it as a computer model. The models can realistically reproduce many of the aspects of cochlear functioning.

Other models suppose that the tectorial membrane, as well as flexing, can undergo compression and expansion vibrations in a direction *radial* across the cochlear duct (Fig. 5.16). The mass of the tectorial membrane as it

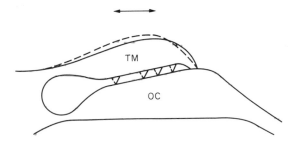

Fig. 5.16   A model of cochlear micromechanics, according to the theory of Zwislocki (1979) and Neely and Kim (1986). The vertical vibration of the cochlear partition sets the tectorial membrane into compression and expansion vibrations in a direction radial across the cochlear duct. TM, tectorial membrane; OC, organ of Corti.

undergoes radial oscillations, together with the stiffness of the stereocilia, form a second resonant system sitting on the cochlear partition (Zwislocki, 1979; Neely and Kim, 1986). The second system would be set into resonance by the motion of the basilar membrane, and would then increase the stimulus to the outer hair cells. The outer hair cells would, by means of their motile properties, contribute to the pressure difference across the cochlear partition, so amplifying the travelling wave. This model differs from the one described previously, in that the resonance of the second system serves to confine the feedback to one region of the travelling wave, set to be just basal to the peak of the travelling wave. The active process in this region would amplify the travelling wave, the wave continuing along the cochlear duct to peak at its own characteristic place. Tuning curves produced by such a model can be made to match the observed neural and basilar membrane tuning curves with a high degree of precision (Fig. 5.17). The model also has the

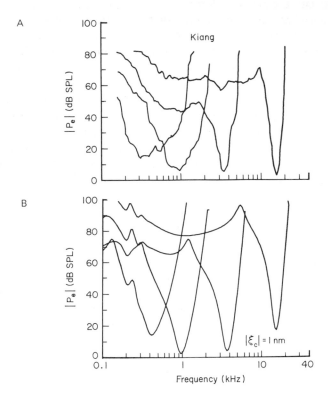

Fig. 5.17 (A) Sample cat auditory nerve fibre tuning curves, compared with predictions (B) from the active mechanical model of Neely and Kim (1986). Note the notches in the tuning curves (B), often seen in auditory nerve fibre tuning curves at these points. From Kiang (1980, Fig. 3) and Neely and Kim (1986, Fig. 12).

advantage of explaining the various odd notches which can be seen in auditory nerve fibre tuning curves and in their rate-intensity functions, as well as certain shifts which occur in their phase of activation as the stimulus frequency and intensity are changed (Liberman and Kiang, 1984; Neely and Kim, 1986; Zwislocki, 1986).

There are two main questions to be discussed in analysing the general validity of these models. First, is there any evidence that outer hair cells are necessary for the development of low thresholds and sharp tuning? Secondly, is there evidence that outer hair cells can act as active mechanical amplifiers?

3. Outer Hair Cells: Role in Low Thresholds and Sharp Tuning

Since nearly all (i.e. 90–95%) auditory nerve afferents synapse with inner hair cells, it must be supposed that the job of inner hair cells is to signal the movements of the basilar membrane to the central nervous system (see Chapter 3). What then is the function of outer hair cells?

The importance of outer hair cells for the sharp tuning and low thresholds of auditory nerve fibres, and hence by extension of the basilar membrane, was initially shown by Kiang *et al.* (1970) in cats poisoned with the ototoxic antibiotic, kanamycin. This agent can selectively damage outer hair cells with much less apparent effect on inner hair cells. Kiang *et al.* showed that in auditory nerve fibres recorded from areas of outer hair cell damage, the sharply-tuned, low-threshold tip of the neural tuning curve was selectively raised, while the high-threshold, low-frequency tail was unaffected. The result was a tuning curve which had a high threshold and was broadly tuned (Fig. 5.18A).

A second way that outer hair cells can be selectively affected, is by stimulation of the crossed olivocochlear bundle (COCB). This is the centrifugal pathway running from the brainstem to the cochlea, and will be discussed in more detail in Chapter 8. The crossed component of the bundle, running from the contralateral superior olivary complex, ends mainly on outer hair cells. Stimulation of the bundle can raise and broaden the tip region of the neural tuning curve (Fig. 5.18B). The result has since been repeated for inner hair cell receptor potentials (Fig. 5.18C). Since the COCB predominantly innervates outer hair cells, this result suggests that normal *outer* hair cell function is in some way necessary for the sharp tuning and sensitivity of *inner* hair cells.

4. Active Mechanical Processes in the Cochlea

There is considerable evidence that active mechanical processes occur in the cochlea, and there is evidence that outer hair cells can generate movement. The most direct evidence for the existence of active mechanical processes comes from the study of cochlear emissions, which show that under certain circumstances the cochlea can produce sound, and that this appears to be related to the activity of the hair cells themselves.

(a) Cochlear emissions

Kemp (1976) sealed both a microphone and a speaker into the ear canal of human subjects. An acoustic click presented through the speaker produced

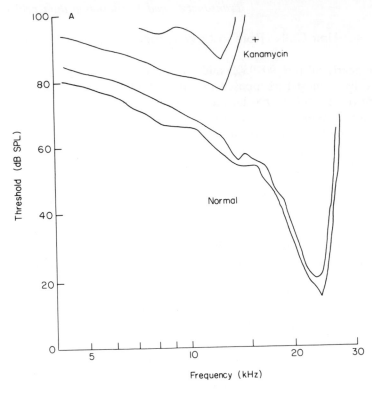

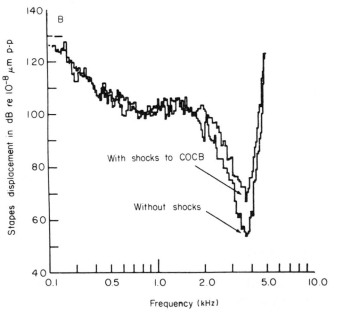

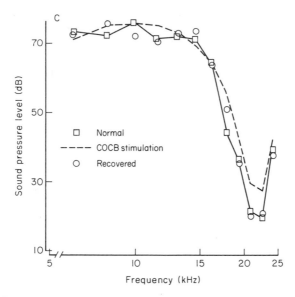

Fig. 5.18   The effects on auditory nerve or inner hair cell tuning, of three manipulations which are thought to have their primary effect on outer hair cells.
(A) Kanamycin reduces the sensitivity and frequency selectivity of auditory nerve
fibres, and lowers the best frequency. In this experiment, single neurones of the
spiral ganglion (auditory nerve cell bodies) were recorded in the guinea-pig at a site
2.2 mm from the base of the cochlea. Adapted from Robertson and Johnstone (1979,
Fig. 1). (B) Effect of crossed olivocochlear bundle (COCB) stimulation on the tuning
curve of an auditory nerve fibre. From Kiang *et al.* (1970, Fig. 16). (C) Effect of
COCB stimulation on the tuning curve of an inner hair cell. From Brown *et al.*
(1983b, Fig. 2). Copyright 1983 by the AAAS.

a brief wave of pressure in the ear canal. However, a second much smaller
sound wave, delayed by 5–15 ms, could also be recorded. The results of a
replication of his experiment by Wilson (1980b) are shown in Fig. 5.19.
The "echo" was strongest relative to the input for low-intensity clicks. The
suggestion was made that the mechanical impulse travelling up the basilar
membrane at some point met a discontinuity in the impedance of the
membrane. This caused a certain proportion of the energy to be reflected,
setting up a pressure wave which travelled back to the base of the cochlea.
The fact that the delayed response could be affected by ototoxic agents and
was absent in sensorineurally deaf subjects indicated a cochlear origin for
the effect. With suitable signal retrieval techniques it could be measured
for stimuli well below the sensory threshold. Moreover, it preserved many
details of the stimulus waveform, and inverted its waveform when the
stimulus was inverted. These suggest that it was generated by a stage before

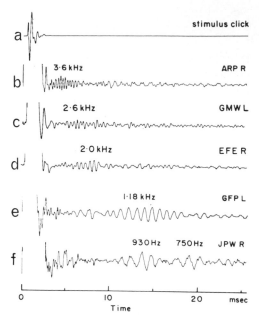

Fig. 5.19   When the ear is stimulated with a click, the cochlea returns an acoustic echo to the external auditory meatus. The form of the echo is different for each subject. The original stimulus in the meatus is shown in trace a. In traces b – f, shown with a much magnified vertical scale, the stimulus has clipped but the waveform of the echo is visible. Each trace is the average of many responses. From Wilson (1980b, Fig. 2).

the synapse, and suggest that this was not the result of the middle ear muscle reflex (Wilson, 1980a; Anderson, 1980).

Analogous echoes can be seen with other types of stimuli. If, for instance, the ear is stimulated with a continuous tone, the cochlea reflects a tone back into the ear canal. With two-tone stimulation, acoustic distortion products produced by intermodulation between the stimulus tones can be detected in the ear canal. The amplitude of the reflected intermodulation tones can be affected by stimulation of the olivocochlear bundle, suggesting that the outer hair cells are involved in its generation (Mountain, 1980; Siegel and Kim, 1982). Further experiments suggest that the reflections occur from a sharply-tuned late stage of the transduction process. For instance, it is possible to mask the echo to a click by a continuous tone. The frequency and intensity relations of the tone necessary to mask any particular frequency component of the echo show that the generators behind the echo must be very sharply tuned.

These observations do not in themselves prove that the echo results from a mechanically active process. However, an observation with more radical implications is shown in Fig. 5.20. It was found that in certain subjects with a tendency to subjective tinnitus of cochlear origin, a short tone burst was able to trigger a long chain of sound pressure fluctuations in the ear canal. The subjects at this point were able to hear an augmentation of their tinnitus. This is very strong evidence that not only are intracochlear events able to affect the sound pressure in the ear canal, but that an amplifying and *mechanically active* physiological process must be involved. The necessity for an amplifying stage was also shown by Kemp (1978), who calculated that even when a click did not evoke tinnitus, more energy could be produced by the cochlea than was originally introduced. The obvious but revolutionary hypothesis that must be put forward is that when the hair cells are stimulated, the cochlear partition is actively moved in return.

(b) Motility in hair cells

Motility of a variety of forms has been demonstrated in hair cells. Ashmore (1987) showed that outer hair cells isolated from the guinea-pig cochlea were able to change length in response to current introduced through an electrode. The cells shortened when the cell was depolarized, and elongated when the cell was hyperpolarized. Analogous motility has been found by Brownell *et al.* (1985) and Kachar *et al.* (1986). In response to step stimuli, the length change occurred with an exponential time course, having a latency as low as 120 μS and a time-constant of 240 μS, suggesting the possibility that the motility might be effective on a cycle-by-cycle basis in the kHz range of frequencies (Ashmore, 1987).

A slower motile process, with a time-constant of several seconds, was shown by Zenner *et al.* (1985), also in outer hair cells isolated from the guinea-pig cochlea. Changing the cell membrane potential by changing the external K^+ concentration produced a slow contraction or extension of the cell body. In other experiments, contraction could be produced by the application of ATP or inositol triphosphate to the cell (Zenner, 1986; Schacht and Zenner, 1987). In these experiments, contraction was accompanied by a tilting of the cuticular plate.

A different type of motility was shown in hair cells of the turtle cochlea, where Crawford and Fettiplace (1985) showed that the bundle of stereocilia was able to generate an oscillatory to-and-fro deflection when stimulated mechanically with a force step. The mechanical oscillations were in time with an electrical oscillation produced inside the cell by the step stimulus (Fig. 5.21).

The mechanisms of the various forms of motility are unclear. Motility of

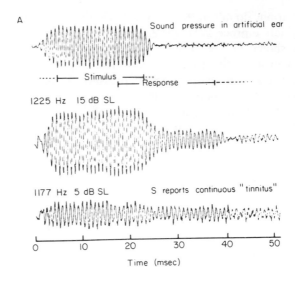

A

Sound pressure in artificial ear

|----- Stimulus -----|
          |---- Response ----|

1225 Hz   15 dB SL

1177 Hz   5 dB SL   S reports continuous "tinnitus"

0   10   20   30   40   50

Time (msec)

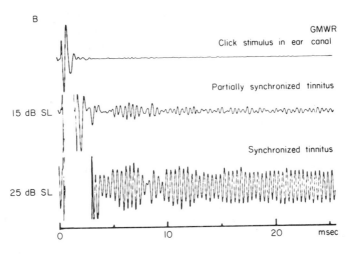

B

GMWR

Click stimulus in ear canal

Partially synchronized tinnitus

15 dB SL

Synchronized tinnitus

25 dB SL

0   10   20   msec

Fig. 5.20 Acoustic waveforms, recorded in the external auditory meatus, of tinnitus of cochlear origin. (A) A tone pip at 1225 Hz and 15 dB SL evoked an echo that outlasted the stimulus. But at 1177 Hz and 5 dB SL the tone pip induced prolonged ringing, and the subject heard continuous tinnitus. (B) Continuous tinnitus was detectable in the external auditory meatus of this subject, although because it was not phase-locked to the stimulus it did not appear in the averaged trace. Clicks of increasing intensity phase-locked the tinnitus to some extent at 15 dB SL and completely at 25 dB SL. From Wilson (1980b, Fig. 5).

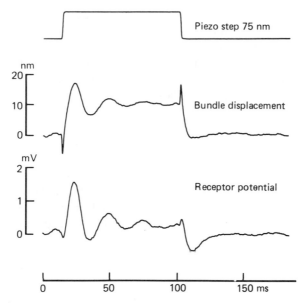

Fig. 5.21 Motility in a hair cell of the turtle cochlea. A flexible quartz fibre was brought up to the bundle of stereocilia, and deflected with a force step (top trace). The movement induced oscillations in the cell membrane potential (bottom trace), and at the same time oscillatory deflections of the bundle were visible (middle trace). Frequency of oscillations: 39 Hz.

The oscillations in the membrane potential are thought to be produced by Ca^{2+}-sensitive K^+ channels in the hair cell membranes, and underlie the frequency selectivity of turtle hair cells (Crawford and Fettiplace, 1981). While there is no evidence at present that similar electrical tuning occurs in the mammalian cochlea, it is possible that it occurs in low-frequency mammalian hair cells. From Crawford and Fettiplace (1985, Fig. 8).

first type is not affected by changes in the Ca^{2+} or ATP concentration, nor by the application of anti-actin or anti-tubulin drugs such as phalloidin or colchicine (Ashmore, 1987; Holley and Ashmore, 1988). These findings, together with the speed of the response, suggests that the motility is produced by a direct physical mechanism, rather than processes such as actin–myosin interactions. Suggested mechanisms involve the movement of charged components of the cell in the electric field (Kachar *et al.*, 1986; Ashmore, 1987). Holley and Ashmore (1988) found that the motility persisted in cells which had been artificially swollen into a spherical shape by direct pressure injection, and suggested that the mechanism of motility resided in structures closely connected to the plasma membrane. Certain specializations are known here, since outer hair cells have numerous cisternae lining their lateral walls, connected to the outer membrane by a regular arrangement of pillars (Flock *et al.*, 1986).

The slower motility of the second type could possibly involve actin–myosin interactions of the sort seen in muscle cells, perhaps involving the actin filaments which circle the cuticular plate, or those which descend from the cuticular plate through the cell body. There is also evidence for other processes involving structural changes in the cytoskeleton, such as the polymerization of actin filaments (Zenner, 1986). The slower systems may use phosphatidyl inositol biphosphate as a second messenger within the cell (Schacht and Zenner, 1987).

In the absence of direct evidence, it is only possible to speculate on the way that motility could affect the travelling wave. Since the travelling wave produces deflection of the stereocilia, active deflection of the stereocilia as shown by Crawford and Fettiplace (1985) could obviously in turn feed back into the travelling wave. Since we do not know the basis of the motility shown by Crawford and Fettiplace, we do not know if it is likely to be capable of acting on a cycle-by-cycle basis in the upper frequency range of mammalian perception.

Brownell *et al.* (1985) and Ashmore (1987) suggested that length changes in the outer hair cells would lead to a dimensional change in the organ of Corti. An upwards movement of the basilar membrane moves the stereocilia in the excitatory direction (Fig. 3.3), and so produces intracellular depolarization. Intracellular depolarization shortens the hair cells, flattening the organ of Corti. It was suggested that this would serve to enhance the upwards movement of the basilar membrane (Ashmore, 1987). However, we do not at the moment know whether the mechanical properties of the organ of Corti would permit this to feed back into the travelling wave.

The slower forms of motility are unlikely to be involved directly in the amplification of the travelling wave, since they are too slow to feed back on a cycle-by-cycle basis. However, it is quite possible that they are involved in adjusting the mechanical state of the cochlea over the long term, keeping it in the optimal mechanical state to detect auditory stimuli. This hypothesis will be elaborated further when discussing the functions of the olivocochlear bundle.

E. Cochlear Nonlinearity

One of the most intriguing aspects of cochlear function is its nonlinearity. On current ideas, the nonlinearity of basilar membrane vibration is produced by the process generating sharp tuning. That process, likely to be the active amplification of the travelling wave by the outer hair cells, derives at least some of its nonlinearity from the intrinsic nonlinearity of hair cell transduction. Cochlear nonlinearity is revealed in four main ways: (1) by the nonlinear growth of cochlear responses with stimulus intensity, (2) by the appearance of d.c. offsets in the position of the basilar membrane during acoustic stimulation, (3) by the reduction in the response to one stimulus by a second stimulus ("two-tone suppression"), and (4) by the generation of combination tones, or in other words intermodulation distortion products.

1. The Nonlinear Growth of Cochlear Responses

Figure 3.10C (p. 43) shows that basilar membrane responses grow nonlinearly with intensity, for stimulus frequencies around the characteristic frequency. The figure shows, for stimulus frequencies of 16–20 kHz, and for stimulus amplitudes greater than 30 dB SPL, that the slope of the amplitude function has a value of 0.2 on log–log scales, indicating that basilar membrane vibration grows in proportion to the stimulus amplitude raised to the power of 0.2.

The most parsimonious current explanation suggests that the nonlinearity arises from the active mechanical amplification of the travelling wave, and that the active process saturates, or reaches its maximum output, by 30–40 dB SPL. Such a function is shown by the solid line in Fig. 5.22. On this theory, the overall amplitude function of the basilar membrane response can be separated into three parts: (i) a nearly linear growth up to about 30 dB SPL, dependent on an approximately linear growth of the active mechanical process, (ii) a saturating function up to 80–100 dB SPL, dependent on saturation of the active process, and (iii) a linear growth at the highest intensities, the linear passive component having now become larger than the active component (Fig. 5.22). The nonlinearity is only seen around the characteristic frequency, because it is only here that the active process is dominant. An electrical analogue of the cochlea has been built, incorporating just such a saturating active process, and can simulate the above phenomena (Zwicker, 1986a,b,c).

Noise trauma might be expected to impair a physiologically active process, and amplitude functions made before and after the induction of noise trauma show a change from the initial three-part function, to one that more closely approximates a simple linear growth (Fig. 5.22).

Why does the active contribution saturate at such a low intensity? On the hypothesis that transducer currents through the outer hair cells drive a motile mechanism, the saturation could occur at two possible stages: (i) saturation of the transducer current as a function of stereociliar displace-

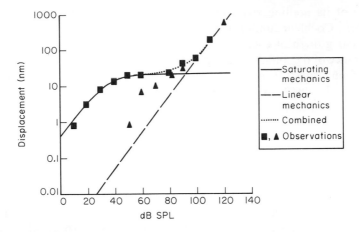

Fig. 5.22 Amplitude functions of the basilar membrane, showing theoretical contributions from the active process (solid line), and the passive process (long dashes), making a net function shown by the short-dashed line. ■, Measured basilar membrane response when cochlea in good condition. ▲, measured response after acoustic trauma. From Johnstone *et al.* (1986, Fig. 5).

ment, and (ii) saturation of the motile mechanism as a function of transducer current.

In the absence of knowledge about the mechanism behind the motile process, it has not been possible to get further evidence on the second possibility. Discussion has therefore concentrated an the possible contribution from saturation of transduction. As described above, manipulation of the stereocilia on vestibular hair cells has shown that transducer current is limited at large deflections (Fig. 5.11). Functions similar to the asymmetrical sigmoidal function of Fig. 5.11B have also been shown in hair cells of the mammalian cochlea, by direct manipulation of the stereocilia on hair cells grown in organ culture (Russell *et al.*, 1986b). The hypothesis therefore suggests that the motile process might saturate, and basilar membrane responses become nonlinear, because outer hair cell transduction saturates (Russell *et al.*, 1986a,b).

2. D.C. Responses of the Basilar Membrane

Acoustic stimulation produces d.c. offsets in the position of the basilar membrane (LePage, 1987). This is further evidence that the basilar membrane moves nonlinearly (Fig. 5.23). The possible generation of the d.c.

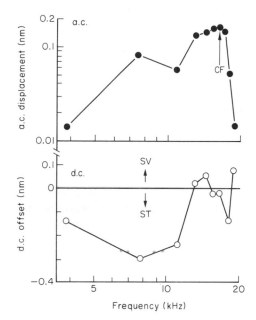

Fig. 5.23   d.c. and a.c changes in the position of the basilar membrane during acoustic stimulation. The d.c. component (lower graph) shows a trimodal change as a function of stimulus frequency, with a small scala-tympani displacement just above the characteristic frequency (CF), a small scala-vestibuli displacement near and just below the characteristic frequency, and a large scala-vestibuli displacement well below the characteristic frequency. Scala-vestibuli displacement would be expected to be the more effective at inhibiting the outer hair cells. Stimulus amplitude: 70 dB SPL. Guinea-pig. From LePage (1987, Fig. 2).

bias can also be understood from nonlinear hair cell input–output functions. Since the functions for cochlear hair cells are asymmetric in the excitatory and inhibitory directions, we would expect the motile process to show a corresponding asymmetry. It is tempting to suggest that this asymmetric motile process generates at least some of the d.c. bias in the position of the basilar membrane. Moreover, we would expect such a bias to feed back round the loop, and in turn deflect the stereocilia of the outer hair cells, since they are coupled directly to the tectorial membrane. The bias towards scala tympani seen for stimuli below the CF in the basal turn may therefore be responsible for turning the predominantly depolarizing asymmetry seen in isolated hair cells into the predominantly hyperpolarizing asymmetry seen *in vivo* in outer hair cells with acoustic stimulation (compare Figs 5.8 and 3.21 B; Russell *et al.*, 1986a). On the other hand, inner hair cell stereocilia would not be biased, since they are not directly coupled to the tectorial membrane.

The full theory of the relation between hair cell and basilar membrane nonlinearity has however not yet been worked out, and would have to take into account the influence of cochlear micromechanics, possibly themselves nonlinear, on the feedback.

3. Two-tone Suppression

It was described in Chapter 4 how the presence of one stimulus can reduce the responsiveness of the cochlea to other stimuli. Figure 4.15 shows the tuning curve for the suppressive effect, when the suppressed tone is set to the characteristic frequency and the suppressing tone is moved in frequency. The explanation given in Chapter 4 for two-tone suppression was the following: since the travelling wave shows a compressive nonlinearity near its peak, suppression will be produced when the travelling waves of the two stimuli overlap.

Further evidence on the possible mechanism of two-tone suppression was produced by Patuzzi *et al.* (1984b), in the basal turn of the guinea-pig cochlea. They measured the mechanical response of the basilar membrane to a high-frequency stimulus at the characteristic frequency (CF) of the region, at the same time as presenting an intense tone of very low frequency. When the response to the CF-tone was measured as a function of the phase of the low-frequency tone, it was found that the response to the CF-tone underwent two phases of suppression for every cycle of the low-frequency tone. The major phase of suppression was seen while the basilar membrane was displaced towards the scala tympani, with a second, minor phase of suppression during displacements towards the scala vestibuli. Inner hair cell tuning curves made during different phases of the intense low-frequency tone showed that the tuning curve moved upwards and increased in width during the suppressive phases, with the greatest effect being seen during scala tympani deflections (Fig. 5.24; Patuzzi and Sellick, 1984). The effect was not simply due to the low-frequency tone moving the basilar membrane to the mechanical limits of its response in either direction, because the effect was greatest for CF-test tones, and completely absent for test tones of much lower frequency. Rather, it looks as though the suppressing tone had interfered with the mechanism for sharp tuning on the basilar membrane – reminiscent of the changes shown in Fig. 5.18. The most reasonable explanation is that the suppressing tone had biased the outer hair cells towards the more saturating part of their input-output functions, with the result that a superimposed test stimulus would be unable to produce as much active mechanical feedback.

Indeed, since outer hair cells normally operate around the steepest part

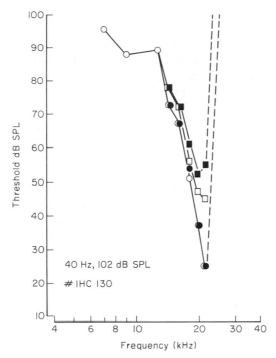

Fig. 5.24   Inner hair cell tuning curves, measured during the different phases of a simultaneous low-frequency stimulus. The tuning curve is sharp at the moments at which the low-frequency stimulus is going through its zero crossings (○●) and broader and of higher threshold during the extremes of movement towards the scala tympani (■) and the scala vestibuli (□). The tuning curve is for the d.c. receptor potential at the 0.9-mV level. From Patuzzi and Sellick (1984, Fig. 3).

of their input–output function, any bias in the operating point can be expected to reduce the amount of active mechanical feedback, and reduce the response around the characteristic frequency. One possible mechanism of two-tone suppression may therefore be related to the nonlinear input-output functions of outer hair cells (Zwicker, 1986c; Lumer, 1987b). Some of the bias will be produced by the a.c. vibration of the basilar membrane in response to the suppressing stimulus. The bias will also be contributed to by the d.c. shifts in the position of the basilar membrane (Fig. 5.23), itself a product of the nonlinear response to the suppressor, and perhaps also originating in outer hair cell nonlinearity. Note that in Fig. 5.23 the basilar membrane is displaced towards the scala tympani for frequencies on either side of the best frequency. This is the direction which would be expected to deflect the stereocilia towards the modiolus (Fig. 3.3), and so towards the shorter stereocilia (Fig. 3.1D). This is the hyperpolarizing and more sharply-saturating direction in isolated hair cells.

The above theory suggests that in order to produce suppression, the travelling wave of the suppressing stimulus must overlap the active region in the travelling wave of the suppressed stimulus. While this mechanism forms one attractive way of explaining two-tone suppression, it does not form a complete explanation. In particular, it cannot cope with the following finding.

Robertson and Johnstone (1981) produced a region of local acoustic trauma in the cochlea, basal to the region innervated by a fibre from which they were recording. They showed that the trauma could reduce suppression by a higher-frequency tone *without* affecting the threshold or tuning of the fibre. In other words, the region producing suppression could be separate from the region producing sharp tuning. This result is in accordance with the explanation which was given earlier in Chapter 4 (Section B.4.a), suggesting how a higher-frequency tone could suppress. It was suggested that, as the basilar membrane moves nonlinearly at the characteristic place of the suppressing tone, the travelling wave to the suppressed tone would be reduced in amplitude as it passed through the region. The result of Robertson and Johnstone suggests that in order to produce suppression, the cochlea has to be functioning normally in the region of the *suppressing* tone, and not only in the region of the suppressed tone, as required by the mechanism described in the previous paragraphs.

The full explanation of two-tone suppression is therefore as yet uncertain.

4. Combination Tones

Combination tones provide powerful evidence for cochlear nonlinearity. If a system is entirely linear, its output waveform will contain only the same frequency components as the input. If a system is nonlinear, single tones will also produce harmonics. A pair of input tones will, in addition, produce combination tones, that is, tones whose frequencies depend on the frequencies of *both* the input tones. The pattern of such harmonics and combination tones immediately tells us a great deal about the nonlinearity. If the input–output function through the nonlinearity is symmetric around the line at input$=0$ (otherwise called even-order), we shall have even harmonics ($2f_1$, $4f_1$, etc.) and sum and difference tones of the form $f_2 \pm f_1$, $2f_2 \pm 2f_1$, etc. (Fig. 5.25). If the input-output function through the nonlinearity is antisymmetric, or odd-order, we shall, in addition to the original frequencies, have odd harmonics and combination tones of the form $2f_1 \pm f_2$, $3f_1 \pm 2f_2$, etc. Combination tones of the form $f_2 - f_1$ (the difference

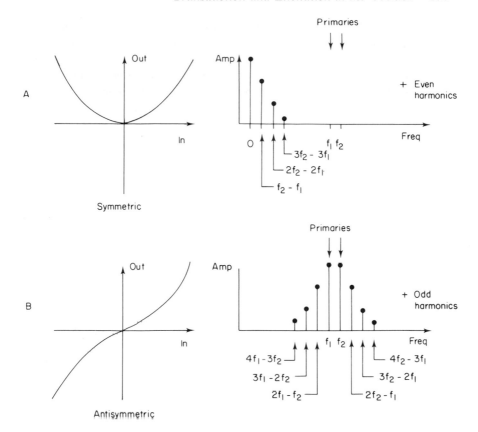

Fig. 5.25 Combination tones produced by different nonlinearities. (A) A nonlinearity with an input–output function symmetrical around the line: input$=0$ (even order function) produces an output with terms of the form n($f_1 + f_2$), and even harmonics. (B) An antisymmetric input–output function (odd-order function) produces the primaries, combination tones of the form $nf_1 \pm mf_2$, and odd harmonics. All realizable input–output functions can be made by combinations of even-order and odd-order functions.

tone), and $2f_1 - f_2$ (the cubic distortion tone) can be demonstrated both electrophysiologically and psychophysically (Goldstein, 1967; Hall, 1972; Kim *et al.*, 1980). This suggests that the input–output function through the nonlinear stage is a sum of both symmetric and antisymmetric components.

Kim *et al.* (1980) recorded the responses of a large number of fibres per animal (up to 418), in response to a two-tone stimulus. For each fibre, they measured the degree of phase-locking in the action potential train to the primary tones f_1 and f_2, and to the combination tone $2f_1 - f_2$. The results demonstrated the extent to which each fibre was driven by the primaries and by the combination tone. They then plotted the results as a function of fibre characteristic frequency to produce a "neurogram", or picture of activity in the whole auditory nerve fibre array. Figure 5.26 shows the neurograms for the primaries and for the combination tone. Note that the two primaries produced separate peaks of activity at their characteristic

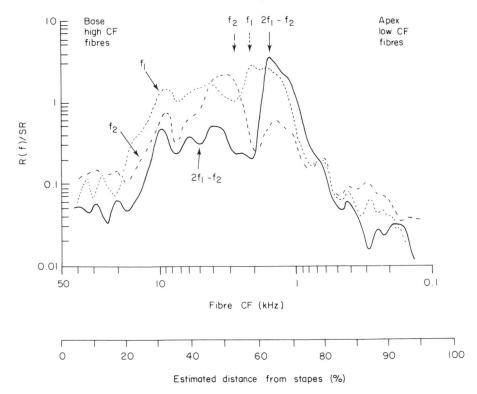

Fig. 5.26   Kim and his colleagues presented a picture of the activity evoked in the auditory nerve fibre array by a two-tone stimulus. For each of a large number of nerve fibres, they calculated the number of spikes phase-locked to the stimulus tone or combination tone of interest, divided by the number of spontaneous spikes. The results were then sideways averaged over the fibres of a small range of characteristic frequencies, to produce the running averages illustrated here.

The activity phase-locked to the primaries (f_1 and f_2) was most prominent in fibres of those characteristic frequencies (arrows). Activity phase-locked to $2f_1 - f_2$ was most prominent in fibres tuned to the frequency $2f_1 - f_2$. (There was also phase-locking to $f_2 - f_1$ which was deleted for the purposes of the illustration.) The frequencies of f_1, f_2 and $2f_1 - f_2$ are indicated by the arrows. Note that fibres of high characteristic frequency are plotted to the left of the figure, so that points on the left refer to the base of the cochlea. From Kim *et al.* (1980, Fig. 4).

place in the cochlea, and that the combination tone produced a peak at *its* characteristic place.

The results suggest that the $2f_1 - f_2$ combination tone produces its own travelling wave in the cochlea, which moves apically to peak at the $2f_1 - f_2$ place, just as though it were an externally-introduced tone. The phase data obtained in the same experiments give very strong support to this idea, since the phase of the responses increased along the cochlea in exactly the same way as did the phase of the response to single external tones. The results also suggest that the combination tone is produced where the primaries overlap, because if the cochlea was noise-damaged in only that region, the response to the combination tone was reduced in fibres tuned to the combination tone (Siegel *et al.*, 1982). The most economical hypothesis to explain the results supposes that the nonlinear active process generates, among its other outputs, a mechanical vibration tuned to $2f_1 - f_2$ when stimulated with f_1 and f_2 together. The vibration sets up its own travelling wave in the cochlear duct. The hypothesis also explains why the combination tone $2f_2 - f_1$ is not large. It is of higher frequency than the primaries, and to reach its resonant point would have to produce a wave travelling towards the base of the cochlea.

The $2f_1 - f_2$ combination tone can also be detected acoustically in the ear canal, again suggesting that an active process is involved (Brown and Kemp, 1984). The active process seems to be similar to the process producing mechanical amplification of the wave, in that it is affected by noise trauma, and by activation of the crossed olivocochlear bundle (Siegel and Kim, 1982). The last observation suggests that the outer hair cells are involved.

F. Hair Cells and Neural Excitation

1. Stimulus Coupling to Inner and Outer Hair Cells

The tips of the tallest stereocilia on outer hair cells are embedded in the tectorial membrane, while the stereocilia of inner hair cells fit loosely into a groove called Hensen's stripe without any evidence that the tips are embedded. This has suggested that, while the stereocilia of outer hair cells are moved directly by displacements of the tectorial membrane, the stereocilia of inner hair cells are moved by viscous drag of the fluid in the subtectorial space. Since viscous drag increases as the *velocity* of the flow, we would expect that for low frequencies of stimulation, inner hair cells would respond to the velocity of basilar membrane movement, while outer hair cells would respond directly to the displacement.

Sellick and Russell (1980) stimulated the cochlea with a low-frequency triangular acoustic wave, which is calculated to give a trapezoidal pattern of displacement on the basilar membrane. They showed that the above conjecture was indeed the case, with inner hair cells giving a large response in phase with the velocity of the movement, and the cochlear microphonic, generated by the outer hair cells, following the displacement (Fig. 5.27). In

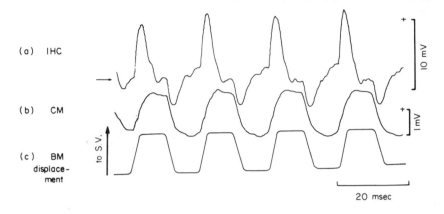

Fig. 5.27   The cochlear microphonic (CM: trace b), dominated by outer hair cells, follows the *displacement* of the basilar membrane (trace c), whereas the inner hair cells (trace a) follow the *velocity*. In both cases, movement towards the scala vestibuli caused positivity below the reticular lamina, indicating a similar functional polarization for inner and outer hair cells.

The cochlear microphonic was here recorded extracellularly in the organ of Corti. The displacement of the basilar membrane was calculated from the rate of change of sound pressure. Adapted from Sellick and Russell (1980, Fig. 1).

accordance with the functional polarization of hair cells, inner hair cell depolarization was produced during movements towards the scala vestibuli, which can be expected to displace the stereocilia in the direction of the tallest (Figs 3.1 and 3.3). Such a velocity response can only be expected for low frequencies of stimulation: at high frequencies, the viscous drag will become strong enough to convert the inner hair cell response to a displacement response. The low-frequency velocity coupling to inner hair cells suggests that the inner hair cells would not be directly sensitive to any d.c. bias in the position of the basilar membrane, but would only pick up the a.c. component in the vibration. On the other hand, outer hair cells would be directly affected by d.c. biases in the position of the membrane, and this may be important in affecting the way that they produce the hypothesized active mechanical feedback. The two types of coupling can therefore be associated with the different roles of the two types of hair cell in cochlear function, inner hair cells detecting the movement of the membrane, and the outer hair cells helping to generate it.

The phase of activation of the auditory nerve afferents by the inner hair cells is a complex issue, and one that has not been resolved. On the basis of the above analysis, we would expect auditory nerve fibres to be activated in phase with the velocity of basilar membrane displacements towards scala vestibuli. While that seems to be true for fibres tuned to low frequencies, it does not seem to be the case for fibres tuned to high frequencies (Ruggero and Rich, 1987). Possible complexities include direct stimulation of the synapse by extracellular currents generated in outer hair cells (Russell and Sellick, 1983), and the influence of cochlear micromechanics (Neely and Kim, 1986). However, certain aspects of the data are still apparently unreconcilable (Ruggero and Rich, 1987).

In spite of these problems, the *degree* of phase-locking in auditory nerve afferents seems to fit with the results of inner hair cell recording. As the stimulus frequency is raised, the a.c. component of the receptor current will tend to leak through the capacitance of the inner hair cell walls to a greater and greater extent, reducing the intracellular a.c. voltage. As described in Chapter 4, the amount of phase-locking declines directly in parallel with the reduction in intracellular a.c. voltage (Palmer and Russell, 1986).

The function of the 5–10% of auditory nerve afferents which innervate outer hair cells is still a puzzle. The only identified outer hair cell afferent to have been recorded from was silent (Robertson, 1984). On the basis of Cody and Russell's (1987) recordings, basal turn outer hair cells do not appear to give a d.c. response to acoustic stimulation near their characteristic frequency, unless the intensity is rather high. Since the membrane capacitance means that they do not generate significant a.c. voltage responses at this frequency either, Robertson's result is understandable. On the other hand, outer hair cells would be expected to respond to any d.c. biases in the position of the basilar membrane, and it is possible that their function is to signal this to the central nervous system.

G. Summary

1. Stereocilia are composed of tightly-packed actin filaments, which give them considerable rigidity. Their rootlets insert into the cuticular plate, which is also composed of actin filaments, although with a less regular organization. The stereocilia and cuticular plate also contain the proteins fimbrin, α-actinin, myosin, tropomyosin and spectrin.

2. The stereocilia on a hair cell are cross-linked, so that all stereocilia tend

to move when some are pushed. Specialized links also emerge from the tips of the shorter stereocilia. These tip links may be involved in sensory transduction.

3. Deflection of the stereocilia opens ion channels in the apical membrane of the hair cell. The cells are polarized, so that deflection in the direction of the tallest stereocilia opens the channels, and deflection in the direction of the shortest closes them. Measurement of extracellular current flow suggests that the ion channels are near the tips of the stereocilia. The best hypothesis at the moment suggests that stretch of the tip links between the stereocilia opens ion channels at one of both ends of the link.

4. There are only a few (1–4) ion channels per stereocilium. The channels appear relatively nonspecific among the cations, although the channel requires the presence of Ca^{2+} as a cofactor for its operation. The size limit of the channel is approximately 0.7 nm. The channel opens and closes with a very short latency (40 μs at 22°C). The operation of the channels is probably very straightforward, with opening being produced by a direct mechanical input to the channel. Results suggest that the channel opens and closes spontaneously under the influence of thermal energy, the mechanical input determining the relative probabilities of the open and closed states. On this theory, therefore, the channel has no absolute threshold.

5. Sharp tuning in the cochlea probably requires an active mechanical process, additional to the passive mechanical system. The evidence, which is not yet conclusive, comes from five main sources.

    (i) It is difficult to produce tuning curves of the type observed experimentally with only passive mechanical models (although they can get near to it).

    (ii) The sharp tuning is very vulnerable, and it is difficult to see how this should be so if it did not depend on a physiologically active process.

    (iii) Manipulations such as stimulation of the crossed olivocochlear bundle, which have their greatest effects on the outer hair cells, affect the sharp tuning.

    (iv) Active mechanical processes occur in the cochlea, as shown for instance by the cochlear echo. Here, if the cochlea is stimulated with sound, it returns an acoustic echo to the ear canal, which can have more energy than was originally introduced.

(v) Outer hair cells can generate movements when stimulated, although it is not yet known if they are the right form to feed back and amplify the mechanical travelling wave.

6. The resonance on the cochlear partition may be more complex than originally thought, with the tectorial membrane acting as a second tuned resonator, driven by the travelling wave on the basilar membrane. Thus, the stimulus to the outer hair cells, and hence perhaps the mechanical feedback to the travelling wave, may depend on the oscillation of the second resonant system.

7. The travelling wave in the cochlea behaves nonlinearly near its peak, with a saturating nonlinearity such that the output grows more and more slowly as the input amplitude is increased. The nonlinearity may well depend on a nonlinearity of the active mechanical process, which probably reaches its maximum contribution by about 40 dB SPL. As well as nonlinear growth of the responses, the nonlinearity produces d.c. shifts in the position of the basilar membrane during acoustic stimulation.

8. The nonlinearity produces two-tone suppression, so that one stimulus can reduce the response to another. This may occur because the first stimulus reduces the extent to which the active process amplifies the travelling wave to the other. That may well depend on the nonlinear responses of the outer hair cells, with the suppressing stimulus driving the hair cells away from the most responsive points of their input-output functions. However, the mechanism is not yet known.

9. The nonlinearity also means that the cochlea generates combination tones in response to two-tone stimuli. The most prominent is the cubic distortion tone at the frequency $2f_1 - f_2$, where the input tones are at frequencies f_1 and f_2 ($f_2 > f_1$). The cubic distortion tone is produced in the cochlea where the travelling waves to f_1 and f_2 overlap, and produces its own travelling wave to peak at its own characteristic place in the cochlea, as well as producing an emitted acoustic distortion tone in the ear canal.

10. The stereocilia of inner hair cells do not contact the tectorial membrane directly, and so are probably moved by viscous drag of the surrounding fluid. At low frequencies, therefore, inner hair cells respond to the *velocity* of the basilar membrane. The stereocilia of outer hair cells, on the other hand, make contact with the tectorial membrane, and outer

hair cells respond to the *displacement* of the membrane. Inner hair cells therefore seem suited to detecting the a.c. component in the response of the basilar membrane, which they signal to the central nervous system. Outer hair cells respond to the d.c. as well as the a.c. component. It is possible that the outer hair cell afferents signal the d.c. component in the response of the basilar membrane.

H. Further Reading

Cochlear function has been reviewed by several authors in *Hearing Research* 1986, volume **22**, and by Pickles (1985a). *Auditory Biochemistry* (ed. D.G. Drescher, C.C. Thomas, Springfield, 1985), *Neurobiology of Hearing: The Cochlea* (eds R.A. Altschuler *et al.*, Raven Press, New York, 1986), and *Peripheral Auditory Mechanisms* (eds J.B. Allen *et al.*, Springer, Berlin, 1986) have many useful chapters.

6. *The Brainstem Nuclei*

The responses of the brainstem auditory nuclei will be described in terms of the neural temporal firing patterns, neural frequency resolution, excitatory–inhibitory interactions, response to complex stimuli and, where appropriate, binaural interactions. Auditory brainstem reflexes and what little information we have on the involvement of brainstem auditory nuclei in learning will be described.

A. Considerations in Studying the Central Nervous System

Many experiments have shown that single auditory nerve fibres have qualitatively uniform properties, although the fibres may vary quantitatively in factors such as bandwidth, spontaneous firing rate, and threshold. It is therefore comparatively easy (but still difficult!) to describe the properties of the whole population from a small number of experiments. In the cochlear nucleus, however, there are many different cell types and regions. Therefore, giving a complete description is already very difficult, and it is much easier to undertake experiments and analyse the results, if we have some theories as to the function of the system.

Three themes can be discerned in the analysis of sensory systems. One concerns "feature detection". In such an analysis, we suppose that certain features of the sensory environment are selectively extracted. In the visual system, the scheme of Hubel and Wiesel (1962) has had great appeal. Here, cortical cells were described as responding selectively to lines and edges in various orientations. Unfortunately, in the auditory system, it has been difficult to describe any critical features beyond the rather elementary ones, either from psychological experiments aimed at finding important features to look for, or from electrophysiological experiments in which neuronal responses to complex stimuli were analysed. For instance, at a simple

163

level, lateral inhibition seems to emphasize the contrast in the neuronal representation of a spectral pattern. This could be said to be one example of feature detection. At a slightly more advanced level, cells in the dorsal cochlear nucleus have been found that give particularly strong responses to stimuli that are amplitude- or frequency-modulated. But even here, there seems to be a continuum in the complexity of neuronal responses, and it is very difficult to decide the extent to which such features are preferentially extracted. It is therefore difficult to decide whether we are entitled to think of such modulated stimuli as forming specific "features" of particular significance for the nervous system. These problems are compounded further, when the analysis of complex sounds such as those of speech is considered in a high-level structure such as the auditory cortex.

A second theme is the localization of functions to the activity of individual cells. In the context of feature detection, it involves finding cells that respond to specific features, so that the detection of a feature can be defined by the activity of single cells studied in isolation. At the other end of a continuum, detection might only be defined by the pattern of activity over many cells. In a common analogy, the first case might be compared to a photograph, in which each point on the photograph represents one point in space, whereas the opposite end of the continuum might be compared to a hologram, in which each point on the hologram represents many points in space, and in which individual points in space can be reconstructed only by the integration of information from many points on the hologram. Undoubtedly, many of the more complex features will only be represented in the second form, and at any level of the auditory system we might expect to find many coexisting stages in between the two extremes. In the context of feature detection, features that are not represented by the activity of individual cells can only be represented by the pattern of activity over many cells. A simple example is seen in the auditory nerve. The fibres of the auditory nerve, by their sharp tuning, appear to be specialized for the detection of specific frequencies. Sharp tuning in such a quasilinear system is necessarily correlated with poor temporal resolution. Yet sound localization experiments show the auditory system is able to detect temporal disparities of the order of 10 μs, and it is hypothesized that such accuracy is achieved by the integration of activity over many fibres.

A third theme is that of hierarchical processing, in which successively more complex analyses are performed at ascending levels of the nervous system. If the only points of interest in an acoustic environment are, say, vowel formants, then it is obviously economical to extract the formants at an early stage in the system, reject all other information, and perform further processing on the information given by the formants.

Schemes for sensory analysis based on the logical extremes of each of the three themes – that is, on the extraction of specific features, on the representation of the features in the activity of single cells, and on hierarchical analysis – naturally spring to mind. But it is likely that the auditory system operates far from these logical extremes on all three points. Such a mode of operation has contributed greatly to the difficulty of the electrophysiological analysis of the central auditory nervous system.

A successful scheme for describing the organization of the auditory central nervous system is based on quite different criteria, being anatomical. The neurones are tonotopically organized in all specific central auditory nuclei. That is, they are arranged in a regular order of best frequency along one dimension across the nucleus. Searches have been made for complementary schemes of organization, with a gradation orthogonal to the frequency axis, based on other criteria. Possible complementary schemes are based on properties such as sharpness of tuning, or binaural dominance.

B. The Cochlear Nuclei

1. Anatomy

In view of the diversity of the properties of cochlear nucleus neurones, anatomical studies are vital in aiding the physiologist in his analysis of the functions of the nuclei.

Each fibre of the auditory nerve branches on entering the nucleus, sending one branch rostrally and the other caudally. The rostral branch innervates the division known as the anteroventral cochlear nucleus, whereas the caudal branch innervates both the posteroventral division of the nucleus and the dorsal cochlear nucleus (Fig. 6.1A). We might expect the orderly arrangement of the incoming fibres to be reflected in an orderly arrangement of characteristic frequencies of the neurones they innervate, leading to a "tonotopic" frequency map. Such maps are indeed found (Rose *et al.*, 1960). But instead of two maps, one for each branch of the auditory nerve, there are in fact three, one corresponding to each of the above-named divisions of the nucleus. One map is supplied by the rostral branch of the auditory nerve, and the other two by the caudal branch. Figure 6.1C shows two of the tonotopic maps, encountered as an electrode was moved from the dorsal to the anteroventral cochlear nuclei.

These three divisions of the cochlear nucleus show broadly different response properties, and it is very likely that they have correspondingly different functions. In general, neurones of the anteroventral cochlear nucleus have properties rather similar to those of auditory nerve fibres, and may well function much as a simple relay for afferent information. Cells of

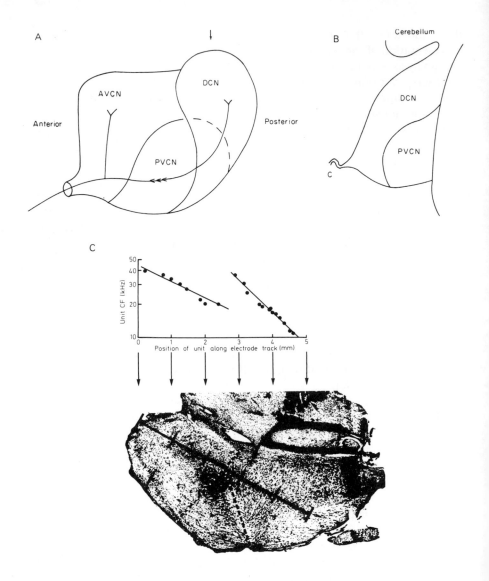

Fig. 6.1(A) A sagittal section of the cat cochlear nucleus shows the three divisions of the nucleus, innervated by a branching auditory nerve fibre. AVCN, anteroventral cochlear nucleus; PVCN, posteroventral cochlear nucleus; DCN, dorsal cochlear nucleus. (B) Transverse section of the cochlear nucleus, at the point marked by the arrow in A. C, choroid plexus. (C) The tonotopic organization of the cochlear nucleus. Two separate high–low sequences were seen as a recording electrode was moved from the dorsal to the anteroventral cochlear nucleus. Sagittal section. From Evans (1975a. Fig. 28).

the dorsal cochlear nucleus, on the other hand, have very much more complex response properties, and may therefore contribute to complex signal analysis. Their output axons bypass the next nucleus in the auditory pathway, the superior olivary complex, and end in the nuclei of the lateral lemniscus and the inferior colliculus. The properties of many neurones of the posteroventral nucleus are intermediate to those of the other two. Interestingly, the dorsal cochlear nucleus is comparatively small in primates.

The detailed study of the cells of the cochlear nucleus has led to the hope that different functional characteristics can be associated with the different cell types. The mapping of cell types is due to Osen (1969) and Brawer *et al.* (1974), both in the cat. The schemes are generally similar; the terminology used here is that of Osen. Cant and Morest (1984) discuss the relation between the different schemes of neuronal organization.

Certain areas can be defined as most obviously being occupied by certain cells. In the anterior pole of the anteroventral cochlear nucleus there is an area of large spherical cells (Fig. 6.2), although there are other cells among them. Auditory nerve fibres contact the large spherical cells by means of particularly large synaptic endings known as end-bulbs of Held, as well as by smaller endings. Caudal to this area there is an area of smaller spherical cells, and then one of globular cells (Fig. 6.2). All these types were classed as bushy cells by Brawer *et al.* (1974). Octopus cells, known as such from the pattern of their dendrites, although Morest *et al.* (1973) remark that they look more like ostrich cells (Fig. 6.3), occupy a region of the posteroventral cochlear nucleus called the octopus cell area. The area consists almost entirely of octopus cells. The other areas of the posteroventral cochlear nucleus contain a variety of cells. The dorsal cochlear nucleus caps the posteroventral nucleus both dorsally and caudally. It contains a striking layer of cells with double processes, one oriented towards the surface of the nucleus and one towards the centre. The cells have been called fusiform cells or pyramidal cells in different terminologies. There are also "giant" cells deep in the dorsal nucleus. Many other smaller cells are distributed throughout the whole cochlear nucleus, some of which are likely to be interneurones. Young (1984) has suggested that the cells of these different types form separate, parallel, systems for relaying the auditory information.

In man, many of the cell groups associated with interneurones are reduced in size, while the number of relay neurones is increased (Moore, 1987). Thus, it appears as though the amount of signal processing at the lower levels of the auditory system is reduced in man, compared with the cat.

2. Neurotransmitters

The neurotransmitter, or neurotransmitters, released by the primary auditory neurones are not certain: excitatory amino acids such as glutamate and

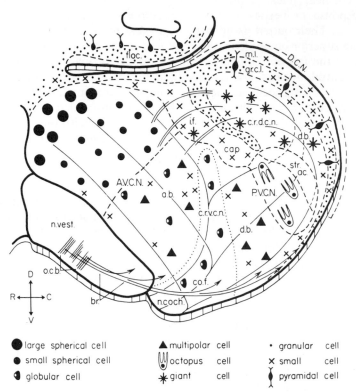

● large spherical cell	▲ multipolar cell	· granular cell
● small spherical cell	Ⓦ octopus cell	× small cell
◐ globular cell	✳ giant cell	⧎ pyramidal cell

Fig. 6.2   A cytoarchitectural map of the cochlear nucleus is shown in sagittal section. The predominant cell types in each region are represented. AVCN, antero-ventral nucleus; cap, peripheral cap of small cells; crdcn, central region of DCN; crcvn, central region of ventral nucleus; DCN, dorsal nucleus; floc, flocculus (cerebellum); gcl, granular cell layer; if, intrinsic fibres; ml, molecular layer; PVCN, posteroventral nucleus; strac, dorsal and intermediate acoustic striae. From Osen and Roth (1969, Fig. 1).

aspartate are often suggested (e.g. Godfrey *et al.*, 1984; Martin, 1985). The evidence is derived from high levels of the amino acids and their associated enzymes in the auditory nerve root, and the effects of the acids and their agonists and antagonists on cells of the cochlear nucleus. The evidence has been reviewed by Wenthold and Martin (1984).

GABA (γ-amino butyric acid) and glycine may be inhibitory transmitters, particularly in the DCN, and associated with interneurones (Godfrey *et al.*, 1978; Peyret *et al.*, 1987). Glycine has also been suggested as the transmitter of fibres which run directly between the cochlear nuclei of the two sides

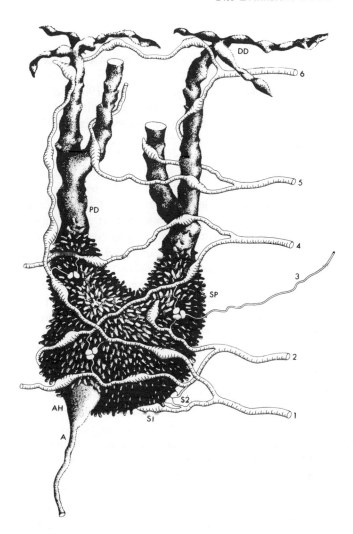

Fig. 6.3 An octopus cell. The thick auditory nerve fibres (1,2,4,5,6) give rise to large endings (S1) on the cell. They may also branch to give thin fibres, which, together with thin afferent axons (3) give rise to small secondary endings (S2). The cell is covered with stubby appendages (SP). AH, axon hillock; A, axon. From Morest *et al.* (1973, Fig. 2).

(Wenthold, 1987). Acetylcholine and noradrenaline in the cochlear nuclei are thought to be mainly transmitters of the centrifugal, or "descending", innervation arising from the central nervous system. These will be discussed further in Chapter 8.

3. Physiology

(a) Classification on the basis of response in time

In an electrophysiological experiment, Pfeiffer (1966a) classified cells of the cochlear nucleus by the apparently arbitrary, but in fact useful, criterion of the time pattern of the response to short tone bursts, delivered just above threshold at the neurone's characteristic frequency.

(*i*) *Primary-like cells.* These have post-stimulus time histograms (PSTHs) to tones similar to those of auditory nerve fibres, with an initial peak at the onset, declining gradually to lower levels (Fig. 6.4A). Such units are found throughout the ventral cochlear nucleus. In particular, those in the antero-ventral cochlear nucleus resemble auditory nerve fibres in other ways, for instance in the shape of the tuning curve, the degree of phase-locking to low-frequency stimuli, monotonic rate-intensity functions, lack of inhibitory sidebands, and relative independence of response classification on intensity. There is evidence that some of the spherical (bushy) cells of the anteroventral nucleus form at least some of the primary-like neurones. For instance, primary-like responses are obtained from the spherical cell area. Further indirect evidence comes from the waveform of the extracellular action potential. Many such recordings show a positive deflection just before the usual monophasic or diphasic waveform recorded from a cell body, and Pfeiffer (1966b) suggested that this corresponded to the depolarization of the large end-bulbs of Held, the presynaptic endings on the auditory nerve fibres on the cells. The correspondence was supported by the intracellular labelling of primary-like cells by Rhode *et al.* (1983b) and Rouiller and Ryugo (1984). The primary-like responses, together with the short synaptic delay on these cells, as well as the time pattern of the spontaneous activity, suggests the existence of what have been called "secure" synaptic connections, in which each afferent action potential produces an action potential in the output. This suggests that the cells act to relay the activity of auditory nerve fibres to the higher centres in a straightforward manner.

(*ii*) *Onset responses.* Cells showing onset responses produce a sharp peak in the PSTH at the beginning of a tone burst, and then either no activity, or a low level of sustained activity (Fig. 6.4B). Such cells are found throughout the cochlear nucleus. However, one region has proved of particular interest. The octopus cell area produces only onset responses, and the area consists almost entirely of octopus cells. Intracellular labelling confirms that octopus cells in this region generate onset responses (Rhode *et al.*, 1983b; Rouiller and Ryugo, 1984). We might suppose that there is an excitatory input, and

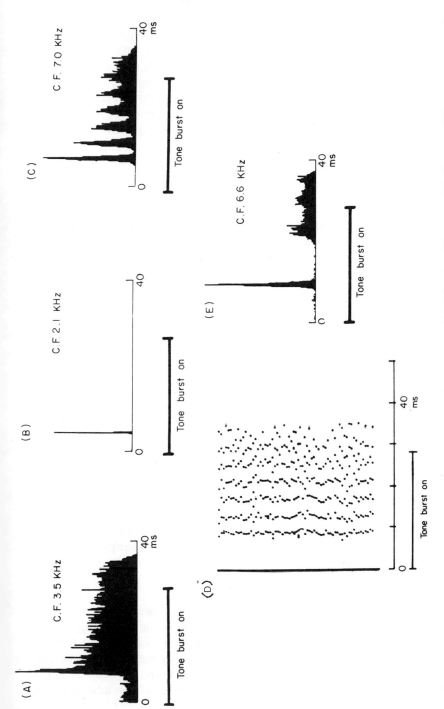

Fig. 6.4 Post-stimulus–time patterns of cells in the cochlear nucleus. (A) Primary; (B) onset; (C) chopper; (D) raster pattern of firing for chopper cell. The trace was swept horizontally after each tone pip, and successive traces are shown one above the other. The occurrence of each spike is shown by a dot. (E) Pauser. From Pfeiffer (1966a, Figs. 1 and 2).

then a delayed inhibitory input. The origin of the inhibitory input is not known. Kane (1973) showed that there were two types of synapse on octopus cells. There are large, primary endings covering much of the cell surface arising from the auditory nerve. There are also finer axons ending in smaller boutons (Fig. 6.3), and Kane suggested that the inhibitory phase might be produced by a delayed inhibition produced by the smaller endings. However, the position is uncertain, because intracellular recordings show that many types of onset cell maintain a sustained *depolarization* during the stimulus, and this is inconsistent with synaptic inhibition (Britt and Starr, 1976a). Ritz and Brownell (1982) have tentatively suggested that the reduction in firing may arise from a depolarization block following intense synaptic activation.

Presumably the inhibition following the onset response will inhibit a response to the next stimulus if the stimuli are presented rapidly enough. Such units will follow every click in a rapid train of clicks up to a certain click rate, beyond which the response drops precipitously (Godfrey *et al.*, 1975a; Rhode and Smith, 1986a). They may therefore respond to the period of complex stimuli. The cells have wide tuning curves, likely to be a correlate of the great degree of synaptic convergence in their inputs (e.g. Rhode and Smith, 1986a).

(*iii*) *Chopper responses.* Chopper units tend to fire repetitively during a sustained tone burst at a rate that is unrelated to the period of the stimulus waveform. The PSTH therefore shows a series of peaks, which, because the timing of the spikes becomes rather ragged during the latter part of the tone burst, declines towards the end (Fig. 6.4C). The increasing raggedness is better seen in the raster diagram showing the timing of the spikes during each tone burst (Fig. 6.4D). Presumably, such cells receive a large number of synaptic inputs, which summate to produce a smooth depolarizing membrane potential, with firing and resetting whenever it reaches threshold. Identification of chopper responses with any particular cell type is not possible, since the responses are found throughout the cochlear nucleus. However, they are strongly represented in some regions of the posteroventral nucleus and the deep layers of the dorsal nucleus (Godfrey *et al.*, 1975a,b; Rhode and Smith, 1986b).

(*iv*) *Pauser and buildup responses.* Pauser cells show an initial onset response, a silent period, and then a gradual resumption of activity (Fig. 6.4E). Buildup units were identified by Rose *et al.* (1959) as those that did not show the initial onset component, but whose activity increased slowly with time. Cells with these two patterns of response are found particularly in the fusiform layer of the dorsal cochlear nucleus, and intracellular labelling with horseradish peroxidase shows that at least some of the pauser and buildup cells are fusiform cells (Godfrey *et al.*, 1975b; Rhode *et al.*, 1983a). The response properties change markedly with changes in stimulus parameters, and it is likely that the temporal pattern is an indication of the complex excitatory and inhibitory inputs playing on the cells (Rhode and Smith, 1986b). It is therefore possible that such cells may be extracting certain complex features from the auditory stimulus.

(b) Patterns of excitation and inhibition

No neural inhibitory responses are seen in single fibres of the auditory nerve. All suppressive phenomena arise from the nonlinearity of the excitatory transduction process, or from rebounds following a period of excitation. However, cells of the cochlear nucleus show strong inhibition arising from inhibitory synapses. In contrast to the auditory nerve, spontaneous as well as stimulus-evoked activity can be reduced. In general, least inhibition is found in the anteroventral division of the cochlear nucleus, where the cells seem to be closest to auditory nerve fibres in their response characteristics, and increasing degrees of inhibition are found as the posteroventral, and then the dorsal cochlear nuclei, are approached.

Extensive investigations of the excitatory–inhibitory properties of cells in the cat cochlear nucleus have been carried out by Evans and Nelson (1973a) and Young (e.g. Young, 1983; Shofner and Young, 1985). At one extreme, cells are found with properties very similar to those of auditory nerve fibres, with similar response areas and no inhibitory responses beyond those arising from suppression in the auditory nerve (Fig. 6.5A). Young classed these cells as Type I cells. It is reasonable to suppose that they correspond most closely to the primary-like cells of Pfeiffer (1966a). Intermediate types of cell, called Type II and Type III cells, show excitatory tuning curves surrounded by inhibitory sidebands (Fig. 6.5B and C). Type II cells do not have spontaneous activity, and therefore their inhibitory sidebands cannot be detected directly. There is, however, evidence that the strength of inhibition is greater in Types II cells (Young, 1984). In some cells, the inhibitory sidebands can overlap the excitatory response area at high intensities, and serve to narrow down the region of excitation. Thus lateral inhibition can at high intensities increase the frequency resolving power of the cochlear nucleus, at least for tones. This only applies to the response well above threshold. It seems that the *tips* of the tuning curves are only a little narrower than in the auditory nerve (Young and Voigt, 1982). The increasing degrees of inhibition seen at high intensities are often associated with nonmonotonic rate-intensity functions (Fig. 6.6). Types II

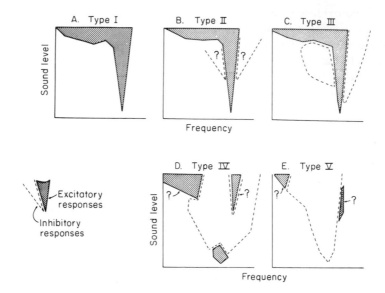

Fig. 6.5   Tuning curves of excitation and inhibition in the cat cochlear nucleus are shown in order of increasing amounts of inhibition (A–E). Purely excitatory responses as in A are predominant in the AVCN. Greater amounts of inhibition are found towards the DCN (D and E). Question marks show variable or uncertain features. From Young (1984, Fig. 12.3).

and III cells are found in all areas of the cochlear nucleus (Young, 1984; Shofner and Young, 1985).

Still stronger inhibitory phenomena are found as the sampling electrode moves towards the dorsal cochlear nucleus. In the dorsal nucleus of chloralose-anaesthetized or decerebrate animals, Evans and Nelson found neurones whose response areas were entirely or almost entirely inhibitory, perhaps possessing narrow islands of excitation (Figs. 6.5D and E). The particularly strong inhibition was blocked by barbiturate anaesthesia and therefore was not seen in earlier experiments. Pfeiffer's classification, performed in experiments under barbiturate anaesthesia, did not include any wholly inhibitory classes, but it is likely that some of his onset, buildup and pauser types resulted from delayed inhibitory inputs. These cells have been classified as Type IV and Type V cells by Young, Type IV cells being distinguished by having an island of excitation at threshold near the CF. Such cells predominate in the fusiform cell layer of the DCN, and fusiform cells make up at least some of the Type IV and Type V cells (Rhode *et al*, 1983a).

Two different sources have been suggested for the inhibitory input to the cells of the DCN. Evans and Nelson (1973b) suggested that it arose from

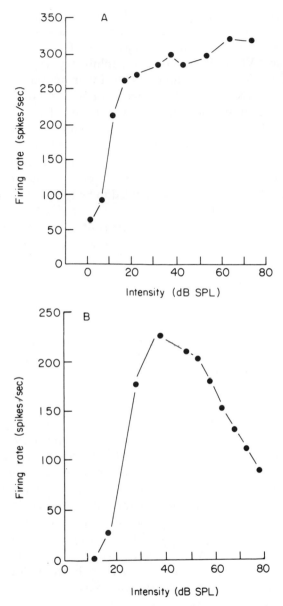

Fig. 6.6 Monotonic (A) and nonmonotonic (B) rate-intensity functions in the cochlear nucleus. A is typical of neurones showing only weak inhibitory sidebands, and B of those with strong ones. Adapted from Greenwood and Goldberg (1970, Figs 1 and 4).

the AVCN, since inhibition in DCN cells could be produced by shocks applied to the AVCN. This has been more recently supported by Shofner and Young (1987), who showed that injections of the local anaesthetic lidocaine into the AVCN, could reduce the inhibitory responses shown by cells in the DCN. Voigt and Young (1980), by an ingenious experiment involving multi-unit recording, have suggested that some of the inhibition could also arise from interneurones in the DCN. They extracted the responses of the different cells from their records, and then cross-correlated the times of firings of the cells. They were able to show that cells with excitatory centres and inhibitory surrounds, which in this experiment they called Type II/III cells, inhibited the Type IV cells in their record. Each action potential in the Type II/III cell tended to be followed by a reduction in firing in the Type IV cell.

Figures 6.7A and B show the response areas of a Type II/III cell and a Type IV cell for which just such an inhibitory relation had been shown by cross-correlation. Fig. 6.7C also shows the response area of the Type IV cell after the activity of the Type II/III cell had ceased following an injury discharge. An injury discharge is a transient high rate of firing, produced just before a cell becomes inactive after it has been damaged by an electrode. After the end of the injury discharge, the inhibitory region of the Type IV cell's response area became smaller, with excitation replacing inhibition in the frequency region of the Type II/III cell's excitatory response area. This is in agreement with the idea that the Type II/III cell had been providing an inhibitory input in this frequency region.

The analysis of excitation and inhibition suggests that a functional division between two components of the auditory pathway has already occurred at the cochlear nucleus. One pathway, arising from the ventral cochlear nucleus, preserves in many ways the response characteristics of primary auditory nerve fibres. That pathway feeds directly by secure, short-latency synapses to the superior olivary complex, where, among other things, spatial information is extracted. The other, arising from the dorsal nucleus, introduces extensive complexity early in the auditory pathway, and by bypassing the superior olivary complex feeds directly to higher centres, such as the nuclei of the lateral lemniscus and the inferior colliculus. There the results of the complex frequency analyses are combined with the results of the spatial analyses performed at the superior olive.

(c) The analysis of complex stimuli

We might expect the complex response areas of the dorsal cochlear nucleus to be appropriate for the analysis of complex stimuli. It is unfortunate that

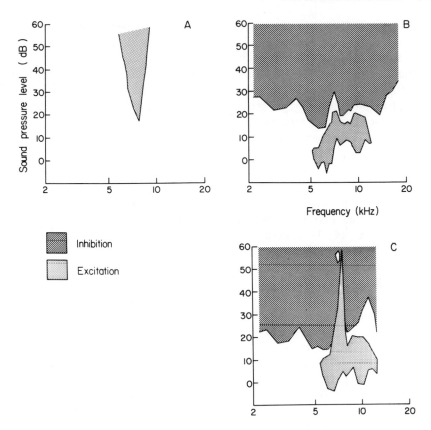

Fig. 6.7 An inhibitory relation was shown between the type II/III cell (excitatory response area in A) and the type IV cell (B). After the response of the type II/III cell had been lost following an injury discharge, the response area of the type IV cell changed (C). The original published response areas have been slightly simplified for clarity. From Pickles (1985b, Fig. 13.10), adapted from Voigt and Young (1980, Figs 4 and 6).

most of the experiments with complex stimuli have been performed under barbiturate anaesthesia, which reduced the amount of inhibition.

Stimuli which spread onto the inhibitory sidebands of neurones possessing them will, of course, reduce the firing rate. Therefore, if a band of noise is centred on the characteristic frequency of such a neurone, and increased in width, the firing rate will at first increase, as more noise falls into the excitatory response area, and then decrease, as some of the noise falls in the inhibitory sidebands. Such effects, which are analogous to those occurring in the auditory nerve as a result of suppression, were described by Greenwood and Goldberg (1970) in the cochlear nucleus. They are stronger in the nucleus than the nerve, because in the nucleus the effect of inhibition is superimposed on the effect of suppression.

When a tone is presented in wideband noise, we might expect the inhibitory sidebands to suppress the weaker parts of the stimulus pattern, so emphasizing the stronger parts. We would therefore expect them to enhance the response, relative to the background, of a tone in wideband masking noise. Gibson *et al.* (1985) showed that this was indeed the case in cells with inhibitory sidebands. We would also expect that under such circumstances the firing would depend on the net *contrast* in the stimulus pattern rather than on the overall intensity; such a point was confirmed in cells of the dorsal cochlear nucleus by Palmer and Evans (1982).

More detailed investigation shows that, although this picture may be representative for those cells with an excitatory centre and inhibitory sidebands, it may not be so for cells with more complicated response areas, similar for instance to those shown in Fig. 6.5D. Young and Brownell (1976) in unanaesthetized cats showed that broadband noise was able to drive such cells more strongly than any tone. In some cases, both tones and noise were excitatory near threshold, but at higher intensities tones became inhibitory and noise was excitatory. Such a finding may seem rather paradoxical, since we would expect the interplay of excitation and inhibition in a complex response area to *reduce* the response to broadband stimuli in comparison with that to tones. However, the responses can be understood simply in terms of the circuitry worked out by Voigt and Young (1980). As described above, they showed that the Type IV cells were themselves inhibited by those cells with the simpler organization of an excitatory centre and an inhibitory surround. If the inhibitory surround was strong, as it is in Type II cells, broadband backgrounds would completely inhibit the responses of the Type II cells. Broadband stimuli therefore inhibited an otherwise inhibitory input to the Type IV cells, leading to excitation. Such cells may, therefore, be specialized for the detection of broadband stimuli. However, it is clear that we do not have the evidence to assess the role of cells with complex response areas in the detection of complex features.

In view of the interplay of excitation and inhibition in the responses of single cells, and the likelihood of different latencies for the different inputs, it is not surprising that interesting responses can be obtained with stimuli which vary in time, such as for instance tones swept in either frequency or intensity across the response area. Dynamic factors have been shown to be important in determining the responses of such cells, because the responses to time-varying stimuli cannot be predicted from the responses to static tones. In this, they are in contrast to cells with simpler response areas, or primary nerve fibres (Britt and Starr, 1976b). Moreover, neurones with either complex and asymmetrical response areas, or many with buildup or pauser time patterns, show marked asymmetries as a tone is swept in frequency across the response area (Britt and Starr, 1976b). In some extreme cases, neurones are found which respond to a frequency sweep in one direction and not to one in the other. Britt and Starr (1976b), by intracellular recording, showed that two different types of inhibition were involved, one arising from stimulation of the inhibitory surround, and the other arising from the off-inhibition following a period of excitation.

Møller (1978) has shown that in certain cells of the cochlear nucleus frequency-modulated tones could produce stronger responses than steady tones. Sometimes the cells were responsive to very small changes in the stimulus parameters. Similar, though less dramatic, selectivities were shown for amplitude-modulated tones. Many of these cells responded to the stimulating waveform in a relatively unchanged way over a wide range of stimulus intensities. It has been hypothesized that a regulating mechanism, perhaps negative feedback from inhibitory interneurones, tends to keep the mean firing rate constant while allowing rapid changes to be transmitted (Møller, 1976b).

What is uncertain is the extent to which we are able to think in terms of feature detection in the cochlear nucleus. There seems to be a continuum of response characteristics along every category of response that has been analysed. It is not therefore certain that we are justified in asserting that the degree to which any one complex feature is extracted results from anything other than a random arrangement of excitation and inhibition on the constituent neurones. However, as a provisional hypothesis, we can assume that the cell types that have been demonstrated form the functional basis for a sensory analysis, that lies in between the logical extremes of the completely holistic and the completely nonholistic.

In the alert animal, the situation is even more complicated. Centrifugal connections from the higher centres innervate the nucleus, and the activity of the nucleus is likely to reflect later sensory processing, and the central state of the animal.

C. The Superior Olivary Complex

1. Innervation and Anatomy

There are three main outflows from the cochlear nucleus, as shown in Fig. 6.8. The fibres in the dorsal acoustic stria arise in the dorsal cochlear nucleus, and those in the intermediate acoustic stria in the posteroventral cochlear nucleus. However, the greatest outflow runs in the ventral acoustic stria or trapezoid body, and arises in both the anteroventral and posteroven-

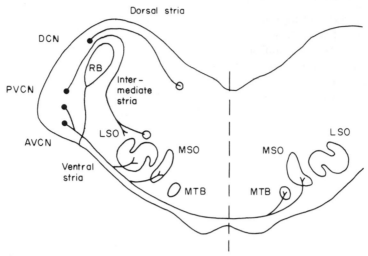

Fig. 6.8   The three main outflows of the cochlear nucleus are shown on a transverse section of the cat brainstem. The dorsal and intermediate acoustic striae pass dorsally around the restiform body (RB), or inferior cerebellar peduncle. The largest outflow is in the ventral acoustic stria or trapezoid body. AVCN, anteroventral cochlear nucleus; DCN, dorsal cochlear nucleus; MSO, medial nucleus of the superior olive; LSO, lateral nucleus of the superior olive; MTB, medial nucleus of the trapezoid body; PVCN, posteroventral cochlear nucleus. The small circles indicate fibres passing to higher levels. The fibres are represented diagrammatically, and do not necessarily branch or join as indicated. Some fibres (not shown) also run directly to the cochlear nucleus of the opposite side (Cant and Gaston, 1982).

tral nuclei. The pathways of the dorsal acoustic stria bypass the superior olivary complex and end in the next highest nuclei, the nuclei of the lateral lemniscus and the inferior colliculus, predominantly on the opposite side (Fig. 6.9). The outflows in the ventral and intermediate striae end in the superior olivary complex of both sides, as well as to a lesser extent in the nuclei of the lateral lemniscus. Within the superior olivary complex itself, several subnuclei can be distinguished (Fig. 6.10), all receiving different distributions of fibres from the different regions of the cochlear nuclei. The main nuclei associated with the ascending auditory system are the lateral and medial nuclei of the superior olive (LSO and MSO), and the medial nucleus of the trapezoid body (MTB). The nuclei are surrounded by other nuclei, known as the pre-olivary and peri-olivary nuclei, which are mainly associated with the centrifugal auditory system, although they receive an ascending input as well. They will be discussed in Chapter 8. As far as the ascending system goes, the MTB is a relay carrying information from the opposite cochlear nucleus to the ipsilateral LSO. The MSO is a disc-shaped

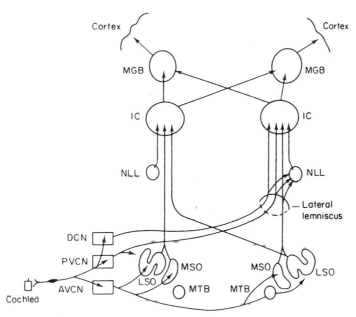

Fig. 6.9 The main ascending auditory pathways of the brainstem. Many minor pathways are not shown. IC, inferior colliculus; MGB, medial geniculate body; NLL, nucleus of the lateral lemniscus. For other abbreviations see Fig. 6.8. The branching and joining of arrows does not mean that the fibres branch or join.

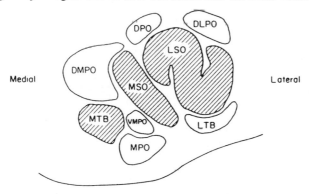

Fig. 6.10 The nuclei of the superior olivary complex are shown in a transverse section in the cat. The main nuclei associated with the ascending system are shaded. DLPO, dorsolateral peri-olivary nucleus; DMPO, dorsomedial peri-olivary nucleus; DPO, dorsal peri-olivary nucleus; LSO, lateral superior olivary nucleus; LTB, lateral nucleus of the trapezoid body; MPO, medial pre-olivary nucleus; MSO, medial superior olivary nucleus; MTB, medial nucleus of the trapezoid body; VMPO, ventromedial peri-olivary nucleus. From Harrison and Howe (1974b, Fig. 8).

structure, and receives direct fibres from the cochlear nuclei of both sides. It predominantly represents low-frequency stimuli, and is involved in detecting the direction of a sound source by means of the main cue available in low-frequency stimuli, namely temporal disparities in the waveforms at the two ears.The LSO is the largest of the component nuclei, and has a characteristic structure of a folded sheet, which in the cat is S-shaped in transverse sections, but which in other species appears more like a boxing glove. It receives direct connections from the ipsilateral cochlear nucleus, and indirect ones from the contralateral nucleus via the MTB. The LSO predominantly represents high-frequency stimuli, and in so far as it plays a role in sound localization, detects the main cues available in high frequency stimuli, namely disparities in interaural intensity, and to a lesser extent ·disparities in the onset time of the stimulus envelopes. The MSO in man is relatively large, and the LSO relatively small (Moore, 1987).

2. Physiology and Function

(a) Introduction

Analysis of the dorsal cochlear nucleus has demonstrated that it may be difficult to assess the functional significance of a nucleus from the response characteristics of its neurones. However, the superior olive shows an advance in sensory processing, that of receiving information from the two ears, which surely must be of functional significance. It is reasonable to suppose that the nucleus plays a part in sound localization.

(b) The lateral superior olive (S-segment)

The principal cells of the LSO have dendritic trees joining the two surfaces of the folded sheet of the nucleus, with afferents from the ipsilateral cochlear nucleus and the ipsilateral MTB contacting the two branches (Fig. 6.11A). The cell characteristic frequencies are arranged tonotopically (Fig. 6.11B), although for a long time this was difficult to detect because of the complex folding of the structure (Tsuchitani and Boudreau, 1966; Glendenning *et al.* 1985). In the anaesthetized preparation, the responses to ipsilateral stimuli are entirely excitatory with tuning curves similar to those of primary fibres, best thresholds in the range 10–20 dB SPL, and chopper time patterns (Tsuchitani, 1977). Brownell *et al.* (1979) have shown that when unanaesthetized cats are used, inhibitory sidebands can be seen to ipsilateral stimuli, and the chopper pattern disappears. Figure 6.12 shows an example of the response to a stimulus at the excitatory centre, and in the inhibitory sidebands. The organization of the excitatory response is more complex

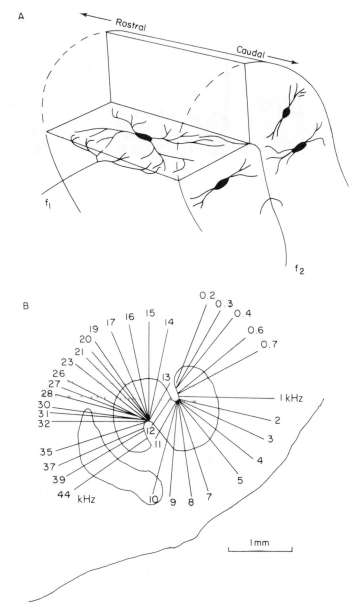

Fig. 6.11   (A) The neural organization of the LSO. f_1, f_2, afferent fibres. Adapted from Scheibel and Scheibel (1974). (B) The tonotopic organization of the LSO. A high proportion of the nucleus is devoted to high frequencies. The numbers denote the characteristic frequencies of the sectors. From Tsuchitani and Boudreau (1966, Fig. 6).

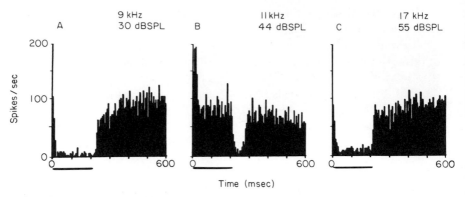

Fig. 6.12   Post-stimulus–time histograms of a cell in the LSO. From Brownell *et al.* (1979, Fig. 1).

than that of the inhibitory one, since it is followed by a complementary rebound. Unlike the responses to ipsilateral stimuli, the responses to contra-lateral stimuli are predominantly inhibitory in both anaesthetized and unanaesthetized animals. Therefore, the majority of binaural neurones in anaesthetized animals are excited by ipsilateral stimuli and inhibited by contralateral ones, forming the so-called "EI" neurones. The threshold and tuning of the ispsilateral excitatory and contralateral inhibitory effects are often comparable (Caird and Klinke, 1983). We might, therefore, imagine that the LSO responds to differences in intensity at the two ears, defined on a spectral basis. Sounds from a source on, say, the left side, will be more intense in the left ear, and the LSO might use this difference as a cue in sound localization. Sounds from a source on the left side will also be delayed in the right ear, and Caird and Klinke (1983) showed that delaying the onset of the stimulus to the inhibitory ear increased the response in EI cells. The LSO therefore appears to perform sound localization on the basis of intensity and timing differences, coding sounds on the ipsilateral side. The intensity differences will be largest at high frequencies, where the degree of diffraction around the head will be small. In accordance with this, most of the LSO is devoted to high frequencies (Fig. 6.11B).

(c) The medial superior olive (accessory olive)

The MSO receives a direct innervation from the anteroventral cochlear nucleus of both sides. The axons innervate opposite sides of the sheet of cells constituting the MSO. It will be remembered that most cells of the anteroventral cochlear nucleus have short-latency, secure, synaptic connections by means of the large end-bulbs of Held, and so the cells of the

MSO are able to receive temporally matched and temporally accurate signals from the two ears.

Single unit studies of the MSO face severe difficulties, because the sheet of cells is thin, and because gross potentials from neighbouring cell groups tend to swamp the single unit action potentials. Such studies as have been successful in identifying cells in the nucleus, found that most were excited by stimuli at both ears, with relatively simple tuning curves and simple temporal patterns of discharge (Guinan *et al.*, 1972; Goldberg and Brown, 1968, 1969). In the dog, which has a particularly large MSO, Goldberg and Brown found that almost all the cells were binaural. Three-quarters of the cells were excited by both ears (EE cells), and the remainder were excited by one and inhibited by the other (EI cells). Many of the low-frequency units with CFs less than 1 kHz were responsive to the relative phases of the stimulating sinusoids at the two ears. Figure 6.13A shows the discharge rate of an MSO neurone as a function of interaural time delay. The firing rate showed a cyclic dependence. The period of the cycle was equal to the period of the sound stimulus. The neurone was therefore responsive to the interaural time difference. This interpretation was supported by experiments in which the stimulus frequency was varied. The time disparity for the optimal response was independent of stimulus frequency, showing that the cells were responsive to time disparity, rather than phase disparity (Fig. 6. 13B). We might suppose that this arose from a difference in the speed of transmission of signals from the two ears. Such a position was supported by the timing of the discharges in response to stimuli at the two ears separately (Fig. 6.14). The two delays calculated for ipsilateral and contralateral stimuli separately could be used to predict the optimal interaural phase disparity. In this way, when the relative phases of the two stimuli were adjusted so that their excitatory effects coincided, there was a large binaural response (Goldberg and Brown, 1969; Moushegian *et al.*, 1975). This was true for neurones where both stimuli had a net excitatory effect (EE neurones), as it was for many EI neurones. Such experiments lead to the notion of a *characteristic delay*, different for each cell, and resulting from differences in the speed of transmission of signals from the two ears. The notion of a characteristic delay, and its relation to the mapping of sound source direction, will be discussed further in Chapter 9.

It appears that the stimuli in the two ears cause cyclic phases of inhibition as well as excitation for both EE and EI neurones, because the response to binaural stimuli in the most effective phase relation could be larger than, and in the least effective smaller than, the response to either monaural stimuli alone (Fig. 6.13A). In the excitatory phase relation, the interaction between stimuli was one of facilitation, because binaural stimuli could drive

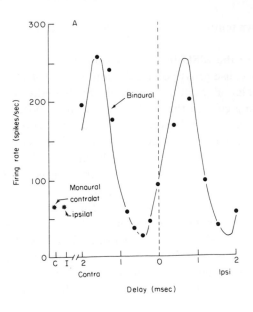

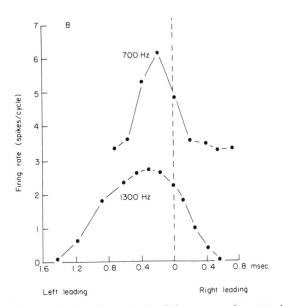

Fig. 6.13   (A) The cyclic function relating firing rate to interaural delay for a cell in the MSO. CF = 444 Hz, period of stimulus = 2.25 ms. From Goldberg and Brown (1969, Fig. 6). (B) Firing rate as a function of interaural delay for sinusoids of two different frequencies. The delay giving the greatest response is independent of stimulus frequency. Kangaroo rat MSO. From Crow *et al.* (1978, Fig. 2).

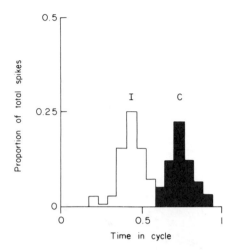

Fig. 6.14   The histograms of latencies of firing to ipsilateral (I) and contralateral (C) stimuli, made with respect to constant phase of the stimulus waveform, indicate different delays in transmission from the two ears. From Goldberg and Brown (1969, Fig. 7).

the neurones far harder than could even more intense monaural stimuli (Fig. 6.15).

Studies in which the uptake of labelled 2-deoxyglucose has been measured, suggest that the MSO is driven most strongly by sounds on the contralateral side (Masterton and Imig, 1984).

We can summarize the role of the LSO and MSO in sound localization as follows. In the LSO most cells are EI cells and so respond to intensity differences between the ears. The majority of the cells are high-frequency cells, and so have characteristic frequencies in the range where a sound source to one side will produce significant interaural intensity differences. In addition, the cells are sensitive to temporal disparities in the time of arrival of the sound waveforms. By contrast, in the MSO the majority of cells are EE cells, and so will not respond to interaural intensity differences. The cells are able to respond to direction only on the basis of interaural temporal cues, and in general are most responsive to low frequencies. Since this is the range in which phase information is preserved, sensitivity to timing differences will be apparent as a sensitivity to interaural phase differences. Anatomical evidence suggests a picture in agreement with this view. Animals with small heads and good high-frequency hearing, who might be expected to favour intensity cues, have large LSO nuclei and small MSO nuclei, whereas animals with large heads and good low-frequency hearing, tend to have the reverse (Masterton and Diamond, 1967).

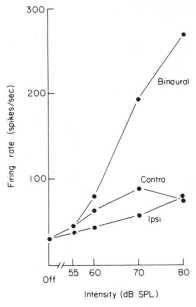

Fig. 6.15   Rate-intensity functions are shown for monaural and binaural stimuli in the dog superior olive. The binaural stimulus was in the optimal phase relation, and could produce a greater firing rate than could either monaural stimulus. From Goldberg and Brown (1969, Fig. 11).

(d) The medial nucleus of the trapezoid body

The medial nucleus of the trapezoid body (MTB) is the relay conveying information from the cochlear nucleus of the opposite side to the lateral superior olive ipsilateral to the MTB. The afferent axons are the largest fibres of the trapezoid body, and end as large end-bulbs of Held. The principal cells therefore receive secure, short-latency, synaptic inputs, as indeed do the cells of the anteroventral cochlear nucleus from which the axons arise, and so are able to influence the LSO with only a short delay. As we might expect, cells of the MTB have simple discharge characteristics, similar to those seen in the anteroventral cochlear nucleus (Goldberg *et al.*, 1964). In addition to the pathways described above, the nucleus receives projections from the fibres projecting to the LSO, and sends axons not only to the ipsilateral LSO but to the surrounding subnuclei as well. Other types of cell and other types of response have also been described. For instance, the olivocochlear bundle, the centrifugal pathway to the cochlea, with its complex response characteristics, also arises partly from the MTB (Warr, 1975). The olivocochlear bundle will be discussed further in Chapter 8.

D. The Ascending Pathways of the Brainstem

The principal ascending pathways were shown diagrammatically in Fig. 6.9.

The main receiving station for the ascending pathways from the superior olivary complex is the inferior colliculus. The LSO projects bilaterally to the inferior colliculus, whereas the MSO projects only ipsilaterally (Adams, 1979; Elverland, 1978). The fibres run in the tract known as the lateral lemniscus. Some send collaterals to the nuclei of the lateral lemniscus. The ventral nucleus of the lateral lemniscus receives an input from the contralateral cochlear nucleus and projects ipsilaterally to the inferior colliculus, whereas the dorsal nucleus of the lateral lemniscus receives a bilateral input and projects bilaterally to the inferior colliculus (Masterton and Imig, 1984). The inferior colliculus also receives direct afferents from the contralateral dorsal cochlear nucleus, from the contralateral posteroventral nucleus, and to a lesser extent from the contralateral anteroventral nucleus (Adams, 1979). Most paths that cross do so at or near the level of the trapezoid body, although the fibres from the dorsal cochlear nucleus cross further rostrally. There is also a smaller uncrossed projection from these nuclei. Interestingly, the direct projection of the cells with simpler response properties, such as the large spherical cells, the globular cells and the octopus cells, to the inferior colliculus seems particularly small. The inferior colliculus is therefore a site of convergence of projections with complex frequency responses but a monaural input from the dorsal cochlear nucleus, and projections with rather simpler frequency responses but a binaural input, from the superior olive. Because the inferior colliculus is tonotopically organized, fibres from different sources but of the same characteristic frequency apparently manage to meet at the appropriate site. The inferior colliculus is an obligatory relay for practically all the auditory input to the medial geniculate body, since injections of horseradish peroxidase into the tracts leading into the medial geniculate give rise to massive labelling in the inferior colliculus, but very little elsewhere (Aitkin and Phillips, 1984).

E. The Inferior Colliculus

1. General Anatomy

The inferior colliculi form the rear pair of a set of four lobes on the dorsal surface of the brainstem. The anterior pair, the superior colliculi, are an important visual reflex centre. The inferior colliculi are an auditory relay and reflex centre. There are three main divisions to the inferior colliculus. The central nucleus (Fig. 6.16A) was defined by Morest and Oliver (1984) in

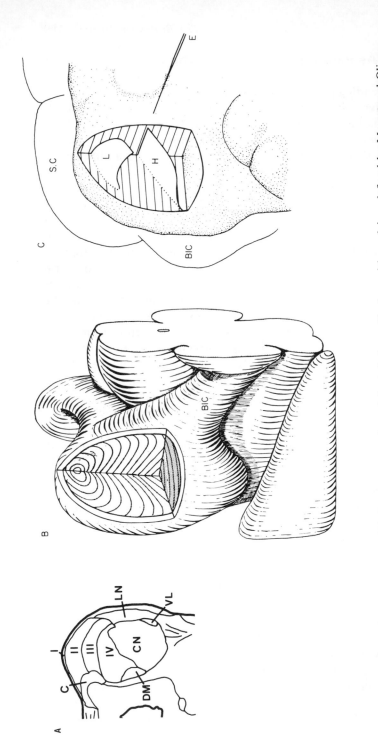

Fig. 6.16 The inferior colliculus. (A) Transverse section of the inferior colliculus, with nuclei as defined by Morest and Oliver (1984). CN: central nucleus; I, II, III, IV: layers I–IV of dorsal cortex. The paracentral nuclei include: LN: lateral nucleus; C: commissural nucleus; DM: dorsomedial nucleus; VL: ventrolateral nucleus. From Morest and Oliver (1984), Fig. 21. (B) Rostrolateral view of the inferior colliculus, showing laminae. It is now recognized that the laminae are not continuous as far laterally as this drawing suggests, and that they do not show the tight curvature dorsally which is shown here. BIC: brachium of the inferior colliculus. From Rockel and Jones (1973a), Fig. 22. (C) Low-frequency (L) and high-frequency (H) iso-frequency sheets in the central nucleus of the inferior colliculus, seen in caudolateral view. BIC: brachium of inferior colliculus; E: electrode; SC: superior colliculus. From Semple and Aitkin (1979), Fig. 5.

Golgi stained sections as being limited to a strongly laminated central region. This definition has superseded an earlier parcellation by Rockel and Jones (1973a), in which the central nucleus stretches more dorsally to include a non-laminated area. Above the central nucleus is a region known as the dorsal cortex, divided into four layers (Morest and Oliver, 1984). This region includes regions which Rockel and Jones described as a pericentral nucleus and the dorsal part of their central nucleus. The lateral nucleus (partly coextensive with the external nucleus of Rockel and Jones) forms part of what Morest and Oliver called the paracentral nuclei. These nuclei are scattered around the central nuclei. In contrast to the rest of the inferior colliculus, the paracentral nuclei are primarily somasthetic and auditory integrative areas rather than an auditory relay (Robards, 1979; Aitkin et al., 1978). The inferior colliculus in man has the same general form as in the cat (Moore, 1987).

2. The Central Nucleus

(a) The spatial organization of the nucleus and its afferents

Oliver and Morest (1984) have given an authoritative account of the structure of the central nucleus. The nucleus has a pronounced laminar structure. The laminae are strongly tilted, so that as one moves laterally in the central nucleus the laminae move ventrally (Fig. 6.16B). At the lateral edge of the central nucleus, the laminae suddenly turn and sweep dorsally. The laminae are formed by layering of the afferent axons and the dendrites of the intrinsic neurones (Rockel and Jones, 1973a). Semple and Aitkin (1979) associated these sheets with iso-frequency sheets of cells (Fig. 6.16C). Low frequencies were found in the dorsal sheets, and high frequencies in the ventral ones.

Interesting overlaps and segregations have been described in the innervation of the central nucleus. Mention has already been made of the wide range of nuclei that project to the inferior colliculus. Roth et al. (1978) showed that any particular injection of horseradish peroxidase (HRP) into a confined area of the central nucleus led to reaction product, transported in a retrograde direction, in only some, but never at once all, of the nuclei labelled by large injections. Aitkin and Shuck (1985) and Maffi and Aitkin (1987) suggested that the different projecting nuclei sent afferents to segregated target zones within the central nucleus. The site of origin of some of the afferents, too, seems to be arranged in a patchy manner (Adams, 1979). Retrograde transport of HRP showed alternate labelled and unlabelled columns of cells in the anteroventral cochlear nucleus. Therefore, while there must be a great deal of convergence onto the inferior colliculi, there seems to be a microstructure in the sites of both the origin and termination which could well be of functional significance.

A similar patchiness was found electrophysiologically in the colliculus by Roth *et al.* (1978). It often appeared that groups of adjacent cells had similar response properties, so that a microelectrode might meet groups of several intensity-sensitive EI cells together, then several time-sensitive EE cells, and so on, even if all were of the same characteristic frequency. A corresponding point was made over a larger scale by Semple and Aitkin (1979), who showed that neurones with different types of binaural interaction were segregated into different parts of the iso-frequency sheet. EE neurones were most common medially, whereas EI neurones were most common rostrally. Monaurally driven units were encountered more caudally, ventrally and laterally. Binaural time-sensitive neurones were almost completely segregated from the ones sensitive to intensity differences, and were encountered rostrally, dorsally and laterally. It seems that there is a considerable segregation of function within each iso-frequency sheet, both generally across the nucleus and on a much smaller scale.

(b) Electrophysiology

In response to single tones, tuning curves and temporal response patterns can be described. Tuning curves show a wide range of bandwidths, some being very broad and some very narrow. Aitkin *et al.* (1975) showed extraordinarily high Q_{10} values of 25–40 for a frequency range around 10 kHz. Such tuning is unrivalled elsewhere in the auditory system. The whole question of a progression to sharper and sharper tuning at higher levels of the auditory system has been controversial. Katsuki *et al.* (1959) suggested that tuning curves became sharper and sharper from the auditory nerve to the medial geniculate body, after which they became broader. Kiang (1965) and Aitkin and Webster (1972) contradicted this progression at the levels of the cochlear nucleus and the medial geniculate body respectively. It should be pointed out that when we analyse frequency resolution it is important to distinguish the frequency resolving power shown by the tip, described by the Q_{10} measure, from that shown well above threshold. Q_{10}'s indicate, except for the above report of Aitkin *et al.* (1975), similar or deteriorating tuning at the higher levels of the nervous system. Any such increase in resolution could be produced only by as yet unknown mechanisms that increased the fundamental resolving power of the auditory system. In contrast, well above threshold, some high-level neurones show narrow bandwidths of excitation to tones, much narrower than for instance those shown by primary auditory nerve fibres. Here it seems that a simple mechanism of lateral inhibition could help preserve at high intensities the resolution seen at low intensities, by narrowing down the response area. Katsuki *et al.* (1959) may have sampled such neurones selectively in their report; other reports indicate that the proportion of broadly tuned neurones increases at higher levels of the nervous system.

Complex excitatory–inhibitory interactions have been found in the inferior colliculus, although they have been described in less detail than in, say, the cochlear nucleus (e.g. Ryan and Miller, 1978). Half of the neurones show the nonmonotonic rate-intensity functions suggestive of complex excitatory–inhibitory interactions. Temporal patterns of response show many onset and pauser types (Rose *et al.*, 1963), with transitions between the types as the stimulus intensity is varied (Ryan and Miller, 1978). For instance, Ryan and Miller describe a common transition from primary-like to pauser, and then to onset, with changes in intensity. Unanaesthetized animals may show sustained excitatory and inhibitory responses rather than onset or pauser ones (Bock *et al.*, 1972), although this has since been disputed (Ryan and Miller, 1978). Some neurones are found to be sensitive to amplitude or frequency modulation and are specifically responsive to a certain speed or direction of modulation (Nelson *et al.*, 1966). As in the dorsal cochlear nucleus, the response to time-varying stimuli cannot necessarily be predicted from the response to tones. But we do not have enough information to decide whether the processing of complex monaural stimuli in the inferior colliculus shows significant advances over that in the dorsal cochlear nucleus.

As in the various nuclei of the superior olive, many neurones are sensitive to interaural timing or intensity differences (Rose *et al.*, 1966; Yin *et al.*, 1986; Caird and Klinke, 1987). As in the MSO, many of the low-frequency time-sensitive neurones seem to have a characteristic delay which is independent of stimulus frequency (Yin and Kuwada, 1983). Most of the neurones are predominantly sensitive to sounds on the contralateral side. This is the same as the predominant representation in the MSO, which sends its outflow mainly to the ipsilateral colliculus. The contribution from the superior olive is more puzzling, since this nucleus, which represents the ipsilateral side, sends bilateral projections to the colliculus (e.g. Roth *et al.*, 1978). A solution was suggested by Glendenning and Masterton (1983), who showed that the main ipsilateral projection from the LSO arose from the lateral side of the LSO, and the main contralateral projection arose from the medial side. Since the medial side codes high frequencies, and since we would expect the high-frequency EI cells of the LSO to code sound direction particularly well, this suggests that the main laterality-specific outflow is crossed, in agreement with the representation in the colliculus.

The sensitivity to interaural timing and intensity differences suggests that neurones should be able to represent the direction of a sound source. By moving a speaker around a cat's head, Aitkin *et al.* (1984, 1985) found that many of the neurones could indeed represent the direction of a real sound source. There seems to be some mapping of source direction, with sources to the midline represented caudally, and those to the side represented rostrally. For high frequency stimuli some of the directional selectivity arose from the directionality of the pinna, since for such stimuli the selectivity could be altered by manipulating the pinna nearest the sound source (Aitkin *et al.*, 1984). However, and in contrast to the position in mammals, an apparently detailed map of auditory space has been found in the barn owl, in its homologue of the inferior colliculus, the lateral dorsal mesencephalic nucleus (Knudsen and Konishi, 1978). Neurones in the lateral rim of the nucleus coded the direction of a real sound source. Points high in space were represented high in the nucleus, and points low were represented low. Points forward were represented anteriorly, and points to the side were represented posteriorly. Each nucleus represented space on the contralateral side, although in front the field crossed 15° over to the ipsilateral side. The owl achieves this map with two specializations which do not appear in the mammal. Firstly, the two ears are set at different heights on the head, and this, combined with the ruff feathers around the head, means that elevation of the source is coded as an interaural intensity difference. Secondly, the owl is sensitive to far smaller interaural time differences than are mammals (Moiseff and Konishi, 1983). The representation of auditory space will be examined in more detail in Chapter 9.

The inferior colliculus therefore seems to combine the complex frequency analysis of the dorsal cochlear nucleus with the sound localizing ability of the superior olive. Very little is known of the details of the interactions between the inputs from the two sources. The convergence may, however, give us a clue to a special function for the inferior colliculus. Localizing sound by interaural time disparities requires the preservation of accurate time relations. These are lost in the dorsal cochlear nucleus by the circuitry needed for complex amplitude and frequency analysis. It is therefore appropriate that the direction of the sound should be extracted separately. The inferior colliculus, by combining information from both sources, might therefore be able to code simultaneously the complexity of sounds and their direction in space.

The inferior colliculus has an important role in many auditory reflexes. They will be discussed below in Section G of this chapter and in Chapter 9.

3. The Dorsal Cortex and Paracentral Nuclei

The dorsal cortex receives a substantial somatosensory input as well as an auditory input (Aitkin *et al.*, 1981). While the cells have broad or complex tuning curves to sound, and show responses to somatosensory stimuli, they are nevertheless tonotopically organized (Aitkin, 1979; Aitkin *et al.*, 1975. On the other hand, the paracentral nuceli can be viewed more as soma- esthetic and integrative areas than as an auditory relay. Tuning curves to auditory stimuli are very broad, so much so that the definition of a best frequency is often arbitrary (Fig. 6.17). Aitkin *et al.* (1978) found that 54%

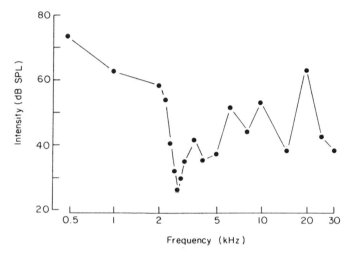

Fig. 6.17 The broad tuning curve of a neurone in a paracentral nucleus of the inferior colliculus. From Aitkin *et al.* (1978, Fig. 5).

of neurones recorded were bimodal, most of these being excited by auditory, and inhibited by somatosensory, stimuli (see also Aitkin *et al.*, 1981). Very little is known about the functions of these nuclei. The nuclei mark the first appearance of a 'diffuse', 'belt', or nonspecific auditory system, which surrounds the specific (or 'core') auditory relay system. This division into specific and diffuse auditory systems is carried up through the medial geniculate body to the auditory cortex.

F. The Medial Geniculate Body

1. Anatomy

The medial geniculate body contains the specific thalamic auditory relay of the auditory system, receiving afferents from the inferior colliculus, and projecting to the cerebral cortex. The medial geniculate body has been divided in different ways by different anatomists. According to the scheme

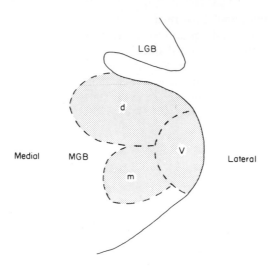

Fig. 6.18   Divisions of the medial geniculate body (MGB). v, ventral (principal) division of MGB, d; dorsal division of MGB; LGB, lateral geniculate body; m, medial division of MGB. From Harrison and Howe (1974a, Fig. 8).

of Morest (1964), there is a ventral (or lateral) medium-celled principal division (Fig. 6.18), a dorsal division, and a medial, large-celled division. Of these, only the ventral division has any claims to being a specific auditory relay. Its afferents run mainly ipsilaterally from the central nucleus of the inferior colliculus (Calford and Aitkin, 1983). The fibres run in the brachium of the inferior colliculus, a bulge on the lateral surface of the brainstem between the inferior colliculus and the geniculate bodies (Fig. 6.16B). The medial and dorsal divisions receive a multiplicity of inputs, the former receiving auditory afferents from the inferior colliculus and the lateral tegmental system running just medial to the brachium of the inferior colliculus, as well as somatosensory afferents, and the latter receiving afferents from the region medial to the brachium, the superior colliculus, the pericentral nucleus of the inferior colliculus, and the somatosensory system (e.g. Harrison and Howe, 1974a; Calford and Aitkin, 1983). They therefore can be viewed as forming a 'diffuse' or nonspecific auditory system surrounding the specific auditory nuclei.

The ventral division projects principally to the AI area of the auditory cortex, and in addition to the anterior, posterior and possibly also the ventral posterior fields (defined in Fig. 7.1). In contrast, the medial division projects to the auditory cortex in a less specific way, and the dorsal division projects to AII, Ep and I-T. Calford and Aitkin (1983) suggest that the different areas of the medial geniculate are parts of three separate projection pathways to the auditory cortex.

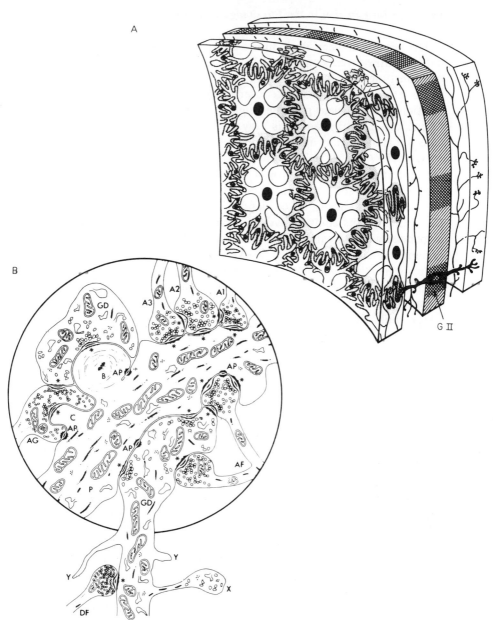

Fig. 6.19 (A) Dendritic trees within the laminae of the medial geniculate body. G II, Golgi type 2 interneurone. From Majorossy and Kiss (1976, Fig. 7). (B) Synaptic contacts in the synaptic nest between the afferent fibre (AF) dendrite of the principal cell (running across the middle of the diagram), and a dendrite of a Golgi type 2 cell (GD). DF, Descending (centrifugal) fibre from cortex. From Morest (1975, Fig. 22).

The ventral division itself shows a laminar structure, the laminae being flat sheet-like layers consisting of both the afferent fibres and dendrites of the constituent neurones (Morest, 1965). Over much of the nucleus the sheets are curved and oriented vertically. Tonotopic organization in this division of the nucleus produces high frequencies located medially and low frequencies laterally, and it is very likely therefore that the sheets of cells detected anatomically are iso-frequency planes (Aitkin and Webster, 1972; Merzenich *et al.*, 1977). The ventral division contains two cell types, namely the principal cells, which project to the auditory cortex, and Golgi type 2 cells, which are short axon interneurones (Morest, 1975). The principal cells have characteristically tufted dendritic trees (Fig. 6.19A; Majorossy and Kiss, 1976). This comparatively simple cellular complement is, however, associated with a great deal of complexity in the local synaptic and dendritic organization, which has been worked out in detail (Majorossy and Kiss, 1976; Winer, 1985a). The interneurones make dendro-dendritic synapses with the principal neurones in terminal clusters called "synaptic nests", containing three-way synaptic contacts between the different cell types. These could lead to gating by descending fibres from the cerebral cortex, as well as to complex transformations of the afferent activity (Fig. 6.19B).

2. Physiology

Neurones in the ventral division, the specific auditory relay, show responses to sound, with best frequencies tonotopically organized over the majority of the division. Tuning curves are relatively sharp, although only a very small proportion seem more sharply tuned than primary auditory nerve fibres, and then by only a small amount (Aitkin and Webster, 1972). Aitkin and Webster did not report any cells with extraordinarily sharp tuning, as in the inferior colliculus. In their temporal firing patterns, many cells are onset or sustained excitatory types in the anaesthetized preparation, although sustained excitation or inhibition were more common in the unanaesthetized preparation (Calford, 1983; Aitkin and Prain, 1974). Inhibitory sidebands existed in the medial geniculate body as in other brainstem nuclei, and as the stimulus frequency was changed, complex excitatory-inhibitory interactions became visible. In general, the neurones with complex temporal properties have nonmonotonic rate-intensity functions and complex frequency response areas, as would be expected for neurones with a multiplicity of excitatory and inhibitory inputs (Aitkin and Prain, 1974).

As in the inferior colliculus, a high proportion of neurones are binaurally sensitive (Aitkin and Webster, 1972). Some, mainly high-frequency units, are predominantly sensitive to interaural intensity differences. Others, mainly low-frequency units, are predominantly sensitive to interaural time differences. The tuning curves for stimuli at the two ears separately are approximately similar, although not exactly so (Fig. 6.20). Units sensitive

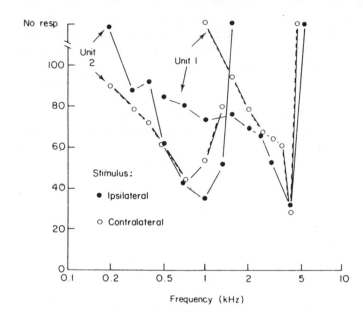

Fig. 6.20   Tuning curves in the ventral division of the medial geniculate body to ipsilateral and contralateral stimuli. From Aitkin and Webster (1972, Fig. 5).

to differences in interaural intensity may show very sharp sensitivity to the intensity differences, sometimes being driven over 80% of their firing range by a change of only 2 dB in interaural disparity.

   Such studies, again, give very little idea of the sensory transformations occurring in the medial geniculate body. Thus attempts have been made to find neurones specifically responsive to the features of complex sounds. Whitfield and Purser (1972) reported that some units responded only to complex sounds and not to tones. Smolders *et al.* (1979) compared the responses of both medial geniculate and cochlear nucleus neurones to tones and the complex sounds produced by cats' vocalizations. They reported that, at least for the neurones studied in the posteroventral cochlear nucleus, the response to complex sounds could be predicted reasonably well from the response to tones. This was, however, not the case in the medial geniculate body. David *et al.* (1977) have shown cells in the medial geniculate of the unanaesthetized cat apparently sensitive only to specific speech parameters or to other complex features. It is obvious that such experiments designed to uncover either specific transformations in a nucleus or specific feature detectors face enormous difficulties. It will be recalled that even as early as the dorsal cochlear nucleus there are neurones that will respond to broadband stimuli but not to tones (Young and Brownell, 1976). Such neurones may have led to reports of neurones in the cochlear nucleus responding only to "complex" stimuli. A great deal of care is needed to distinguish the different types of complex response, and this may be beyond the limits of our present techniques in a nucleus as high in the system as the medial geniculate. It is also obvious that the central state of the animal influences the responses. Thus Whitfield and Purser (1972) noted that in the freely moving animal the responses in the medial geniculate body were labile, with, say, bands of excitation and inhibition appearing and disappearing over time.

In the medial division, part of the "diffuse" auditory system which projects generally to the auditory cortex, Aitkin (1973) in the unanaesthetized cat showed very wide and complex response areas, with many onset responses and much variability and habituation. Three-quarters of the units were binaural. We expect there to be multimodal interactions in the nucleus, and Aitkin suggested that the wide response areas were a reflection of this nonspecificity. Units in the deep division of the dorsal nucleus showed short-latency and sharply-tuned responses to sound in the anaesthetized cat; those situated more dorsally and medially (in the caudodorsal and suprageniculate divisions of the dorsal nucleus) responded more weakly and inconsistently, with much habituation (Calford and Aitkin, 1983).

G. Brainstem Reflexes

The brainstem is the main auditory reflex centre of lower vertebrates, and it would be surprising if some of these functions were not retained in man and other mammals.

1. Unlearned Reflexes

One of the most elementary auditory reflexes is the middle ear muscle reflex. The tensor tympani and stapedius muscle contract reflexly to loud sounds. Borg (1973) showed that an arc of three to four neurones was involved, consisting of a projection from the ventral cochlear nucleus to the MSO and then to the motor nuclei of the facial and trigeminal nerves. But in addition to this short-latency pathway, he presented evidence for a slower pathway, perhaps projecting via the red nucleus or the reticular formation, both of which receive an auditory input.

At a rather higher level, the inferior colliculus has also been implicated in many auditory reflexes. It has been implicated in a reaction known as the auditory startle response, in which a sudden sound produces a characteristic and widespread muscular contraction. Fox (1979) showed that the reaction persisted in rats decerebrated above the inferior colliculus, but that additional lesions of the colliculi abolished the response. While the nervous pathways involved are not known, it was suggested by Willott *et al.* (1979) that the pericentral and external nuclei, rather than the central nucleus were involved, because in unanaesthetized animals the neurones there habituated in the same way as did the startle reflex. Audiogenic seizures also seem to require structures up to the level of the inferior colliculus, but not beyond. In certain susceptible strains of animals, early deprivation of auditory input leads to a hypersensitivity of the central nervous system, so that a later auditory stimulus produces a motor seizure (Saunders *et al.*, 1972). Lesion of the inferior colliculus, or structures below it, reduces the susceptibility to seizure, whereas lesion of higher centres does not (Kesner, 1966). Similarly, auditory influences on spinal reflexes require an intact inferior colliculus (Wright and Barnes, 1972). Wright and Barnes suggested that, of all the brainstem auditory nuclei, the inferior colliculus had the richest projection to the reticular formation. Thompson and Masterton (1978) have also shown that structures in the region of the colliculus were necessary for the initial reflex turning of the head towards a sound source.

The inferior colliculus also seems important for directing the animal's attention to auditory stimuli. Jane *et al.* (1965) trained cats to avoid an electric shock, with a combined tone and light as a warning stimulus. In later testing, the efficacy of the two stimuli separately was measured in unreinforced trials in which only one or the other stimulus was presented; normal animals responded mainly to the tone rather than the light. However, animals in whom the inferior colliculus had been lesioned before training responded primarily to the light. By contrast, lesions in other auditory structures did not have this effect. It seems, therefore, that the inferior colliculus was necessary for establishing the importance of sound in governing the animals' normal behaviour.

2. Learned Reflexes

Modifiability of single-unit auditory evoked responses during training has been seen in the inferior colliculus (e.g. Ryan and Miller, 1977; Birt *et al.*, 1979) and in the medial geniculate body (e.g. Birt and Olds, 1981). While attempts have been made to ascertain the lowest level of the auditory system at which modifiability occurs during learning, the interpretation of these interesting experiments is fraught with difficulties. For instance, there is the difficulty of stimulus control during the measurements. In addition, the responses may be influenced by pathways descending from the higher levels of the nervous system. Thus it is possible that descending pathways from the cortex affect the responses of the earlier stages of the auditory system according to their significance for the animal. Some learning, even if it requires the presence of the auditory cortex, may be stored subcortically under the influence of descending pathways from the cortex.

The role of the brainstem in sound localization behaviour will be further described in Chapter 9.

H. Summary

1. The electrophysiological analysis of the auditory brainstem is faced with difficulties, because we do not have good ideas of the sensory features to which the auditory system is particularly responsive. It is also quite possible that many of the features extracted are represented not in the activity of single cells, but only in a pattern of activity over many cells. A successful scheme of organization is based on tonotopicity, such that neurones are arranged in ascending order of best frequency across one dimension in each nucleus, with complementary schemes of organization along the orthogonal axes.

2. The cochlear nucleus has three divisions, known as the anteroventral, the posteroventral, and the dorsal cochlear nuclei. Each division of the cochlear nucleus is tonotopically organized, and the best frequencies of the neurones make a spatially ordered map. In the anteroventral division, the neuronal responses are rather similar to those of auditory nerve fibres, with simple tuning curves, no inhibitory sidebands, and monotonic rate-intensity functions. In the dorsal cochlear nucleus, the tuning curves are very complex, with strong bands of inhibition, and rate-intensity functions that are nonmonotonic. Responses in the posteroventral cochlear nucleus have an intermediate form. This is correlated with the form of the post-stimulus–time histograms to tone bursts: those in the anteroventral cochlear nucleus are similar to those of auditory nerve fibres, whereas those more dorsally in the nucleus tend to show inhibitory pauses. It is reasonable to suppose that neurones in the anteroventral nucleus relay the auditory information to the next nucleus with very little transformation, whereas those in the dorsal cochlear nucleus have already

begun some complex sensory analysis. Some neurones in the dorsal cochlear nucleus appear particularly responsive to tones which are amplitude- and frequency-modulated, and others respond only to wide-band stimuli. In addition, inhibitory sidebands may serve to extract signals from background noise over a wide range of stimulus intensity.

3. The anteroventral and posteroventral cochlear nuclei project mainly to the superior olivary complex, which receives an input from the cochlear nuclei of both sides. The superior olivary complex has several component nuclei. The largest, known as the lateral superior olivary nucleus (or S-segment) receives an input from both sides, the ipsilateral input being predominantly excitatory, and the contralateral input predominantly inhibitory. The nucleus is mainly a high-frequency nucleus, is responsive to interaural disparities in intensity and timing, and uses these to code the direction of a sound in space. Another component nucleus, the medial superior olivary nucleus, is mainly a low-frequency nucleus. It is responsive to disparities in interaural timing, and therefore can be said to code the direction of a sound in space on the basis of timing differences.

4. The next major nucleus of the auditory pathway is the inferior colliculus, which receives afferents bilaterally from the superior olivary complex and contralaterally from the cochlear nucleus, mainly the dorsal division. The inferior colliculus therefore combines the spatially coded input from the superior olivary complex with the results of the complex sensory analysis of the dorsal cochlear nucleus. The central nucleus of the inferior colliculus is tonotopically organized, with cells arranged in iso-frequency sheets across the nucleus. Within each sheet, there seems some degree of mapping of sound source position, with cells responding to sources near the midline situated caudally, and those responding to sources at the side situated rostrally. The inferior colliculus seems to play an important part in many auditory reflexes.

5. The medial geniculate body receives its input from the inferior colliculus and projects to the auditory cortex. It has three divisions, only one of which, the ventral division, is a specific auditory relay. Neuronal responses are complex, although it is difficult to judge the extent to which they are in advance of those in the inferior colliculus.

6. Many auditory reflexes are established at the brainstem level, although our knowledge is sketchy. One of the most elementary reflexes is that by which the middle ear muscles contract in response to loud sounds. It involves an arc of three to four neurones, running from the ventral

cochlear nucleus to the medial superior olive and then to the motor nuclei of the facial and trigeminal nerves. The inferior colliculus has also been implicated in many auditory reflexes. It seems important for the startle response to loud sounds, and for the development of audiogenic seizures, a motor response resulting from a central hypersensitivity of the auditory system following early deprivation of auditory input. Auditory influences on spinal reflexes also seem to require an intact inferior colliculus. The inferior colliculus further affects the animal's attention to auditory stimuli. Learning has been shown to produce modification of neuronal responses in the inferior colliculus and medial geniculate body, but we do not know if the modification resulted from neuronal plasticity at those levels, or was due to descending influences from, say, the cortex.

I. Further Reading

The anatomy of the cochlear nucleus has been reviewed by Cant and Morest (1984) and J. K. Moore (1986b). The physiology has been reviewed by Young (1984) and Brugge and Geisler (1978). Neurotransmitters of the nerve and nucleus have been reviewed by Wenthold and Martin (1984), Caspary (1986), and in chapters in *Auditory Biochemistry* edited by D.G. Drescher (Charles C. Thomas, Springfield, 1985). The superior olivary complex has been reviewed by Brugge and Geisler (1978). The inferior colliculus has been reviewed by Aitkin (1986), who also gives a general review of the brainstem. The brainstem in relation to sound localization has been reviewed by Masterton and Imig (1984). An excellent general review of the brainstem is given by Irvine (1986).

7. The Auditory Cortex

The anatomical definition of auditory cortex has been simplified by the introduction of axonal transport techniques, which allow a definition of cortical areas on the basis of thalamic connections, and by detailed electrophysiological mapping, which allows a definition on the basis of tonotopic organization. The anatomical and physiological organization of the auditory cortex will be described, together with what is known of the neuronal responses. The auditory cortex has been a favourite target for behavioural scientists: unfortunately the behaviour-ablation method has in recent years proved to be less powerful for analysing the function of the auditory cortex than it seemed to be 20 years ago. The general functions of the auditory cortex are still not certain, and some hypotheses are listed.

A. Organization

1. Anatomy and Projections

The auditory cortex has been most commonly studied in the cat, where it is displayed on the surface of the brain. Until recently, much less work has been done in primates, where the auditory cortex lies on the superior temporal plane hidden in the lateral or Sylvian fissure.

A framework for analysing the areas of the cat auditory cortex can be built on Rose's (1949) delimitation of the cytoarchitectural areas of the temporal cortex (Fig. 7.1). By defining areas with constant cellular characteristics as seen with the Nissl stain, he described primary auditory cortex (AI), secondary cortex (AII), and a further auditory area on the posterior ectosylvian gyrus (Ep). Primary auditory cortex was described as being cytoarchitecturally similar to other primary sensory cortex, with six layers and a high density of pyramidal and granule cells in layers II, III and IV, but with sparse staining in layer V. The high density of granule cells leads to the term koniocortex, or "dust cortex". Additional auditory areas were

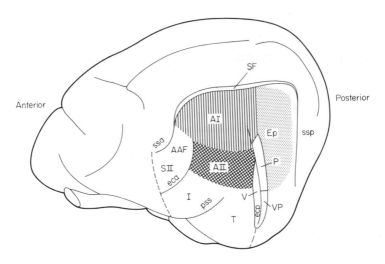

Fig. 7.1   The divisions of the cat's auditory cortex described by Rose (1949) are indicated by shaded areas, together with the other auditory cortical areas now recognized. Cortical areas: AI, primary auditory cortex, AII, secondary auditory cortex; AAF, anterior auditory field; Ep, posterior ectosylvian field; P, posterior field; V, ventral field; VP, ventral posterior field; SII, secondary somatosensory area; I, insular area; T, temporal area; SF, suprasylvian fringe area, buried on the upper surface of the suprasylvian sulcus; Sulci, ssa and ssp, anterior and posterior suprasylvian sulci; eca and ecp, anterior and posterior ectosylvian sulci; pss, pseudosylvian sulcus. Adapted from Rose (1949, Fig. 1), with additions from Brugge and Reale (1985).

defined by other methods. Rose and Woolsey (1958) showed that the secondary somatosensory area (SII) and the insulo-temporal area (I-T) were also auditory areas on the basis of their thalamic connections. Reale and Imig (1980) have more recently used microelectrode mapping to subdivide the previous auditory areas to give in addition the anterior auditory field (AAF), and the ventral (V), posterior (P), and ventral posterior (VP) fields (Fig. 7.1).

The projection from the thalamus to the auditory cortex was worked out with the horseradish peroxidase technique by Winer *et al.* (1977) and Niimi and Matsuoka (1979). The conclusions of these studies are shown in Fig. 7.2, although some of the minor projections have been omitted. The major part of the ventral division of the medial geniculate, the specific auditory relay, projects mainly to AI, with smaller projections to the other areas. The medial division projects to almost all the areas of the auditory cortex, and the dorsal division, as defined in Chapter 6, projects to AII, the insulo-temporal area and Ep. Thus we can define a "core" system, running from

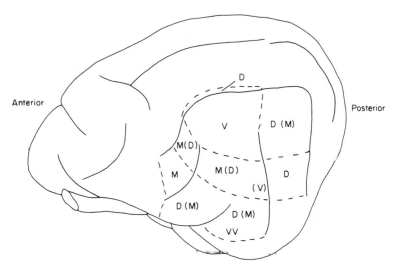

Fig. 7.2 Divisions of the medial geniculate body projecting to the auditory cortex are indicated on a map of the cat auditory cortex. The smaller projections are indicated in brackets. V, from ventral division; M, from medial division; D, from dorsal division; VV, from extreme ventral (non-laminar) division. Many still smaller contributions are not shown. Primary auditory cortex (AI) receives projections only from the ventral division. The surrounding areas receive their projections from the dorsal and medial divisions. Adapted from Ravizza and Belmore (1978, Fig. 3), to incorporate the results of Niimi and Matsuoka (1979).

the specific thalamic auditory relay, namely the ventral division of the medial geniculate body, to the primary auditory cortex. One or more "belt" or "diffuse" systems, surrounding AI, receive projections from the other divisions of the medial geniculate. The belt also receives projections from other thalamic groups, and in particular the posterior group of thalamic nuclei. There are intense reciprocal connections between the cortical areas, and back from the cortex to the brainstem (Diamond *et al.*, 1969; Andersen *et al.*, 1980).

The afferent fibres from the ventral division of the medial geniculate end mainly in layer IV but also in the other layers (Mitani *et al.*, 1985). The constituent cells of AI, especially in layer IV, appear to be organized in vertical columns. In the cat, the cells appear to be situated around the periphery of small vertical cylinders, of 50–60 μm diameter, oriented with their axes at right angles to the cortical surface (Sousa-Pinto, 1973). Vertical columns have also been described in layer IV of human beings and monkeys (Seldon, 1981a; Smith and Moskowitz, 1979). Many of the cells show direct soma-to-soma contacts with other cells in the column. Smith and Moskowitz suggest these may represent gap junctions between the cells in one column.

The cell types of the primary auditory cortex appear to be similar to those of other cortical areas, including for instance pyramidal cells (with axons extending into the white matter) and fusiform cells (with two tufts of dendrites) (e.g. Winer, 1985b; Mitani *et al.*, 1985). It appears that dendrites and axons ramify horizontally more than is typical of sensory cortex (Sousa-Pinto, 1973), especially along the iso-frequency lines running across the cortex (Glaser *et al.*, 1979). There is also a particularly rich ramification vertically within each column of cells. Callosal afferents, from the contralateral cortex, ramify vertically within "callosal columns", i.e. within columns of cells having a particularly rich callosal innervation (Code and Winer, 1986).

2. Tonotopic Organization

The tonotopic organization of the afferent projections was worked out with gross evoked potentials by Woolsey (1960). His plan for the cat (Fig. 7.3A) shows that auditory evoked potentials were recordable in many auditory areas. Tonotopic organization was shown for some of these areas and is indicated by the representation of the base of the cochlea (B) or the apex (A). Tunturi (1952), in the dog, showed that between these two extremes there was a complete representation of stimulus frequency, with areas having the same frequency lying in strips at right angles to the line of frequency progression (Fig. 7.3B).

Merzenich *et al.* (1975) and Reale and Imig (1980) showed that the tonotopic organization of the cortex was preserved at the single cell level (Fig. 7.4). In AI, cells were sufficiently sharply tuned for best frequencies to be clearly definable, and tonotopic organization and iso-frequency strips as described above were found. In AII, by contrast, the degree of tonotopicity appeared to be poor, with cells in the same region having a wide range of characteristic frequencies (Reale and Imig, 1980; Schreiner and Cynader, 1984). The increased detail from single unit studies has also permitted closer parcellation of the different fields. For instance, Knight (1977) and Reale and Imig (1980) suggested that the region anterior to AI, which Woolsey (1960) thought was the low-frequency continuation of the suprasylvian fringe, was in fact a separate area of its own with a tonotopic frequency map in its own right (the anterior auditory field: Fig. 7.1). This technique also enabled them to define the posterior and ventral posterior fields on the anterior edge of the Ep field, buried in the posterior ectosylvian sulcus, and the ventral field on the posterior edge of Woolsey's AII field (see Fig. 7.1).

The map of frequency therefore seems to undergo a series of transformations up the auditory pathway. A sound of one frequency is represented by a single point in the cochlea, by a two-dimensional sheet of cells in each of

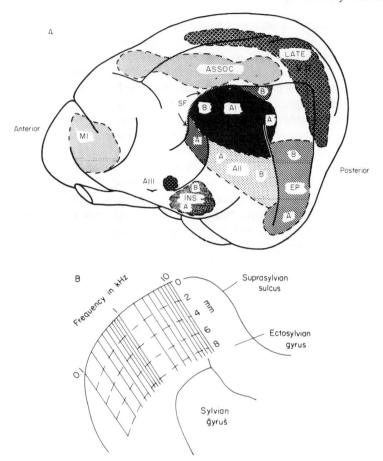

Fig. 7.3 (A) Auditory areas and their tonotopic organization were shown with gross evoked potentials by Woolsey (1960) in the cat. Where the areas are organized tonotopically, the representation of the high-frequency base of the cochlea (B), and of the low-frequency apex (A) are indicated. MI, precentral motor area. AIII is now called SII. Other abbreviations as in Fig. 7.1. The part of "SF" labelled "A" is now called the anterior auditory field, with a tonotopic organization in its own right, AII now contains the ventral field, and Ep now contains the posterior and ventral posterior fields (see Fig. 7.1). From Woolsey (1960), *Neural Mechanisms of the Auditory and Vestibular Systems* (eds G. L. Rasmussen and W. F. Windle). Courtesy of Charles C Thomas, Springfield, Illinois. (B) Iso-frequency strips in the dog auditory cortex. From Tunturi (1952, Fig. 2).

the intervening auditory nuclei, and by a one-dimensional strip of cells in each of the tonotopically-organized fields of the cortex, with multiple representations in the different fields.

Tonotopicity is less obvious in unanaesthetized cats (Evans *et al.*, 1965).

The cortical areas have been investigated less extensively in primates. The primary area is situated on the superior temporal plane within the Sylvian fissure (Fig. 7.5A). Again, there is a tonotopic organization of AI, with low frequencies represented rostrally and high frequencies caudally (Merzenich and Brugge, 1973; Imig *et al.*, 1977). As in the cat, AI is surrounded on all sides by other auditory areas (Fig. 7.5B), some of which are tonotopically organized, although the terminology differs between investigators, and the apparent details of the fields differ between the different species of primate. In man, in which the limits of the auditory cortex have to be defined on cytoarchitectonic rather than electrophysiological grounds, the auditory cortex appears to be in a similar position (Economo and Horn, 1930). Its detailed anatomy has been described more recently by Seldon (1981a,b, 1985). Positron emission tomography in alert human subjects has shown auditory responses to be tonotopically organized

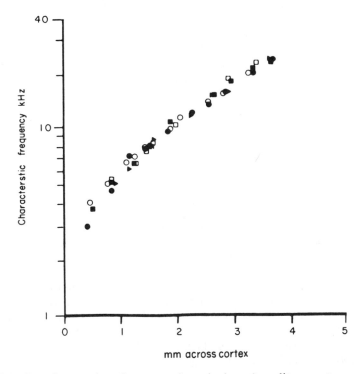

Fig. 7.4   Best frequencies of neurones in a single cat's auditory cortex are plotted as a function of distance across the cortex. The neurones were located on five parallel lines across the cortex, and different symbols are used for each line. From Merzenich *et al.* (1975, Fig. 6).

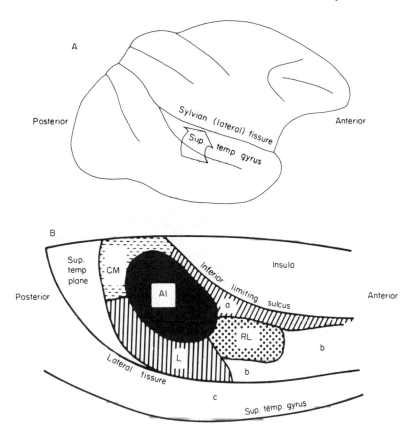

Fig. 7.5 (A) The auditory cortex in the monkey lies buried in the Sylvian or lateral fissure (arrow). If the cortex above the fissure is removed, the areas of cortex on the lower surface of the fissure, the superior temporal plane on the upper surface of the superior temporal gyrus, become visible. (B) The superior temporal plane seen from above, after removal of the overlying cortex. AI, primary auditory area; RL, rostrolateral field; L, lateral field; CM, caudomedian field; a,b,c, additional auditory areas. From Merzenich and Brugge (1973, Fig. 14).

(Lauter *et al.*, 1985). In this technique, positron emission from radioactive ^{15}O-rich H_2O is detected, as modulated by changes in regional blood flow resulting from altered neuronal activity.

3. Columns in the Auditory Cortex

Within AI, many attempts have been made to identify functional columns of cells similar to those described in somatosensory and visual cortices (e.g. Mountcastle, 1957; Hubel and Wiesel, 1963). It had been found in such studies that cells in a single radial column of the cortex had similarities in their optimal stimulus characteristics. Neighbouring columns may have different stimulus characteristics, and there would be a sharp jump in the characteristics measured as a sampling electrode left one column and entered another. In the auditory cortex, the possibility of frequency-specific columns within the overall tonotopic organization was first investigated. Thus Merzenich *et al.* (1975) reported that cells in a single radial electrode penetration of the cortex had the same frequency, and that stepwise changes in frequency were often observed in oblique penetrations. In a systematic analysis, Abeles and Goldstein (1970) showed that units close together had similar best frequencies. However, there was no sign of discrete transitions between adjacent columns. Either there are smooth transitions in frequency across the iso-frequency contours, or the frequency columns are too small (100 μm or less) to be detectable by the technique. Sousa-Pinto's (1973) anatomical cylinders of cells were 50 μm across. He did not think they could be associated with discretely different frequencies, because the separation of the columns was less than the lateral spread of the incoming axonal arborization.

With an analysis of binaural sensitivity, however, there appears to have been more success in identifying functional columns. In AI, Imig and Adrian (1977) showed that cells excited by stimuli in one ear but inhibited by stimuli in the other ear (EI cells) were located in discrete radial columns. They were separate from cells excited by stimuli in both ears (EE cells). The different categories of cells were located in discrete radial columns. In a surface view, the two types of cell formed patches wandering over the surface of the cortex. Middlebrooks *et al.* (1980) suggested that these patches were organized in strips running roughly at right angles to the iso-frequency contours (Fig. 7.6). There was a close relation between the electrophysiological responses and the innervation as shown anatomically, since those areas showing electrophysiological evidence of a strong input from the contralateral half of the brain had a particularly rich innervation from the contralateral auditory cortex via the corpus callosum (Imig and Brugge, 1978; Code and Winer, 1986). Thus the two-dimensional sheet of the auditory cortex seems to be organized in both directions. In one direction, it is organized in frequency, although probably with smooth rather than stepwise transitions in frequency between the iso-frequency strips. At approximately right angles to this, the cortex is organized in terms of binaural dominance; the strips in this direction seem to have discrete borders.

B. The Responses of Single Neurones

1. Response Types

In the early stages of the auditory pathway there seemed to be some hope that an objective classification into discrete response types might be possible,

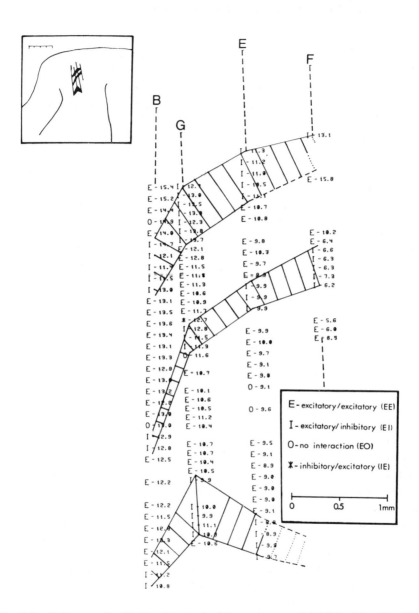

Fig. 7.6 Columns of cells that are excited by one ear and inhibited by the other (EI cells, shown in cross-hatched areas) are segregated from columns of cells that are excited by both ears (EE cells). The cell CFs are indicated by numbers along the electrode tracks. Inset shows the cortical area sampled. From Middlebrooks *et al.* (1980, Fig. 4).

and that the classes might correspond to anatomically-separable classes of cells. However, in the auditory cortex, there seems such a diversity of response types, and such a degree of dependence on the behavioural state of the animal, that this does not at the moment seem possible. In addition, even in the unanaesthetized, paralysed animal, a certain proportion of the units (20% according to Goldstein and Abeles, 1975) do not seem to respond in a determinate way.

Only a certain proportion of neurones seem responsive to sound at all. In the anaesthetized cat, Erulkar *et al.* (1956) found that only 66% of neurones in AI were responsive to sound. In the unanaesthetized and freely moving, or unanaesthetized and paralysed, cat this proportion seems higher, being 77% according to Evans and Whitfield (1964), and 95% according to Goldstein *et al.* (1968).

Oonishi and Katsuki (1965) described a wide variety of tuning curve shapes in AI of the barbiturate-anaesthetized cat. Some cells were sharply tuned, with the V-shaped tuning curves of units in the lower stages of the auditory system. Others had more than one dip, and were termed multi-peaked units. Some had broad tuning curves (Fig. 7.7). The thresholds at the tip of the tuning curve have been reported by some to be as low as in earlier stages of the auditory system (e.g. Goldstein *et al.*, 1968), although others have found best thresholds of 50–60 dB SPL (Evans and Whitfield, 1964).

Different temporal patterns of response can be seen for different cells. For the cells which respond in a determinate way, Abeles and Goldstein (1972) in the unanaesthetized paralysed cat described "through" (i.e. sustained), "on", "on-off" and "off" responses (Fig. 7.8). The same unit may show different temporal patterns for different frequencies of stimulation. A similar variety of response patterns has been described earlier in the auditory system, and it is by no means certain that such an analysis indicates any significant increase in response complexity.

Many cells show wide bands of lateral inhibition in unanaesthetized preparations (Shamma and Symmes, 1985). In multipeaked units, responses to excitatory stimuli in one range can be inhibited only by stimuli in the same frequency range, and in broadly tuned units the inhibitory range of frequencies varies with the frequency of the exciting tone (Abeles and Goldstein, 1972). Some of the latter properties can be explained by the convergence, on to single units, of the projections of cells with different best frequencies, where each of the projecting cells has inhibitory sidebands.

Many neurones show very sharp non-monotonicity in their rate-intensity functions, with the firing rate falling by perhaps 50% for deviations of stimulus intensity by 10 dB or so from the optimum (Fig. 7.9; Brugge and Merzenich, 1973; Benson and Teas, 1976).

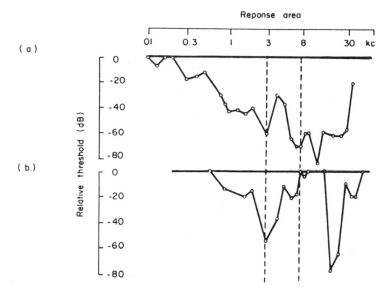

Fig. 7.7 Broad (a) and multipeaked (b) tuning curves are seen in the auditory cortex. Single peaked tuning curves are also present. From Oonishi and Katsuki (1965, Fig. 1).

In the posterior field (Fig. 7.1), most neurones have sharp V-shaped tuning curves. Nearly 90% of rate-intensity functions are reported to be strongly non-monotonic in the cat, as against less than 40% for AI (Phillips and Orman, 1984). The neurones respond with very long latencies, with minimum values of 20–50 ms, compared with 20 ms or less for AI. In AII, tuning curves are broad, and often have complex dips on the low-frequency side. There may well be a continuum in breadth of tuning across both AI and AII, such that the sharpest tuning curves are found in the dorsal part of AI, with a gradual transition to broader tuning across AII (Schreiner and Cynader, 1984).

2. Sound Localization

Many neurones in AI show binaural interactions, being sensitive to interaural phase or intensity differences (e.g. Brugge and Merzenich, 1973; Reale and Kettner, 1986). Many are also selective for the direction of sound sources (Middlebrooks and Pettigrew, 1981). For unilateral stimuli, stimulation of the contralateral ear is generally the more potent, and this has a correlate in the binaural case where the interaural intensity and phase responses are

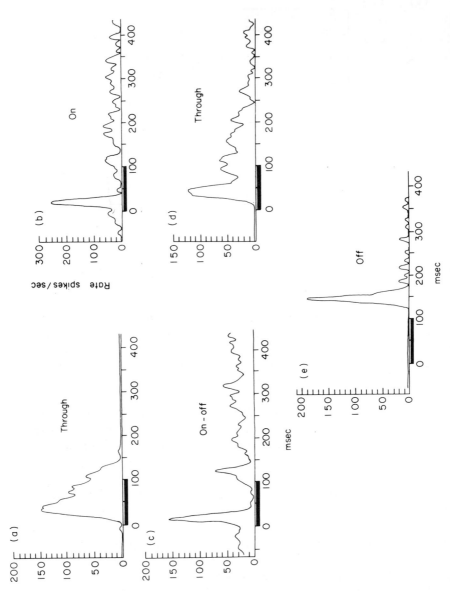

Fig. 7.8 Temporal response patterns in primary auditory cortex, according to Abeles and Goldstein (1972, Fig. 2).

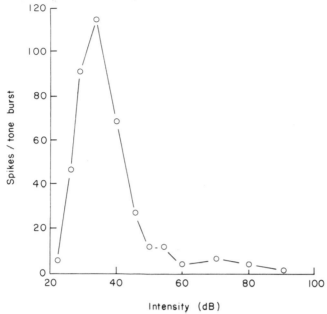

Fig. 7.9   Neurones in the auditory cortex can have very sharply nonmonotonic rate-intensity functions. From Brugge and Merzenich (1973, Fig. 8A).

such that each cortex predominantly responds to stimuli on the contralateral side. It seems that the cells share the directional selectivity of cells in the brainstem, modified by the callosal interactions seen in the cortex. It further seems that different cells in a single cortical binaural interaction column share the same directional selectivity, although they may be optimally responsive to different types of sound (Brugge and Merzenich, 1973).

In one type of experiment, stimuli are presented separately to the two ears, and the relative timing or intensity of the stimuli is varied. When interaural phase is varied, the firing of many units shows a cyclic function similar to those obtained at the lower stages of the auditory pathway (Fig. 7.10). In some cases the time disparity for maximum response is independent of stimulus frequency (Fig. 7.10A), indicating that it is appropriate to think in terms of a characteristic delay. In others, this is not the case (Fig. 7.10B). For such a unit, if it is appropriate to think in terms of sound localization at all, we would expect the optimal location of the sound source to vary with stimulus frequency. In these studies continuous tones were presented to the two ears, and the interaural phase delay varied. However, it is also possible to vary the interaural disparity in the onset or temporal envelope of a stimulus. In one group of neurones studied by Kitzes *et al.* (1980), a very sharp peak of sensitivity was found for zero or near-zero disparities in

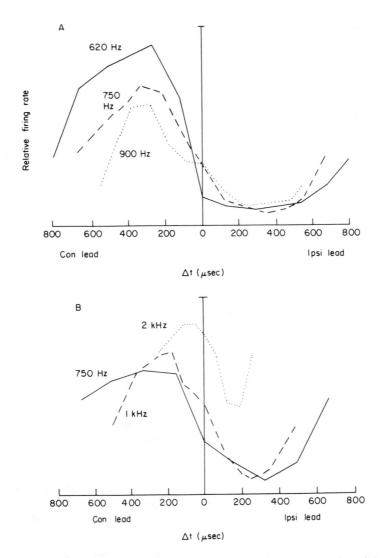

Fig. 7.10 The cyclic dependence of firing rate on interaural time delay indicates in A a neurone with an optimal time disparity which is independent of stimulus frequency, and in B one with disparity which is dependent on stimulus frequency. From Benson and Teas (1976, Fig. 4).

onset time, indicating that the cells could code sound sources directly, or nearly directly, ahead. Time disparities for click stimuli have been studied as well (Benson and Teas, 1976; Brugge and Merzenich, 1973). In some cells the functions for time disparities measured with clicks matched the functions for interaural phase delays measured with tones. In other cells, however, they clearly did not. For these cells, a simple explanation in terms of sound localizing ability does not seem appropriate.

A sound source in space can give rise to differences in intensity at the two ears, as well as to differences in phase. In a corresponding way, some neurones give their optimal response for specific intensity differences between the two ears (Brugge and Merzenich, 1973; Phillips and Irvine, 1983). There is a tendency for cells with the more strongly contralaterally-dominated responses to be situated more ventrally along each iso-frequency line (Reale and Kettner, 1986).

In the experiments described so far, directionality of the sound source was simulated by varying the interaural phase and intensity differences. However, some units (50%) are able to code the real direction of sound sources, measured by moving a speaker around a cat's head (Middlebrooks and Pettigrew, 1981). As in lower auditory nuclei, some of the cells have small, circumscribed receptive fields centred on the acoustic axis of the pinna nearest to the sound source. These cells tend to have high characteristic frequencies (>12 kHz), and tend to be sensitive to interaural intensity differences. Middlebrooks and Pettigrew presented evidence that the cells obtained their directional selectivity from the strong directionally-selective amplification produced by the pinna in this frequency range, maintained and enhanced by the cells' sensitivity to interaural intensity differences (Middlebrooks and Pettigrew, 1987). Other cells responded to sound sources over the whole of the contralateral hemifield. They tended to have characteristic frequencies below 12 kHz, and it was suggested that they obtained their selectivity from the lesser directional selectivity of the pinna in this frequency range, again maintained and enhanced by the cells' sensitivity to interaural intensity differences.

Sovijärvi and Hyvärinen (1974) have described units specifically responsive to the direction of movement of a sound source. In response to stationary stimuli such cells gave a complex on–off response, indicating that this particular sensitivity arose from multiple excitatory and inhibitory inputs to the cell.

3. The Detection of Other Features

Whitfield and Evans (1965) in the unanaesthetized, unrestrained cat described cells apparently specifically responsive to frequency-modulated

(FM) tones. Many seem to be specifically sensitive to one particular direction of frequency modulation, and some are tuned to a certain rate of frequency modulation (Mendelson and Cynader, 1985). Other cells are strongly driven by amplitude-modulated stimuli, again being tuned to a specific frequency of modulation (Schreiner and Urbas, 1986). In some of these units, and particularly those with complex temporal patterns of discharge, the response to modulated tones cannot be predicted from the response to steady tones (see also Funkenstein and Winter, 1973). As was described above, some such units have been described as early as the dorsal cochlear nucleus (Britt and Starr, 1976b). However, the FM-sensitivity of cells in the cortex shows a significant advance over that in the cochlear nucleus. Cells in the cortex tend to show a greater specificity in their sensitivity to sweep direction. Moreover, the effective range of frequency modulation found by Whitfield and Evans, sometimes ± 2.5% or less, is much smaller than that necessary in the dorsal cochlear nucleus. In some cases they found responses to small modulations when the modulated frequency range lay entirely within the steady tone response area. In other cases they found responses to frequency-modulated stimuli with stimuli entirely outside the steady tone response area. This was in accordance with the observation that frequency-modulated stimuli were in general the more effective stimuli for cortical cells, as was the observation that some units could be driven by frequency-modulated stimuli but not steady tones at all.

Such studies show that a degree of response specificity for feature extraction exists in the cortex. But the extent to which we are able to divide such cells into separate classes of specific feature detectors, analogous to, say, simple or complex cells in the visual system, is doubtful. Thus, while Goldstein and Abeles (1975) report that FM stimuli were very effective stimuli for cortical cells, the FM-sensitive cells were situated on a continuum of cells responding to a range of complex features.

If we wanted to pursue the idea of a hierarchy of specific feature detectors, we might expect neurones at the highest levels of the nervous system to respond only to stimuli of particular significance for the animal concerned. Attempts have therefore been made to measure the responses of cortical cells to the vocalizations of the species, or to other sounds of presumed biological significance. However, by comparing the responses to normal and time-reversed species-specific vocalizations in unanaesthetized squirrel monkeys, Glass and Wollberg (1983) concluded that there was no particular evidence that cells were generally more responsive to the normal (non-reversed) vocalizations. Instead, the cells seem to be responsive to the discrete stimulus components in the vocalizations, and in particular to their transients. Similarly, in response to human speech sounds, the cells appear to be particularly affected by transients in the stimuli (Steinschneider *et al.*,

1982). Although these experiments suggest that the cortex is indeed involved in the handling of complex sounds, these results cannot be used to suggest that animal calls, or indeed other stimuli of presumed biological significance when not incorporated in a behavioural task, form special classes which are specifically represented in the cortex.

Although some cells respond to auditory stimuli in a repeatable, if complex, way, there are others which do not. For instance, Manley and Müller-Preuss (1978) found that 50% of cells in AI, and 62% of cells in AII, spontaneously varied their response to a constant vocalization. Evans and Whitfield (1964) noted that the responses of many units habituated rapidly. In these units, the apparent novelty of the stimulus was an important factor in governing their response. Some cells responded only when the attention of the animal was drawn to the source of the sound, perhaps by visual means, and in some cases the response to sound disappeared when the animal was induced to shut its eyes (Evans, 1968). These may be the "attention" units described in auditory cortex by Hubel *et al.* (1959). Such units responded only to novel stimuli, and once the response to one stimulus had habituated, a new stimulus would again evoke a response. It is apparent that the animal's behavioural relation with the stimulus is an important factor in governing the response of such neurones. Some studies have attempted to control this by recording responses under different conditions of arousal, or by using the auditory stimulus in a behavioural task.

Thus a greater number of action potentials are found to auditory stimuli in the awake than in the drowsy or sleeping animal (Brugge and Merzenich, 1973; Pfingst *et al.*, 1977). Responses may be enhanced by using the stimulus in a conditioning task; some cells even change the tuning of their response areas in such cases (Weinberger and Diamond, 1987). These studies suggest that the function of neurones in the auditory cortex might be more satisfactorily assessed by using the stimuli as cues in a behavioural task. It further implies that the most useful data might be obtained if the task is one for which the cortex has been shown to be necessary. In studying the function of a high-level structure such as the auditory cortex, behavioural studies therefore become paramount.

C. Behavioural Studies of the Auditory Cortex

1. Introduction

Considerable effort has been put into analysing the function of the auditory cortex, by testing performance on various auditory tasks before and after cortical lesions. In spite of a great deal of progress, and some intriguing leads, we are still uncertain about the cortical function, or functions, underlying many of the discovered deficits. For instance, it is still controversial as to whether experimental ablation of the auditory cortex leads to raised absolute thresholds. Although the commonly quoted position is that there are no, or only small, changes in absolute threshold (Kryter and Ades, 1943; Neff *et al.* 1975), some experimenters have found substantial losses (e.g. Heffner and Heffner, 1986a). Tests have generally been aimed at higher-level functions for the auditory cortex. In spite of many attempts to suggest a unifying function for the auditory cortex, it is most likely that there will be several different functions underlying the deficits seen in the different tasks. And it is unfortunately likely too, that apparently small differences in the training and testing procedures, perhaps so small that they were unreported in the original papers, will turn out to have had a decisive influence on the results. Thus, in frequency discrimination, for instance, we are still not sure of the role of the auditory cortex even after more than 40 years of work.

2. Frequency Discrimination

The earliest experiment was performed by Allen (1945), who trained dogs to lift a foreleg to a sound of one frequency, produced by tapping a bell, but not to a sound of another frequency, produced by tapping a tin cup. He showed that the discrimination was lost after large lesions of the auditory cortex. Later, Meyer and Woolsey (1952) trained cats in a rotating cage to remain still to a short series of 1.0-kHz tone pips but to rotate the cage when the series was terminated by a pip at 1.1 kHz. The discrimination could be relearned if any portion of the auditory field (i.e. AI, AII, Ep, SII of Fig. 7.3A) remained intact, but not if all areas were ablated bilaterally. Thus at this time it seemed that the auditory cortex was necessary for frequency discrimination, and that any sector of the auditory cortex remaining could mediate the discrimination. There followed an interesting series of experiments which showed that under some circumstances cats were indeed able to make frequency discriminations after cortical lesions. Two concepts emerged from these experiments: (i) after cortical lesions, cats were only able to detect stimuli, and not identify them on an absolute basis, and (ii) after cortical lesions, cats had difficulty withholding responses to irrelevant stimuli (e.g. Thompson, 1960; Elliot and Trahiotis, 1970; Neff *et al.*, 1975). While these hypotheses undoubtedly summarize many aspects of the animals' behaviour, they unfortunately are not true in an absolute way: Cranford *et al.* (1976a) showed that with appropriate methods of training cats are able to perform satisfactorily on both counts after cortical lesions.

Task difficulty is a confounding factor here. It seems that when the method of training was chosen such that the task was easy for the animal, performance survived the lesion, but with methods of training that made the task more difficult, it did not. This suggests that lesions may interfere with complex strategies; alternatively they may just produce a generalized interference with behaviour, which is revealed only with difficult tasks.

3. Sound Localization

A very different task, with important implications for the function of the auditory cortex, is that of sound localization. Many experiments have found changes in the ability to localize sounds after cortical ablation. For instance, Neff (1968) trained cats to approach the source of a sound, as in Fig. 7.11. He showed that after bilateral lesions of AI, AII, Ep, I-T, SII and the

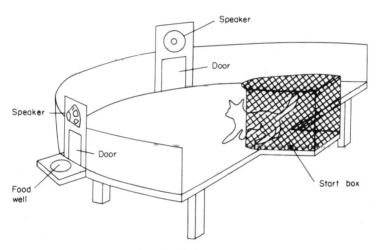

Fig. 7.11   In an apparatus for testing sound localization, one speaker sounds, the cat is released, and has to push open the door under the correct speaker.

suprasylvian gyrus (defined as in Fig. 7.3A), cats were unable to approach the correct box. If the suprasylvian gyrus was spared, performance was still poor post-operatively, although better than chance (Neff, 1968; Strominger, 1969). Wegener (1973) found analogous effects in the monkey, and some effects have also been reported in man (e.g. Jerger *et al.*, 1969).

More recent data shows that the position is not entirely clear-cut, because with different methods of testing normal performance can be found after these lesions (Cranford, 1979a).

The results of sound localization experiments become clarified if two conditions are observed: (i) the sound signals are brief (i.e. <40 ms), so that the subject cannot orient or explore within the sound field while the stimulus is sounding, and (ii) there must be several speakers, rather than two as in Fig. 7.11, so that the subject has to make a genuine choice of direction, rather than a simple left–right decision (Jenkins and Masterton, 1982).

Using these techniques, Jenkins and Merzenich (1984) showed that cats had profound deficits in sound localization after unilateral lesions confined to AI. The deficits were confined to the hemifield contralateral to the lesion, with performance in the ipsilateral hemifield being unaffected. If lesions were confined to a single iso-frequency strip in AI, deficits in localization were found for tone pips of only the corresponding frequency. If the complementary experiment was performed, and a narrow iso-frequency strip was left in AI while the rest of AI was removed, sound localization was possible only for the frequencies represented by the strip.

The results of Jenkins and Merzenich's experiment strongly indicates that AI is required for sound localization, and that sound locus is coded in a frequency-specific way. The results are in entire agreement with the electrophysiological evidence. Electrophysiological experiments show that cortical cells respond preferentially to sound sources on the contralateral side of the head. Moreover, they suggest that information about sound location is coded according to stimulus frequency, since cells responding to binaural information have best frequencies for stimulation and are tonotopically organized (see Chapter 6 and sections A.3 and B.2 in this Chapter).

For the result to hold, the two above conditions had to apply: Firstly, if longer (200 ms) stimuli were used, no effect was found. Jenkins and Merzenich suggested that with long stimuli the animal would be able to scan by means of head movements, and perform the task on the basis of monaural cues. However, an additional effect is possible, since there is evidence that animals have general difficulty with short stimuli after cortical lesions (see below). Secondly, if a simple left–right decision is needed, sound localization is possible even with brief stimuli (Jenkins and Masterton, 1982). Jenkins and Merzenich suggested that it would not be possible to measure unilateral deficits in this case, because sound sources on the unaffected side could still be localized, and the subject would be able to achieve good detection of sound direction by categorizing all sounds as localizable (i.e. from the unaffected side), or as unlocalizable (i.e. from the affected side). However, while this may account for the effects of unilateral lesions, it does not account for the lack of effects of bilateral lesions on certain simple left–right lateralization tasks.

In such left-right lateralization tasks, the animal did not have to locomote to the source of the sound, but had to indicate its direction by making another response. For instance, in the experiments of Ravizza and Masterton (1972), opossums were trained to drink from a water spout when sounds came from the right, but to stop when sounds came from the left. They were still able to localize after nearly complete removal of the neocortex. In the experiments of Cranford (1979a), binaural signals differing in phase and intensity were presented to the two ears through headphones. In these conditions, in man at least, the sound appears to be coming from one side or the other. Cranford (1979a) showed that after bilateral ablations of AI, AII, Ep, SII and I-T, cats could be trained to use a change in binaural intensity or phase disparity as a signal to move across a shuttle box. Moreover, the threshold disparities were little greater than those of normal animals. The negative effects of bilateral lesions with these tasks has an interesting implication. The results suggest that locomotion towards the source of the sound might be essential if a deficit is to appear. That is, the animals might have had difficulty in orienting in auditory space, or in locomotion, or in relating the two.

Experiments by Heffner (1978) addressed these possibilities. He trained dogs to approach one of two goal boxes not on the basis of the position of a sound source, but on the basis of the rate of a brief train of clicks presented through a central speaker. After bilateral removal of primary and secondary auditory cortices the dogs were able to use this cue to approach one goal box or the other. But if the very same stimuli, presented now through one of two speakers over the goal boxes, were used in the localization task, the animals failed. Thus under these conditions, neither making a motor reaction, nor remembering the correct response, nor attending to a brief stimulus, were the critical factors. It appears that the dogs had a specific deficit in connecting the location of a sound source with the necessary movement towards it. As Neff *et al.* (1975) say: "Perhaps, in the absence of auditory cortex, organization of a 'spatial world' based on acoustic information is no longer possible."

The auditory cortex is also necessary for the discrimination of binaural signals simulating moving sound sources, when signals are presented to the two ears with a varying time delay between them (Altman and Kalmykova, 1986). Again, unilateral lesions affect performance only for sound sources contralateral to the lesion.

We should end with a word of caution. Although hypotheses of complex function for the cortex are attractive, we should not be seduced away from simpler, if less interesting, hypotheses. Thus Jerger *et al.* (1969) found that a patient with unilateral damage to the temporal lobe had deficits in sound localization and also abnormal thresholds for short stimuli. Plainly, abnormal temporal integration of brief stimuli, if unilateral, could lead to a distortion of auditory space, with the result that the subject might not be able to walk to or point to the source of a sound, while still being able to discriminate changes in source position. He would still have the concept of auditory space; it merely no longer matches real space, due to a comparatively simple sensory deficit.

4. Ear Selection

Following the complex interpretations of the previous section, we now turn to experiments where a simpler explanation of one aspect of auditory cortex function is at least possible. It appears that if a cat is trained to respond to auditory signals in one ear, and to ignore competing signals in the other ear, performance is reduced by lesions of the auditory cortex contralateral to the attended ear. Kaas *et al.* (1967) trained cats to respond to changes in the pattern of tone pulses, when the elements of the pattern were presented separately to the two ears through headphones. Figure 7.12 shows the

	Neutral stimulus	Avoidance stimulus
Attending ear	L – L – L – L – L – L	L – H – L – H – L – H –
Ignoring ear	– L – H – L – H – L –	– L – H – L – H – L – H
Binaural pattern	L L L H L L L H L L L	L L H H L L H H L L H H

Fig. 7.12   In a test of ear selection used by Kaas *et al.* (1967), binaural stimuli were presented through headphones. Cats learned the task by responding to the stimuli in the "attending ear". Attention to one ear was upset by lesions of the contralateral cortex.

paradigm. Although the task can be learned on the basis of the binaural pattern, it is simpler to attend to one ear only, the one in which the pattern changes from L–L–L–L for the neutral stimulus to L–H–L–H for the warning stimulus. Transfer tests showed that this is in fact what the cats did. When the auditory cortex (AI, AII, Ep, SII, I-T) was ablated unilaterally, there were severe initial deficits with lesions contralateral to the attended ear, but not ones ipsilateral to it. It appears, therefore, that each cortex relates specifically to the contralateral ear. It will be recalled that this was also suggested by the electrophysiological evidence; neurones in one auditory cortex are most strongly excited by stimuli in the contralateral ear.

Analogous deficits can be found if the cat has to detect tone pips in one ear in the face of a particular effective masker in the other. Cranford

(1975) used a continuous train of noise bursts as a contralateral masker, synchronized with the tone pips in the signal ear. He showed that unilateral lesions of the cortex, contralateral to the signal ear, increased the amount of contralateral masking. The effect, as we might expect, only appeared with lesions contralateral to the attended ear. It appears, therefore, that unilateral cortical lesions can alter the effective balance of excitation arriving from the two ears. Such an interpretation was supported by the further ablation of the auditory cortex on the other side. Once the cortices on both sides were ablated, the degree of contralateral masking returned to normal.

The above tasks are reminiscent of those that have been used in the analysis of interaural attention in man. A stimulus, such as spoken text, might be presented to one ear, and a competing stimulus to the opposite ear (e.g. Cherry, 1953). The subjects have to respond to one stimulus or the other. Not surprisingly, such tests reveal deficits with unilateral cortical damage (e.g. Berlin and McNeil, 1976).

5. Patterns, Memory and Time

Diamond and Neff (1957) suggested that the auditory cortex might be important for the analysis of auditory patterns, just as the visual cortex is necessary for the analysis of visual patterns. They tested the ability of cats to discriminate changes in an ongoing pattern of 800-Hz and 1-kHz tone pips, when the temporal ordering of the tone pips was changed, but the sequences were otherwise identical. They found that the pattern could be relearned if any part of AI were preserved, but there was complete loss and no relearning if the lesions included AI, AII, Ep and I-T (as defined in Fig. 7.3A). These areas therefore seem necessary for distinguishing the temporal order of auditory stimuli. The I-T area seems to have a particularly important function in this task, since Colativa *et al.* (1974) have shown that such discriminations are lost after lesions of the I-T area alone. It seems probable that the I-T area in the cat has a supramodal role in temporal pattern perception, since Colativa has shown that lesions of I-T upset the perception of visual and somatosensory as well as auditory temporal patterns (Colativa, 1972, 1974). In man, there seems a correlate, since subjects with temporal lobe lesions have difficulty in perceiving the temporal pattern of auditory stimuli (e.g. Karaseva, 1972).

Lesions of the auditory cortex also upset tasks in animals where the subject has to indicate whether two successive sounds are the same or different by having to make a response when they are the same but not when they are different. In cats, dogs and monkeys, lesions of the primary auditory cortex or the belt area prevent the task from being relearned (e.g.

Cornwell, 1967). These animals, therefore, had a difficulty in relating the trace left by one auditory stimulus with the next auditory stimulus, or translating this into response terms, or both.

The same point was investigated by Dewson *et al.* (1970) who trained rhesus monkeys to reproduce a two-element pattern of tone and noise bursts on two panels, one corresponding to each stimulus. The intervals between the two stimuli and the duration of the stimuli were systematically varied by means of a rule that maintained performance around 79% correct. After lesions of the auditory cortex, performance was possible only for a much more limited set of time relations than before, either for shorter stimuli or for shorter silent intervals between them. This suggests that the monkeys had difficulty relating one stimulus to another if the two were separated in time. However, the deficit is not one of short-term memory in general. Forcing the monkeys to wait before making a response did not produce further deficits (Cowey and Weiskrantz, 1976). These subjects, therefore, had a specific difficulty in relating one auditory stimulus to a later one, and could be said to have a deficit in "auditory memory".

Such a deficit may be one aspect of a general deficit in coding or utilizing the temporal dimension of auditory stimuli. One of the most elementary of such deficits was described by Gershuni *et al.* (1967) and Baru and Karaseva (1972), who showed that in man and dogs unilateral lesions of the auditory cortex resulted in a loss of sensitivity to short tones in the opposite ear, but not to long ones, nor to either type of stimulus in the ipsilateral ear (Fig. 7.13). Presumably the cortex plays some role in extending the effect of brief stimuli so that they can influence other neural events. Note, however, that the deficit only appeared for very short stimuli, lasting 10 ms or so. This is an order of magnitude less than the time intervals over which the deficit appeared in the pattern discrimination task of Dewson *et al.* (1970), and so presumably reflects a different mechanism. The subjects appeared to have a general difficulty with short stimuli, because frequency discrimination limens were raised for short but not long stimuli (Gershuni *et al.*, 1967), as has also been found in the cat (Cranford, 1979b).

An example of another temporal task, which used time intervals in the range of the temporal pattern tests described above, was that of duration discrimination. Scharlock *et al.* (1965) trained cats to respond when the duration of tone pips increased from 1 to 4 s. Lesions of AI, AII, Ep and I-T allowed relearning, but if SII was included relearning was not possible.

D. Hypotheses as to the Function of the Auditory Cortex

1. The Analysis of Complex Sounds

Electrophysiological experiments have shown the stimuli to which neurones in the auditory cortex are specially responsive. However, it is very difficult

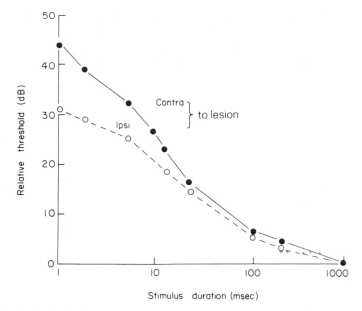

Fig. 7.13 Lesions of the auditory cortex selectively affect the thresholds of short stimuli in the contralateral ear. From Gershuni *et al.* (1967, Fig. 2).

to decide any specific role for the cortex from such experiments. The stimuli to which neurones in the cortex seem specifically responsive, which may include frequency-modulated tones and vocalizations, are in themselves so complex, and on close analysis produce such complex responses, that it is generally very difficult to decide the extent to which the cortex shows any significant advance in processing over the lower stages of the auditory system. Lesion experiments have at least the virtue of indicating whether or not a structure is essential for a particular task; but again pinning down the exact role in complex tasks is also often difficult. Leaving those difficulties aside, there are possible links between the electrophysiological observations of responses to complex sounds (whether or not this represents specific feature detection) and behavioural experiments. The discrimination of complex sounds, such as speech sounds, is upset after cortical lesions. This has been shown in the cat (Dewson, 1964) and the monkey (Dewson *et al.*, 1969), and agrees with the extensive literature in man showing deficits in speech perception after lesions of the temporal cortex.

2. Sound Localization

AI subserves sound localization, representing sounds on the contralateral side. Sound locus is coded in a frequency-specific way in the individual frequency channels (Jenkins and Merzenich, 1984). It has been suggested that the deficit is in relating the direction of the source to a spatial schema (e.g. Cranford, 1979a). The deficit may be in the formation of "auditory space", and the cortex may be necessary for performing manipulations within such a space.

3. Ear Selection

The cortex may govern interaural attention, the cortex on one side potentiating the effect of stimuli in the opposite ear (Kaas *et al.*, 1967; Cranford, 1975). This may be related to a selective attention to sound sources on the basis of source position.

4. Response Inhibition

A hypothesis which related the auditory cortex to a motor function rather than a sensory function was that of Thompson (1960). He suggested that the auditory cortex was necessary for the inhibition of inappropriate responses. This can no longer be held in its simple form because in some experiments lesioned animals tend to make too few rather than too many responses (e.g. Cranford *et al.*, 1976a,b). The appearance of this particular deficit depends on the method of training, although the hypothesis still seems valid under certain circumstances.

5. Identification Versus Detection

A hypothesis, derived from Neff's (1961) hypothesis of the abilities that survive cortical damage, is that the cortex is necessary for the identification of stimuli on an absolute basis, but not for their detection, nor for the detection of change (Elliot and Trahiotis, 1970). Again, this hypothesis cannot be held in its simple form, because in some experiments lesioned animals could still identify stimuli by their absolute attributes, rather than only by their relative ones (e.g. Cranford *et al.*, 1976a,b; Cranford, 1979a).

6. Discrimination of Temporal Patterns

Deficits have been found in many tasks where animals have to relate one stimulus to another, when the stimuli are separated in time (e.g. Diamond and Neff, 1957; Dewson *et al.*, 1970. The I-T cortex may play a particularly important role in this, and may govern the utilization of the temporal relations of stimuli in the visual and somatosensory as well as the auditory modalities (Colativa, 1972, 1974). When the time dimension is critical, many auditory tasks have been disrupted, and the cortex may well have a role in prolonging and utilizing the trace left by an auditory stimulus. The deficit may not be in memory generally, but in relating one element of an auditory stimulus to a later one.

7. Concept Formation

Whitfield (1979) suggested that the auditory cortex had a role in forming auditory concepts. More specifically, he suggested that the auditory cortex posits the real objects to which auditory stimuli relate. These posited real objects form the concepts unifying the different auditory stimuli. Transfer tests, in which dichotic click pairs were treated as equivalent to stimuli on one side or the other before and not after cortical lesions, form part of the evidence (Masterton and Diamond, 1964).

8. Task Difficulty

It has been a general finding that difficult tasks tend to be most frequently disrupted by cortical lesions. The difficulty of a task may depend on the details of the training technique used by the experimenter, and will vary with the strategy used by the animal in finding a solution. The function of the auditory cortex should therefore be thought of in terms of strategies rather than the tasks themselves. It has been suggested, for instance, that where the training procedure makes absolute frequency discrimination difficult for cats, cortical lesions produce deficits, but where they are such as to make the task easy, cortical lesions do not (Cranford *et al.*, 1976a). This may occur because cortical lesions produce a generalized interference with behaviour, which is revealed most clearly with difficult tasks. It is also possible that where the task is arranged so as to be difficult, the auditory cortex becomes involved in establishing or maintaining performance. The function of the cortex would therefore be that of helping to store or utilize strategies. The analysis of the strategies actually used by the individual animals in behavioural experiments would therefore be an essential, though forbidding, requirement.

It is obvious that present electrophysiological techniques would have great difficulty in uncovering the single neurone correlates of the more complex of these hypotheses.

E. Summary

1. The auditory cortex consists of a "core" area, surrounded by a "belt". The core, which is the primary auditory cortex or AI, receives its input from the main specific auditory relay of the thalamus, the ventral division of the medial geniculate body. The belt receives its input mainly from the other divisions of the medial geniculate.

2. The primary auditory cortex, and some of the divisions of the belt area, are tonotopically organized. Iso-frequency strips lie at right angles to the line of frequency progression.

3. A discrete columnar organization of frequency is not obvious in the auditory cortex; although cells in the same radial direction have similar characteristic frequencies, there do not appear to be sudden jumps in frequency as an electrode is moved tangentially in the cortex. Binaural dominance, on the other hand, does seem to be related to the existence of discrete columns. Cells of the same binaural dominance (e.g. one ear excitatory, the other ear inhibitory) lie in the same radial direction in the cortex, and are segregated into discrete strips, running along the cortical surface at roughly right angles to the iso-frequency strips.

4. Not all neurones in the primary auditory cortex show responses to sound. In those that do, a variety of shapes of tuning curve, including broad and multipeaked ones, can be found. Many neurones show complex temporal patterns of response. Many neurones show binaural interactions suggesting that they code sound direction. Each cortex predominantly represents sound sources on the contralateral side.

5. Many neurones show particular sensitivity to the features of complex sounds. Some cells seem specifically responsive to frequency-modulated stimuli. Others respond only to complex sounds such as animal calls. However, there is no evidence that such cells can be regarded as specific detectors for those features; rather the cells respond to the basic acoustic elements of the stimuli.

6. Behavioural studies of the auditory cortex, in which auditory performance is tested before and after cortical lesions, have shown that the

auditory cortex is implicated in many tasks. However, it is often very difficult to work out the functions underlying the deficits. Frequency discrimination, for instance, was once thought to be impossible after complete lesions of the auditory cortex. Later it was shown that frequency discrimination was possible after cortical lesions if the animals had to detect changes in the frequency of an ongoing series of tone pips, and this led to hypotheses either that lesioned animals were only able to respond to stimuli on the basis of change, or that they had difficulty in inhibiting inappropriate responses. Now however, lesioned animals have been shown to respond in ways which contradict both theories, and we have to resort to explaining the results in terms of task difficulty. Cortical lesions upset performance on these tasks only if the initial learning was difficult.

7. The auditory cortex seems to be necessary for normal sound localization. AI is necessary for the localization of sounds on the contralateral side of the head. Sound locus is coded in a frequency-specific way, such that each iso-frequency strip is involved in coding the source locus for sounds of that frequency.

8. Cortical lesions upset tasks where the animals have to utilize the temporal dimension of auditory stimuli, and when they have to detect or discriminate very short stimuli. This suggests that the auditory cortex may be necessary for auditory short-term memory, and for prolonging the effects of short stimuli.

9. The auditory cortex seems to affect the ability to attend to sounds in the contralateral ear.

10. Hypotheses as to the function of the auditory cortex suggest:

    (i)   that it may be necessary for the analysis of complex sounds;

    (ii)   that it subserves sound localization and the representation of "auditory space";

    (iii) that it is necessary for selective attention to auditory stimuli on the basis of source position;

    (iv) that it serves to inhibit inappropriate motor responses;

    (v)   that it serves to identify stimuli on an absolute basis;

    (vi) that it is necessary for the discrimination of auditory temporal patterns;

    (vii) that it is necessary for short-term memory when one auditory stimulus has to be related to another later in time;

(viii) that it is necessary for auditory tasks that are difficult.

F. Further Reading

The anatomy of the auditory cortex is reviewed by Brugge and Reale (1985) and Seldon (1985). Neuronal responses are reviewed by Goldstein and Abeles (1975) and Brugge and Reale (1985). Earlier behavioural studies are reviewed by Neff *et al.* (1975) and Ravizza and Belmore (1978), and some of the more recent ones by Pickles (1985c). Sound localization is reviewed by Phillips and Brugge (1985). Auditory cortical function in man is reviewed by Pinheiro and Musiek (1985).

8. *The Centrifugal Pathways*

The centrifugal auditory pathways run from the higher stages of the auditory system to the lower. One pathway, the olivocochlear bundle, runs from the superior olivary complex to the hair cells of the cochlea. The central auditory nuclei are targets for other centrifugal pathways. It has been suggested that the pathways are organized into a chain, running from the cortex to the cochlea. In this chapter, electrophysiological and behavioural experiments on centrifugal pathways will be described, and some hypotheses as to the function of the pathways discussed.

A. Introduction

So far we have considered the auditory pathway as one in which information is handed exclusively from the lower to the higher levels of the nervous system. Such a view is, however, far from that of the whole picture. In particular, the auditory system possesses a large number of nerve fibres running in the reverse direction, from the higher levels of the nervous system to the lower. The fibres run close to, but not generally within, the tracts carrying the ascending information. In this way the activity of the lower levels of the nervous system can be influenced by the complex responses of the highest. We might also expect the central state of the animal to affect the sensory responses of the early stages of the auditory pathway. Centrifugal pathways have been known since the end of the nineteenth century (e.g. Held, 1893); however, the more recent interest in centrifugal pathways was triggered by Rasmussen's description in 1946 of the olivocochlear bundle, running from the superior olive to the hair cells. Interest was also triggered by the possibility that the centrifugal pathways could modify the sensory input during processes such as attention.

235

B. The Olivocochlear Bundle

1. Anatomy

The cochlea receives a centrifugal, commonly called "efferent", innervation from the superior olivary complex. The innervation is bilateral, the fibres from the opposite side running over the dorsal surface of the brainstem just below the floor of the fourth ventricle (Rasmussen, 1946; Fig. 8.1). The

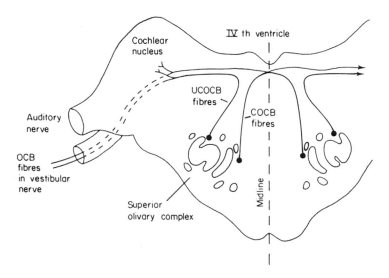

Fig. 8.1   The paths of the uncrossed olivocochlear bundle (UCOCB) and crossed olivocochlear bundle (COCB) are shown on a schematic cross-section of the cat's brainstem.

fibres are then joined by ipsilateral fibres, and a few branch off to enter the cochlear nucleus. The others leave the brain stem by way of the vestibular nerve, cross over into the auditory nerve, and enter the cochlea. Within the cochlea, the fibres terminate in two ways. Some fibres terminate with large, granulated, synaptic terminals around the lower ends of the outer hair cells. They appear to envelope both the base of the outer hair cells and the afferent terminals (Fig. 3.5C). They therefore appear to be able to control not only the state of the hair cells but possibly also synaptic transmission to the afferent pathway. A rather greater proportion of the fibres end in the region of the inner hair cells; they make axodendritic synapses *en passant* with the afferent fibres under the inner hair cells and also make contact with the afferent terminals on the base of the inner hair cells. Only rarely, however, do they make contact with the inner hair cells themselves (e.g. Brown, 1987b). The density of efferent terminals is greatest towards the basal or high-frequency end of the cochlea, although they are missing at the extreme base.

The details of the sites of origin in the brainstem have been worked out by means of axonal transport techniques by Warr (1975) and Warr and Guinan (1979). They showed that in the cat there were about 1800 fibres in all, of which about 1200 were uncrossed. The cell bodies lay in the superior olive, not in the divisions associated with the ascending pathway, such as the lateral and medial olivary nuclei, but around their borders and in many of the surrounding pre- and peri-olivary nuclei (Fig. 8.2; see also

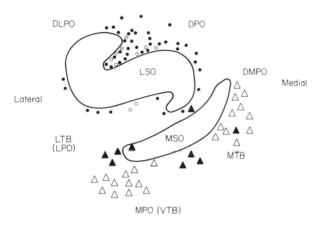

Fig. 8.2 The cells of origin of the crossed and uncrossed olivocochlear bundles were shown by applying horseradish peroxidase to either the contralateral or ipsilateral cochleae. Small cells (●,○) are situated laterally and project to the nerve fibres below the inner hair cells, large cells (▲,△) are situated medially and project mainly to the outer hair cells. Cells projecting ipsilaterally are shown by filled symbols, those projecting contralaterally by open symbols. The cells of origin, here represented schematically, lie in the pre- and peri-olivary cell groups and on the borders of the LSO. LSO, lateral superior olivary nucleus; MSO, medial superior olivary nucleus; DLPO, dorsal peri-olivary nucleus; DMPO, dorsomedial peri-olivary nucleus; DPO, dorsal peri-olivary nucleus; LTB, lateral nucleus of the trapezoid body, or lateral preolivary nucleus (LPO); MPO, medial pre-olivary nucleus, or ventral nucleus of the trapezoid body (VTB), MTB, medial nucleus of the trapezoid body. Data from different levels of the cat brainstem have been projected onto one cross-section. Data from Warr *et al.* (1986, Fig. 1).

Fig. 6.10). Such an association of the centrifugal system with the areas surrounding, but not identical with, the ascending pathway seems to be reproduced at many levels of the auditory system.

The cells of origin can be divided into two types (Fig. 8.2). The cells in a lateral group have bodies tightly clustered around the lateral superior olivary nucleus (LSO). The cells have relatively small bodies, and give rise to small, unmyelinated axons, which project to the region of the inner hair cells, almost entirely on the ipsilateral side (Fig. 8.3). Cells of the medial

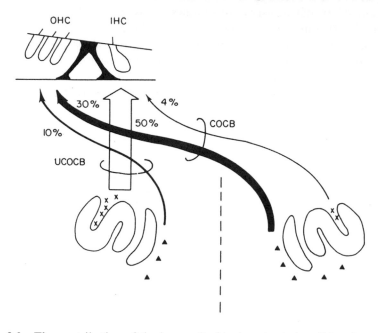

Fig. 8.3   The contribution of the large cells (▲, to outer hair cells) and small cells (×, to region of inner hair cells) to the different components of the olivocochlear bundle in the cat. The indicated proportions do not add up to 100% because other fibres add a few per cent. COCB, crossed olivocochlear bundle; UCOCB, uncrossed bundle. Adapted from Warr (1978) to incorporate the results of Warr and Guinan (1979).

group are scattered around the medial superior olive. They have larger bodies, give rise to large, myelinated axons, and project to the outer hair cells, mostly on the contralateral side (Guinan *et al.*, 1983). This almost complete separation into two systems, one to the region of the inner hair cells, and one to the outer hair cells, may well be associated with a functional separation, associated with the different roles of the inner and outer hair cells in transduction.

2. Neurotransmitters

There is clear evidence that acetylcholine is a transmitter of the olivocochlear bundle (for review, see Klinke and Galley, 1974). Acetylcholine can be collected when the cochlea is perfused during stimulation of the olivocochlear bundle. Furthermore, the efferent terminals, axons, and cell bodies contain acetylcholinesterase and choline acetyltransferase (e.g. Eybalin and Pujol, 1987). In addition, the effects of stimulating the olivocochlear bundle can be reproduced by the infusion of cholinomimetics, and can be blocked by both muscarinic and nicotinic cholinergic blockers. Unusually for a cholinergic system, however, strychnine is also a powerful blocker. Other neurotransmitters seem to be present too: there is evidence for enkephalins, dynorphins and γ-amino-butyric acid (GABA) (Altschuler and Fex, 1986; Abou-Madi et al., 1987). Some may act as co-transmitters; it is also possible that the olivocochlear bundle is not homogeneous and that different fibres use different transmitters.

3. Physiology

(a) Effect on the ascending system

(*i*) *Mechanism of action on hair cell membranes.* Centrifugal effects on hair cell membranes have most clearly been evaluated for hair cells of the turtle cochlea. Art et al. (1984) showed that stimulation of the centrifugal supply to the hair cells caused hair cell hyperpolarization. The potentials could be affected by changing the K^+ concentration of the perilymph, but not the Cl^- concentration. This suggests that centrifugal stimulation had opened K^+, and not Cl^-, channels in the hair cell membrane. This result disagrees with results obtained by gross recording in the mammalian (cat) cochlea, where the potentials produced by centrifugal stimulation were reduced by perfusing the cochlea with solutions of altered Cl^-, but not altered K^+, concentration (Desmedt and Robertson, 1975). It would appear therefore that centrifugal fibres open K^+ or Cl^- channels in the hair cells, reducing the cell membrane resistance, and in some cases hyperpolarizing the cells, with perhaps species variation.

(*ii*) *Effect on the afferent fibres.* The medial group of cells sends its axons to the outer hair cells, mostly on the contralateral side. These axons, which make up most of the crossed olivocochlear bundle (COCB), can be stimulated electrically where they cross the floor of the fourth ventricle. Galambos (1956) showed that such stimulation reduced the gross neural response of the cochlea, the N_1 potential. The effect has a comparatively long latency, the inhibitory effect appearing some 15 ms after the onset of stimulation, and increasing over a further 50 ms. The reduction was greatest at low intensities. Under the most favourable circumstances the effect was equivalent to reducing the stimulus intensity by 20–25 dB, although 15 dB was a more common figure. At the same time, stimulation increases the magnitude of the cochlear microphonic by a few dB (Fex, 1959). By discrete stimulation of the cells of origin in the superior olive, Gifford and Guinan (1987) showed that the effects could indeed be produced by stimulation of the medial group of olivocochlear cells.

It is now known that stimulation of the crossed olivocochlear bundle (i.e. mainly fibres from the large-celled or medial group, running to the *outer* hair cells) can indeed affect the responses of the *inner* hair cells, and hence the auditory nerve fibres to which they are connected. As shown in Fig. 5.18B and C, the result is to raise the thresholds of inner hair cell and auditory nerve tuning curves at the tip, and to broaden the tuning curves. The inhibition can also be revealed by the rate-intensity functions for tones, which are shifted along the intensity axis (Fig. 8.4). This would explain

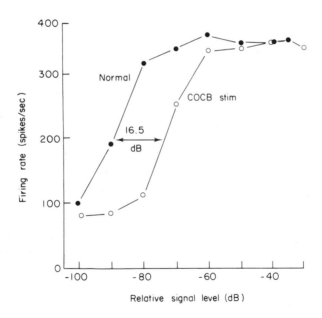

Fig. 8.4   Electrical stimulation of the crossed olivocochlear bundle (COCB) shifts the rate-intensity functions of single auditory nerve fibres. From Wiederhold (1970, Fig. 4C).

how the N_1 potential was reduced, and why the effects were greatest at low intensities. However, there is a problem, since the medial olivocochlear cells send their fibres to the *outer*, rather than the inner, hair cells (Fig. 8.3).

The effects can most easily be explained by supposing that the stimulation influences the sharpness of mechanical tuning on the basilar membrane, although that hypothesis has not yet been tested directly, by for instance measuring basilar membrane mechanics during olivocochlear stimulation. As explained in Chapters 3 and 5, there is now considerable indirect evidence that the sharp tuning of the basilar membrane depends on an active mechanical process, and that the outer hair cells actively put energy into the travelling wave as it moves up the cochlea, increasing its amplitude and sharpness of tuning. Under the hypothesis, stimulation of the olivocochlear bundle interferes with the ability of outer hair cells to do this. The idea that stimulation of the olivocochlear bundle alters cochlear mechanics is supported by the finding that stimulation of the bundle alters the active emission of acoustic energy from the cochlea to the ear canal (Mountain, 1980; Siegel and Kim, 1982; see also Chapter 5)

In the absence of evidence on the nature of the motile process, we can only speculate on the way that olivocochlear stimulation might alter the contribution of outer hair cells to the sharp tuning and sensitivity of the mechanical travelling wave. Changing the membrane potential of the cell might alter the operating point and perhaps shape of the outer hair cell input–output function (Fig. 3.21B). This, and changes in hair cell membrane resistance, might affect the efficacy of the motile process.

Little is known about the function of the lateral group of olivocochlear cells. Stimulation of their axons was once thought to reduce the N_1 potential of the cochlea without any effect on the cochlear microphonic (Sohmer, 1966). However, Gifford and Guinan (1987) stimulated the cells of origin of the medial system directly, and suggested that the effects had been due to spread of current to the medial system. They were unable to find a differential function for the lateral cells.

(b) What normally activates the olivocochlear bundle?

The fibres of the olivocochlear bundle are responsive to sound. It is also likely that the fibres are affected by the more central neural activity of the animal, since the bundle can be activated by the electrical stimulation of certain sites higher in the central nervous system. Fex (1962) showed that fibres of the olivocochlear bundle fired spontaneously with a particularly regular firing pattern. Sound drove the olivocochlear fibres with a latency of 10–30 ms. The fibres have thresholds within 15 dB of the most sensitive afferent nerve fibres, and have as sharp, or nearly as sharp, tuning curves (Robertson and Gummer, 1985; Liberman and Brown, 1986). When recordings are made within the cochlea, just before the fibres enter the organ of Corti, it is possible to relate fibre properties to the site of termination along the cochlear duct. In this case, it can be shown that the fibres are tuned to the frequency of the region they themselves influence (Robertson and Gummer, 1985: Liberman and Brown, 1986). Evidence for such an organized tonotopic projection can also be found anatomically (Robertson *et al.*, 1987).

The tonotopic organization can also be shown functionally with acoustic stimulation, by recording the response of afferent auditory nerve fibres from one ear, while stimulating the opposite ear with sound. If care is taken to exclude crosstalk and contractions of the middle ear muscles, then the effect of one cochlea on the other must be due to the olivocochlear bundle. Buño (1978) showed that stimulation of one ear reduced the response of auditory nerve fibres running from the other ear. When the frequency of the contralateral stimulus was plotted against the effectiveness of its action on the ipsilateral response, the resulting frequency response curves were usually V-shaped and centred on the characteristic frequency of the fibre being inhibited (Fig. 8.5). Occasionally W-shaped crossed inhibitory effects were found, with the dips of the W on either side of the characteristic frequency.

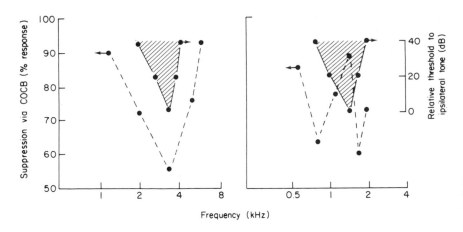

Fig. 8.5   Interactions of ipsilateral and contralateral influences in the auditory nerve reveal the effects of acoustic stimulation of the olivocochlear fibres on afferent activity. Tuning curves for ipsilateral stimuli were obtained first (shaded areas). Then a continuous tone was presented to the ipsilateral ear, 10 dB above threshold. The change in firing rate was plotted as a 30-dB *contralateral* tone was swept in frequency (dotted lines). The contralateral effect (via the OCB) is tuned to the same frequency region as the ipsilateral activation. From Buño (1978, Fig. 4).

The maximum rate at which sound can drive fibres of the olivocochlear bundle is of the order of 100/s even in unanaesthetized decerebrate animals; yet the maximal olivocochlear effect on the cochlea is attained with electrical stimulation at 400/s (Fex, 1962; Liberman and Brown, 1986). This has suggested to many workers that there is an additional central facilitation of the olivocochlear bundle, enhancing the reflex effects when required. One such influence was shown by Desmedt (1975). He showed that the olivocochlear bundle could also be activated by electrical stimulation of the insulo-temporal cortex. The importance of the central state of the animal was shown by Banks *et al.* (1979), who recorded multi-unit activity from the olivocochlear bundle with chronically implanted electrodes. They showed that continuous activity was present in awake animals, but was absent in anaesthetized ones. Tones activated the fibres, but not spontaneous movements or vocalizations, suggesting that the fibres were not involved in reducing the responses to self-produced noise.

(c) Functional significance of the olivocochlear bundle

In spite of a large number of experiments, we are still unsure of the importance of the olivocochlear bundle in hearing. For instance, behavioural experiments have often found no difference in auditory performance after sectioning the crossed olivocochlear bundle. The bundle can be cut where it crosses the midline. This would be expected to interrupt three-quarters of the fibres to the outer hair cells. Of course, it is always possible that the remaining fibres were sufficient to carry on the function. Current hypotheses fall into four main groups.

(*i*) *Improving the detection of signals in noise.* Winslow and Sachs (1987) showed that stimulation of the olivocochlear bundle could reduce masking in single auditory nerve fibres. Figure 8.6 shows the rate-intensity functions of an auditory nerve fibre for tone pulses in silence (A), and for tone pulses heavily masked by constant broadband noise (B). Here, the firing rate is plotted as a function of tone level. With no olivocochlear stimulation, the background noise flattened the tone intensity-function (full line in B: compare with full line in A). Electrical stimulation of the crossed olivocochlear bundle partially reversed the flattening, so that the tone could now produce a greater relative change in firing rates.

Winslow and Sachs showed that this occurred because olivocochlear stimulation reduced the response to the constant background of masking noise. The effect was to lower the firing rate at low tone levels (because the noise drove the fibre less strongly), and to increase the tone-induced rate at high tone levels (because there had been less adaptation of the firing to the continuous background noise). The rate-intensity function therefore steepened.

A role in the detection of signals in noise was also suggested by the behavioural experiments of Dewson (1968). He trained rhesus monkeys to

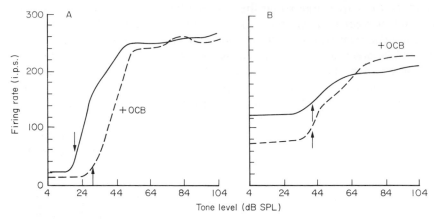

Fig. 8.6   Rate-intensity functions for an auditory nerve fibre, without (——) and with (— — — —) electrical stimulation of the COCB. (A) No masking noise. The fibre here is activated only by the tone pulses, and shows the standard sigmoidal relation between tone intensity and firing rate. Olivocochlear stimulation shifted the function along the intensity axis by 12 dB. (B) With continuous masking noise, the function was flatter (see text). Olivocochlear stimulation reversed the effects, and steepened the function to some extent.

Arrows: point at which the tone increased the firing rate by 20 spikes/s. For clarity, the original published functions have been slightly smoothed. From Winslow and Sachs (1987, Fig. 4).

discriminate between two vowel sounds in the presence of masking noise. Performance was reduced by cutting the crossed olivocochlear bundle. It seems that the *discrimination* between two sounds was the key factor, because in experiments where the animal has to *detect* a single sound in noise, cutting the bundle has no effect (Trahiotis and Elliott, 1970). We can see how this fits with the neural results: olivocochlear stimulation steepens the rate-intensity function in Fig. 8.6B, relatively increasing the response to the suprathreshold stimuli required for a complex discrimination. However, it does not change the position of the foot of the function on the intensity axis (arrows), and so does not affect masked thresholds for a simple detection.

(ii) Protection from noise damage. Cody and Johnstone (1982) showed that simultaneous contralateral acoustic stimulation could reduce the degree of acoustic trauma produced by an intense tone. The protection could be abolished by the application of strychnine, a known blocker of the olivocochlear bundle, suggesting that the protection was mediated via the bundle (Rajan and Johnstone, 1983). Here, the mechanism of the action is not known. It is possible that the contralateral stimulation reduces the size of

the travelling wave in the cochlea by affecting the active mechanical process. However, it is seen for traumatizing tones of over 100 dB SPL, an intensity at which the active process is probably no longer making a contribution (Fig. 5.22). It is possible, therefore, that the olivocochlear bundle protects the hair cells in some other ways. The finding that contralateral stimulation reduces acoustic trauma has important practical applications, suggesting that it might be unwise to protect only one ear in a noisy environment.

(*iii*) *Controlling the mechanical state of the cochlea.* If outer hair cells indeed amplify the mechanical travelling wave by means of an active process, it is possible that the amplification could be critically affected by small degrees of bias in the system. For instance, Johnstone *et al.* (1986) showed that substantial degrees of two-tone suppression could be produced by low-frequency stimuli which generated pressure differences across the cochlear partition equivalent to a 1-mm head of water. Presumably, as described in Chapter 5, the pressure differences deflect the basilar membrane, and can bias the outer hair cells away from the steepest part of their input–output function. Pressure differences of this order might easily arise in life, for instance due to postural changes transmitted passively along the ducts connecting the cochlea to the other fluid spaces of the body. It is therefore possible, as suggested by Johnstone *et al.*, that the olivocochlear bundle acts to keep the cochlea in its optimal mechanical state. The bundle might affect the mechanical properties of the hair cells, changing the stiffness of the stereocilia, or introduce a bias in their displacement by actively deflecting the stereocilia. This might be a function for the slower of the motile processes described in Chapter 5, for instance those depending on actin–myosin interactions in the hair cell (Flock *et al.*, 1986; Zenner, 1986). Such a scheme also requires the cochlea to detect a static pressure difference across the partition. That may be a function of the $5 - 10\%$ of afferents in the auditory nerve which innervate the outer hair cells, since outer hair cells are responsive to d.c. deflections of the cochlear partition (e.g. Russell *et al.*, 1986a).

(*iv*) *Role in attention.* Could the olivocochlear bundle be involved in attention, attenuating the auditory input when it is judged to be irrelevant? Such a hypothesis could explain the common experience of fluctuations in the awareness of auditory stimuli. Although the recent consensus has been that the olivocochlear bundle does not gate the sensory input in attention, there is one positive report, in a nevertheless apparently well-controlled study. Oatman (1976) trained cats, with middle ear muscles cut, to make a visual discrimination for a food reward. Clicks were continuously delivered at 1/s through ear tubes. When the cat was performing visual discrimination, the

N_1 potential of the cochlea was smaller, and the CM larger, than when the cat was relaxed. The effect on N_1 was greatest at low intensities. The pattern of changes therefore exactly paralleled those expected with the olivocochlear bundle. This was important corroborative evidence that the olivocochlear bundle was activated. However, the important control, of showing that the effect disappeared when the olivocochlear bundle was cut, was not done.

C. Centrifugal Pathways to the Cochlear Nuclei

1. Anatomy

The cochlear nuclei receive centrifugal fibres from several sources. By far the largest innervation appears to arise in the superior olivary complex. Centrifugal fibres from the dorsal and ventral nuclei of the lateral lemniscus, the inferior colliculus, and the reticular formation have also been described. Some of the centrifugal fibres from the superior olivary complex consist of branches of the olivocochlear bundle. Others run ipsilaterally by rather more direct paths in the dorsal and intermediate acoustic striae, and from superior olives on both sides, in the trapezoid body (Fig. 8.7). As with the olivocochlear bundle, the centrifugal innervation does not arise in the main nuclei associated with the ascending system, but in some of the surrounding pre- and peri-olivary nuclei (Elverland, 1977; Adams and Warr, 1976; Spangler *et al.*, 1987). These nuclei, of course, receive an auditory input, so the centrifugal pathways can be activated by sound as well as by central influences. The centrifugal innervation from the inferior colliculus arises laterally in the colliculus, and descends and runs along the ventral surface of the brainstem, where it turns and ascends dorsally into the middle layers of the DCN (Rasmussen, 1967). The fibres from the nuclei of the lateral lemniscus have been the least well described and not all authors agree on the details. Both the dorsal and ventral nuclei of the lateral lemniscus on both the ipsilateral and contralateral sides have been suggested as sites of origin. The reticular formation also sends a projection to the cochlear nucleus (Adams and Warr, 1976). All divisions of the cochlear nuclei receive centrifugal fibres in different degrees from the different sources. There is considerable detail in the projections – for instance, Cant and Morest (1978) described six groups of centrifugal axons ending in different ways in the anteroventral cochlear nucleus alone.

2. Neurotransmitters

As in the cochlea, the neurotransmitters associated with the centrifugal system are known rather better than those associated with the centripetal

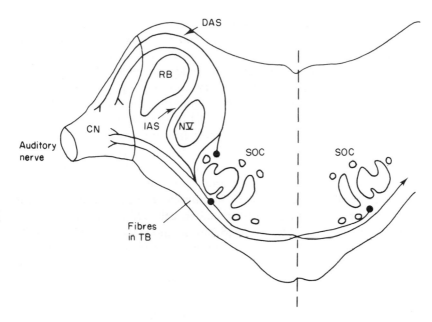

Fig. 8.7 A schematic representation of some of the centrifugal pathways from the superior olivary complex (SOC) to the cochlear nucleus (CN). Branches of the olivocochlear bundle also run from the SOC to the CN but are not shown here. The fibres run by three routes: the dorsal acoustic stria (DAS), the intermediate acoustic stria (IAS) and in the trapezoid body (TB). RB, restiform body (inferior cerebellar peduncle); NV, fifth nerve nucleus. Data from Elverland (1977).

system. Moreover, knowledge of their pharmacology has been useful in discovering the function of the centrifugal pathways.

There is evidence that some of the centrifugal pathways to the cochlear nucleus are cholinergic. The cochlear nucleus contains relatively high levels of choline acetyltransferase, an enzyme used in synthesizing acetylcholine. Large reductions in the choline acetyltransferase levels can be produced by lesion of the superior olivary complex, mainly ipsilaterally (Godfrey et al., 1987b). In addition, many of the terminals in the cochlear nucleus and the cells of origin, particularly in the superior olive, react positively for acetylcholinesterase. Osen and Roth (1969) ascribe the reaction exclusively to the olivocochlear bundle, which of course sends branches to the cochlear nucleus. They disagree with the initial observations of Rasmussen (1964), who described the tracts running directly from the ipsilateral superior olive in the intermediate acoustic stria as also reacting positively. This more ventral route is also supported by the choline acetyltransferase measurements of Godfrey et al. (1987a,b).

Whatever the pathways involved, Comis and Whitfield (1968) produced strong evidence that an excitatory centrifugal pathway running from the superior olive used acetylcholine as a transmitter. The excitation produced by electrical stimulation of the pathway could be blocked by cholinergic blockers such as atropine and dihydro-β-erythroidine applied to target cells of the cochlear nucleus. The effects of stimulating the pathway could be mimicked by applying acetylcholine to cells of the cochlear nucleus. And Comis and Davies (1969) showed that electrical stimulation of the superior olive could release acetylcholine from the cochlear nucleus.

A second centrifugal neurotransmitter in the cochlear nucleus seems to be noradrenaline. Noradrenaline-containing terminals have been demonstrated in the cochlear nucleus by an immunofluorescence technique, and the noradrenaline-containing fibres traced to cell bodies in, at least, the dorsal nucleus of the lateral lemniscus (Swanson and Hartman, 1975). Comis and Whitfield (1968) showed that stimulation of the nuclei of the lateral lemniscus could give inhibition in the cochlear nucleus. The effects were similar to those of noradrenaline applied to single cells of the cochlear nucleus, which was always inhibitory.

It is very likely, of course, that there are centrifugal neurotransmitters to the cochlear nucleus in addition to acetylcholine and noradrenaline.

In contrast to noradrenaline which was associated with inhibitory centrifugal pathways, it appears that at least some of the inhibition intrinsic to the nucleus may be mediated by γ-amino-butyric acid and glycine (Godfrey *et al.*, 1977, 1978; Peyret *et al.*, 1987).

3. Physiology

Comis and Whitfield (1968) electrically stimulated the superior olive and recorded the activity of cells in the cochlear nucleus. Stimulation of most of the superior olive excited cells of the ipsilateral anteroventral cochlear nucleus. There was evidence of "gating" of the auditory input by the centrifugal pathway, since electrical stimulation could lower the threshold of neurones in the nucleus by as much as 15 dB. On the other hand, stimulation of the extreme lateral region of the superior olive, in the region of the dorsolateral peri-olivary and the lateral pre-olivary nuclei, inhibited cells in the cochlear nucleus (Comis, 1970). In some cases the inhibition occurred because the olivocochlear bundle had been activated, reducing the responses of the cochlea. In such cases the inhibition could be released by the application of strychnine, a powerful blocker of the olivocochlear bundle in the cochlea, to the round window. In other cases this could not be done, and it is possible that direct inhibitory pathways were involved. It is not known to what extent such effects can be explained by the activation of the branches of the olivocochlear bundle which run directly to the cochlear nucleus. There is evidence that such branches can be both excitatory and inhibitory. Starr and Wernick (1968) found that with the cochlea destroyed, stimulation of the crossed olivocochlear bundle increased the spontaneous activity of 42%, and decreased the activity of 16%, of cells recorded in the cochlear nucleus. Of course, some of these effects may have been indirect, mediated by the activity of inhibitory interneurones.

A pathway which was generally inhibitory, but sometimes excitatory, ran from the contralateral nuclei of the lateral lemniscus (Comis and Whitfield, 1968).

Some effects of natural stimulation of the centrifugal pathways can be shown by stimulation of the contralateral ear. There are no direct afferent auditory nerve fibres from one cochlea to the contralateral cochlear nucleus: all influences must be by pathways such as the direct nucleus–nucleus connections, or centrifugal pathways via the superior olivary complex, or perhaps via higher nuclei such as the nuclei of the lateral lemniscus and the inferior colliculus.

Mast (1973) showed that some cells in the dorsal cochlear nucleus could be excited or inhibited by contralateral sound. Tuning curves for the direct ipsilateral and the centrifugal contralateral effects were generally very similar in threshold, best frequency, and shape. Klinke *et al.* (1969) showed that the contralateral stimulus could sometimes produce a W-shaped tuning curve, with two inhibitory sidebands. The latter effect is reminiscent of the effects obtained in a minority of cases via the olivocochlear bundle (Buño, 1978).

4. Behavioural Experiments

The behavioural analysis of the olivocochlear bundle is comparatively straightforward, because the crossed component of the bundle can be cut in the midline, without danger of damage to other auditory structures. In contrast, most centrifugal pathways to the cochlear nucleus run intermingled with the afferent pathways, so an analogous approach is not available. However, the centrifugal system uses acetylcholine and noradrenaline as neurotransmitters. Since these are not used by the centripetal system pharmacological methods can be used to affect the centrifugal system selectively.

Pickles and Comis (1973) chronically implanted a cannula over the cochlear nucleus in cats, so that drugs could be applied to the nucleus in unanaesthetized animals. There is, as described above, good evidence that some of the centrifugal pathways from the superior olive to the cochlear nucleus are cholinergic and that their action can be blocked by atropine. The cats were trained to detect tone pips both in silence and against masking noise. In such behavioural tests, atropine applied to the cochlear nucleus raised the absolute thresholds for tone pips by a few dB, but raised masked thresholds significantly more (Fig. 8.8). This suggests that atropine might

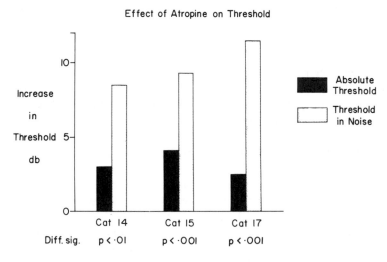

Fig. 8.8   Atropine applied to the cochlear nucleus in unanaesthetized cats raised masked thresholds to a greater extent than absolute thresholds. Thresholds were determined behaviourally. Data from Pickles and Comis (1973).

have been blocking a system whose normal action was to *help* the animal hear signals in masking noise.

The psychophysics of the atropine effect was investigated in further experiments. It was shown that atropine had an effect on masked thresholds only if the signal was masked by noise of a wide, but not narrow, bandwidth (Pickles, 1976a). In itself, this was an important control for the validity of the effect, since it suggested that the atropine effect was related to the properties of the masker. It also suggested a possible functional implication. It is known that, if we are detecting a tone against masking noise, most of the masking is due to noise components which are near to the signal in frequency. The more remote frequency components do not contribute to masking, being removed by the filtering processes of the auditory system (the critical band; see Chapter 9). By systematically varying the frequency relations of the masker and signal, it was possible to show that the application of atropine to the cochlear nucleus abolished this filtering process (Pickles, 1976a). The results therefore suggested that the filtering of signals from the background noise, on the basis of frequency, was under centrifugal control at the cochlear nucleus. It is interesting that both the centrifugal pathways so far investigated in detail, namely the olivocochlear bundle and the centrifugal pathways to the cochlear nucleus, seem to affect the processing of signals in noise, although in different ways.

Noradrenaline, a transmitter in a second centrifugal system to the cochlear nuclei, has also been applied through the cannula to block the normal operation of the noradrenergic system by the blanket activation of the receptors (Pickles, 1976b). Both absolute and masked thresholds were affected. The loss in masked threshold depended on the bandwidth of the masking noise, again suggesting that frequency filtering had been affected.

D. Centrifugal Pathways in Higher Centres

1. Anatomy

Two descending systems have been described as originating in the auditory cortex. First, it appears as though each cortical area sends a descending projection to the division of the medial geniculate from which it receives an ascending projection (Diamond *et al.*, 1969). The descending fibres terminate on the very cells that project back to the cortex (Morest, 1975). Such a terminal of a descending fibre is shown by the ending "DF" in Fig. 6.19B. There seems, therefore, to be a close coupling in the loop of afferent and efferent fibres, suggesting that, within each functional division, the thalamus and cortex act as one unit. Secondly, the cortex also sends descending projections to a wide range of diencephalic and midbrain nuclei, including other nuclei of the thalamus, the tegmentum, the inferior colliculus and two areas connected to the motor system, namely the corpus striatum and the pontine nuclei (Diamond *et al.*, 1969).

The inferior colliculus receives a descending innervation from both the auditory cortex and the medial geniculate body. The axons terminate in the dorsal and external regions of the nucleus (Fig. 8.9; Rockel and Jones, 1973a,b). These areas do not receive the main ascending supply. A similar picture was produced in electrophysiological experiments by Massopust and Ordy (1962). Stimulation of the auditory cortex produced evoked potentials in the external layers of the inferior colliculus, whereas sound produced evoked potentials in the central region. There seems, therefore, some separation between the ascending and descending auditory systems, already noted in the superior olivary complex.

Harrison and Howe (1974b) suggest that the collicular areas receiving a

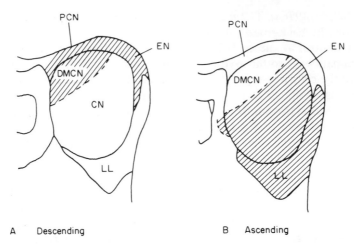

A   Descending   B   Ascending

Fig. 8.9   Descending and ascending fibres to the inferior colliculus end in predominantly different parts of the nucleus. Here, patterns of terminal degeneration are
shown, as they appear after lesions of the descending (A) and ascending (B)
innervation. The terminology is the older terminology of Rockel and Jones (see also
Fig. 6.16A). CN, central nucleus of the IC; DMCN, dorsomedial division of the
central nucleus of the IC; EN, external nucleus of the IC; LL, lateral lemniscus;
PCN, pericentral nucleus of the IC. From Rockel and Jones (1973a, Fig. 21) and
Rockel and Jones (1973b, Fig. 1).

descending centrifugal innervation from the cortex might in turn give rise
to centrifugal axons descending further down the auditory system. These
might end on the pre- and peri-olivary nuclei of the superior olivary complex,
which themselves are known to give rise to centrifugal axons. Other
centrifugal fibres run directly from the inferior colliculus to the dorsal
cochlear nucleus (Rasmussen, 1967). In this way, a complete chain of
descending pathways, running from the cortex to the periphery, could exist.

2. Physiology and Function

There have been very few studies of the physiology of the more central of
the centrifugal pathways. There have been reports that electrical stimulation
of the cortex can produce both excitation and inhibition in single units of
the medial geniculate body (Watanabe *et al.*, 1966; Andersen *et al.*, 1972).
Ryugo and Weinberger (1976) suggested that the closed corticogeniculate
loops could be responsible for the reverberations seen in the firing of
many geniculate neurones. Many such neurones show reverberation after

discharge, lasting 100 ms or more after an auditory stimulus. Cooling the cortex reduced the response.

Dewson *et al.* (1966) detected the influence of the I-T (insulo-temporal) area of the auditory cortex on the cochlear nucleus by recording recovery functions to paired clicks in the cochlear nucleus. In this technique, the responsiveness of the nucleus after one click was tested by measuring the size of the response to a second click. In unanaesthetized cats, after lesion of the I-T area, the responsiveness recovered particularly rapidly after each click, and Dewson *et al.* interpreted this as suggesting that normally the I-T cortex, by means of the centrifugal pathways, prolongs the effect of stimuli in the cochlear nucleus. Beyond that, they did not speculate on the significance of the finding.

Desmedt (1975) presented physiological evidence for a train of centrifugal pathways descending from the cortex to the cochlea. Electrical stimulation of many stages of the brain, generally close to but not within the ascending tracts, altered auditory evoked potentials earlier in the system. Desmedt suggested that the centrifugal pathways formed one system, which he called CERACS (Centrifugal Extrareticular Activating System), the extrareticular to indicate that the system was not part of the reticular formation, which can also have a modifying effect on the sensory input.

We should not, however, think of the centrifugal system as affecting only the periphery. It is very likely that it produces changes in neuronal processing in the intermediate nuclei. Examples have already been given where the cholinergic efferent supply seems to alter the detection of signals in noise in the cochlear nucleus as well as at the periphery. There is also evidence that centrifugal pathways must be involved in more complex behaviour. It was described in the previous chapters how some learned auditory tasks survive complete ablation of the neocortex. The reflexes may, therefore, have been established subcortically. Furthermore, it is likely that in the intact animal the cortex would have been able to influence the reflexes. This influence must therefore occur through the centrifugal pathways. We therefore have a system in which reflexes can be established at many levels, and in which the cortex controls the reflexes through descending influences, and not one in which all sensory activity is relayed to the cerebral cortex, where learned connections are made, and then relayed down the effector pathways. In this way it is possible that the centrifugal pathways from the cortex can be thought of as being some of its principal output pathways.

E. Summary

1. Centrifugal (efferent) auditory pathways parallel the centripetal (afferent) auditory pathways along the entire length of the system, forming a chain

which runs from the cortex to the hair cells. In many stages of the auditory pathway, they run adjacent to, but not actually within, the tracts and nuclei principally associated with the ascending system.

2. The cochlea is innervated by the olivocochlear bundle, which arises bilaterally in the superior olivary complex. Cells of a medial group, situated near the medial nucleus of the superior olive, have relatively large bodies, and give rise to relatively large axons which innervate the outer hair cells, mainly on the contralateral side. Cells of a lateral group are situated around the lateral nucleus of the superior olive, have smaller cell bodies, and give rise to smaller axons which innervate the nerve fibres just below the inner hair cells, almost entirely on the ipsilateral side. The transmitter is acetylcholine. Peptides such as dynorphin and enkephalins are implicated too.

3. Electrical activation of the crossed olivocochlear bundle reduces the response of auditory nerve fibres to sound. This may occur because the efferents affect the contribution of the outer hair cells to the sharp tuning and sensitivity of the mechanical travelling wave on the basilar membrane.

4. Fibres of the olivocochlear bundle are themselves responsive to sound and have sharp tuning curves and low thresholds similar to those of afferent fibres. Olivocochlear fibres terminate in areas of the cochlea corresponding to the frequency of the sound that drives them best, and so make closed frequency-specific feedback loops. Central influences also affect the activity of the olivocochlear fibres.

5. The function of the olivocochlear bundle in auditory performance is uncertain. Hypotheses fall into four main groups:

   (i)   it may improve the detection of signals in masking noise;
   (ii) it may help protect the cochlear from acoustic trauma;
   (iii) it may control the mechanical state of the cochlea, compensating for changes in factors such as the stiffness and static displacement of the basilar membrane; and
   (iv) it may be involved in attention.

6. The cochlear nucleus receives branches of the olivocochlear bundle, together with other centrifugal fibres from the superior olivary complex, and from higher auditory nuclei, including the nuclei of the lateral lemniscus and the inferior colliculus. Some of the transmitters seem to be acetylcholine and noradrenaline.

7. Centrifugal fibres to the cochlear nucleus are both inhibitory and excitatory. The cholinergic innervation affects the animal's ability to detect signals in noise. The innervation affects the bandwidth of noise that contributes to masking.

8. There have been comparatively few studies of the physiology of centrifugal fibres at higher levels of the auditory system. There is evidence that the fibres are organized into a functional chain, so that the cerebral cortex can affect the activity of the lower stages of the auditory system. Their significance is not known: they may for instance serve to control auditory reflexes established at lower levels of the auditory system.

F. Further Reading

Warr *et al.* (1986) describe the organization of the olivocochlear bundle, and Harrison and Howe (1974b) describe the organization of the more central centrifugal fibres. Wiederhold (1985) reviews the physiology of the olivocochlear bundle, and Klinke (1986), Caspary (1986) and Fex and Altschuler (1986) review information on the neurotransmitters. Desmedt (1975) reviews some of the physiology of the central centrifugal auditory pathways. Behavioural experiments are described by Neff *et al.* (1975).

9. *Physiological Correlates of Auditory Psychophysics and Performance*

There has been a tendency for auditory psychophysicists to seek correlates of psychophysical phenomena in the neuronal responses of the earliest stages of the auditory system for which we have definite information. Thus, for instance, the factors governing our handling of frequency and intensity information have been sought in the responses of single fibres of the auditory nerve. The validity of such correlations will be examined here. The relation of the absolute threshold to the best thresholds of auditory nerve fibres, together with the possible aspects of auditory nerve firing used as cues by the subject, will be discussed first. The physiological correlates of psychophysical frequency resolution and discrimination will be examined, followed by the coding of auditory stimuli as a function of intensity and the determination of the sensation of loudness. The relation of sound localization to neuronal binaural interaction will be examined. Finally, neuronal responses to speech will be described. This chapter requires knowledge of the information contained in Chapters 1 to 4. In addition, there are some specific referrals back to Chapters 6 and 7.

A. Introduction

In this chapter we shall try to analyse the extent to which some aspects of auditory performance can be explained in terms of known physiological processes. The chapter by no means attempts to be a balanced review of the psychology or psychophysics of hearing, but to deal with certain phenomena whose explanation has, or seems to have, a close correlate with physiology.

B. The Absolute Threshold

The absolute threshold seems to be a reasonably close match to the minimum thresholds of auditory nerve fibres (Fig. 4.4, p. 84). The behavioural audiogram is near, or is about 10 dB below, the lowest neural thresholds. A small discrepancy is not surprising. Even if the subject uses only mean neural firing rates, he may be able to do better than suggested by the thresholds of individual fibres, because he is able to average activity over many fibres. The discrepancy between the neural and behavioural data is more serious at high frequencies. The reason for this is not clear; the surgical preparation for the electrophysiological experiment may well have had an influence.

Is phase-locking rather than an increase in mean firing rate used as a cue at threshold? In electrophysiological recordings of auditory nerve fibres the first indication of activation by a tone of low frequencies can be a phase-locking of the spontaneous activity to the stimulus, rather than a net increase in firing. The threshold for phase-locking may be 10–20 dB below that for an increase in mean firing rate. How this happens may be understood from the membrane potential of inner hair cells. At low intensities the depolarizing phases of the intracellular potential will add just as many spikes as the hyperpolarizing phase subtracts, and phase-locking will increase without any net increase in firing rate. In order to see whether phase-locking is used as a cue, we might think of seeing whether the behavioural threshold falls below the neural mean rate threshold to a greater extent at low, than at high frequencies, where phase-locking does not occur. Unfortunately, there are so many uncertainties in the comparison that it turns out not to be practicable. There are such large interindividual differences in both the neural and electrophysiological sets of data that comparisons have to be made in the same animals. The anaesthetic and the surgical preparation for recording will produce unknown and uncertain effects. At the moment, the comparison has not been possible at the required level of certainty.

There is, however, some psychophysical evidence that phase-locking is not used as a detection cue at threshold. A beating sensation can be produced between two sinusoids with slightly different frequencies presented to the two ears. The beats are likely to arise from an interaction of phase-locked action potentials in the central nervous system. Groen (1964) showed that binaural beats could be detected when one of the stimuli was as much as 20 dB below its absolute threshold. This suggests that phase information could be transmitted by the auditory nerve when the stimulus was below threshold, and has the corollary that the phase information did not by itself determine the absolute threshold.

C. Frequency Resolution

1. A Review of the Psychophysics of Frequency Resolution

(a) Frequency resolution and frequency discrimination

One of the most fundamental properties of the auditory nerve system is its frequency selectivity. Psychophysically, we distinguish two phenomena of frequency selectivity. One is that known as frequency discrimination. Two tones are presented one after the other, and we have to tell whether there is a difference between them. Frequency difference limens in this case can be very small, perhaps as small as 0.2 or 0.3% of the stimulus frequency. The fine resolution comes from our ability to compare two neural patterns that are separated in time. The other phenomenon is known as frequency resolution, and corresponds more closely to the physiological mechanisms of frequency selectivity studied by the electrophysiologist. In this case the subject has to detect one frequency component of a complex stimulus in the presence of other frequency components, all presented simultaneously. It measures the extent to which the subject is able to filter one stimulus out from others on the basis of frequency. In an analogy, we can think of tuning a radio receiver so that the filter in the input circuit receives the desired station and rejects all others. The resolution of the filter tells us how good it is at passing one station while rejecting others that are close to it in frequency. In the auditory system such resolution bandwidths are a measure of the fundamental frequency filtering properties of the auditory system. These resolution bandwidths are much larger than frequency difference limens, and are perhaps 10–20% of the stimulus frequency. It is the latter case of frequency resolution that will be dealt with in this section; frequency discrimination will be dealt with later.

(b) Masking patterns as an indication of frequency resolution

One of the most immediate demonstrations of psychophysical frequency resolution is provided by the masking pattern produced by a narrowband stimulus, such as a narrow band of noise (Fig. 9.1). The masker is presented continuously, and the threshold of a tonal signal, called the probe, is plotted as a function of probe frequency. It is assumed that the threshold of the probe is a measure of the amount of activity produced by the masker in neurones with characteristic frequencies near to the probe's frequency. The logic is that the probe is detected only if it produces more activity than the masker alone. The masking pattern therefore becomes a correlate of the iso-intensity curves of Figs 4.7B–D. The masked threshold is greatest near

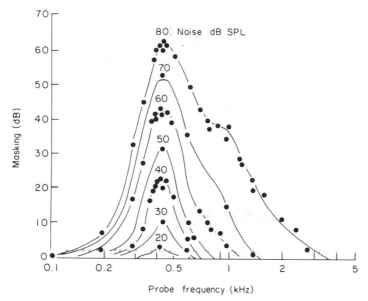

Fig. 9.1 The masking pattern produced by a narrow band of noise. The elevation in threshold of a probe tone was plotted as a function of the probe frequency. At high intensities, the masking pattern spreads more to high than to low frequencies. Masker width 90 Hz, centred at 410 Hz. From Egan and Hake (1950, Fig. 8).

the masker frequency, and at high masker levels the pattern spreads more to high than to low frequencies. The asymmetry has an obvious correlate in the asymmetry of tuning curves, which in the cat for neurones above 1 kHz extend further below than above the characteristic frequency. The reversal of the pattern of asymmetry between the psychophysical and neural cases stems simply from the method of measurement. An electrophysiologist plots the response of *one* neurone to stimuli of many different frequencies. In contrast, the psychophysicist measures the response in *many* different frequency regions to a masker of *one* frequency. Viewed in another way, a masker is able to activate the low-frequency tails of the tuning curves of neurones of much higher characteristic frequency, and so mask probes of much higher frequency. But a masker is comparatively ineffective at activating neurones of lower characteristic frequency, because the high-frequency cutoffs of their tuning curves are so sharp. Maskers are therefore not effective at masking probes of much lower frequency. This accounts for the predominantly upwards spread of masking at higher masking levels.

A technique which might give a close psychophysical correlate of neural tuning curves is that known as the "psychophysical tuning curve" (Zwicker, 1974). In this case, the probe is fixed in frequency and is presented at a low constant intensity, such as 10 dB above threshold. Presumably, such a low-level probe will activate only a few neurones and will provide a near approximation to the electrophysiologist's measurement of single neurones. The masker is varied in frequency, and is adjusted in intensity to keep the probe at threshold. Again we presume that the probe is detected if it produces more activity in any neurones than the masker alone. We might therefore expect the probe to stay at threshold as the masker is moved in frequency and intensity around the edges of the tuning curves of the neurones at the probe frequency. The resulting psychophysical tuning curves indeed appear very similar to the tuning curves of auditory nerve fibres (Fig. 9.2).

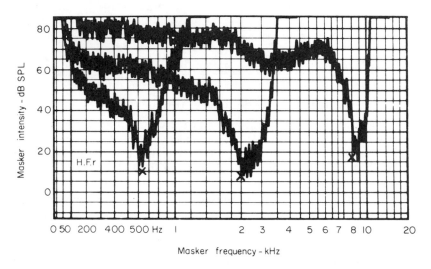

Fig. 9.2 Psychophysical tuning curves are produced by plotting the locus of frequency and intensity necessary to just mask a constant low-level probe (shown by crosses). The curves were determined with a Békésy audiometer, in which the subject continuously adjusted the masker intensity as its frequency was swept, in order to keep the probe at threshold. From Zwicker (1974, Fig. 2).

The masking patterns of Figs 9.1 and 9.2 appear to be similar to the frequency responses of bandpass filters. This suggests that we can think of the frequency resolving power of the auditory system as being due to a set of bandpass filters. Such filters have been measured extensively, and have given us the phenomenon known as the critical band.

(c) Critical bands and auditory filters

Fletcher (1940) introduced the concept of the critical band to deal with the masking of a pure-tone signal by wideband noise. He found that he could electronically filter out noise components remote in frequency from the signal without affecting the signal's threshold. However, there was a critical frequency region around the signal in which removing noise did affect masking. He called this range the critical bandwidth. We can think of critical bands as a series of bandpass filters situated early in the auditory system. Only noise which falls in the same critical band filter as the signal will mask it. The filters correspond to the filters underlying the masking patterns of Figs 9.1 and 9.2. The bandwidth of the filters is known as the critical bandwidth, and has a value of 10–20% of the stimulus frequency. Stimuli will interact to different extents depending on whether or not they lie within the same critical band filter. Critical bands in this way affect a wide variety of auditory tasks, including masked thresholds, frequency discrimination, sensitivity to phase relations, and judgements of loudness, tonal dissonance and roughness (e.g. Scharf, 1970; Moore and Glasberg, 1986).

Current techniques for measuring the critical band filter generally depend on masking the response to one signal, usually a tone, by noise of variable spectral characteristics. In one method, the tone is masked by noise which has a notch in its spectrum. The notch is centred on the signal frequency, and the signal threshold is measured as a function of noise notch width (Patterson *et al.*, 1982). The resulting calculated auditory filter has a form with a rounded top, the so-called "rounded-exponential" shape (Patterson and Moore, 1986). Thus, it has a form similar to the tuning curves of auditory nerve fibres. The filter is slightly asymmetric, being wider on the low-frequency side for high-intensity maskers (Fig. 9.3). The asymmetry at high masker levels is related to the asymmetry of the masking patterns in Figs 9.1 and 9.2.

(d) Non-simultaneous masking techniques

A technique for measuring frequency resolution that has produced a great deal of interest depends on non-simultaneous rather than direct or simultaneous masking (Houtgast, 1972, 1977). One example is forward masking. The masker is pulsed, each pulse being followed by a brief probe, lasting only 10 or 20 ms. The stimuli are ramped on and off to reduce the effects of spectral splatter. The masker will, of course, raise the threshold of the probe. If the critical band measurements are repeated with non-simultaneous masking a surprising difference emerges: the calculated psychophysical filters necessary to explain the results are commonly rather narrower, and surrounded, particularly on the high-frequency side, by areas of *negative* transmission. The negative areas can be explained by supposing that they are areas of lateral inhibition or suppression (Fig. 9.4; Houtgast, 1977).

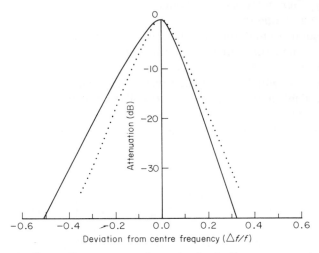

Fig. 9.3. Auditory filters measured psychophysically at two masker levels. The attenuation of the filter was plotted as a function of relative deviation from the centre frequency. The filter becomes asymmetric towards low frequencies with increases in masker level. Modified from Patterson and Moore (1986, Fig. 3.13). Noise spectrum level: ——, 50 dB; ....., 35 dB.

Lateral inhibition or suppression is of course widespread in the auditory system, but had not hitherto been demonstrated convincingly by simultaneous masking techniques. Houtgast pointed out an obvious reason. We calculate the internal representation of a masker by measuring the threshold of a probe superimposed on the masker. If the elements of a complex masker interact to suppress the masker in some frequency regions, a simultaneously-presented probe will be suppressed as well and to a similar extent. Because the signal-to-noise ratio is in effect unchanged, the probe threshold will be unchanged, and the suppression areas will not be reflected in the probe threshold. Forward masking acts rather differently. A complex masker will produce regions of high activity and, as a result of lateral suppression or inhibition, some regions of low activity. When the masker is turned off, its after-effects will raise the thresholds of neurones of some later stage in the auditory system, to an extent which depends on the amount of previous activity. If a probe is now presented, its threshold will reflect the influence of the inhibitory bands as well as the excitatory ones.

Non-simultaneous masking techniques have generated interest because they reveal the effects of lateral inhibition. This has suggested that non-simultaneous masking might provide a more accurate picture of the neural representation of auditory stimuli than does simultaneous masking. On the

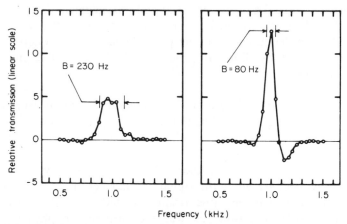

Fig. 9.4   Calculated psychophysical filter shapes derived by direct masking techniques (left) are broader than those determined by non-simultaneous techniques
(right). That determined by non-simultaneous masking also has an inhibitory
sideband. "B" indicates the effective bandwidth of the filter, i.e. the bandwidth of
the equivalent rectangular filter. From Houtgast (1974, Fig. 9.5).

other hand, the detection cues used by the subject have their own effect on
the measured thresholds, and this can change the apparent filter shape
(Moore and Glasberg, 1982a).

2. Relating Psychophysics to Physiology in Frequency Resolution

Having accumulated the information needed on the psychophysics of frequency resolution, we are now in a position to see how psychophysical
frequency resolution relates to the physiology of the auditory system.

Fletcher (1940) suggested that the critical bandwidth was equal to the
resolution bandwidth of the pattern of mechanical excitation on the basilar
membrane. On current evidence, his hypothesis would also mean that the
critical bandwidth was equal to the resolution bandwidths of single auditory
nerve fibres.

It is obvious that the masking patterns and filter functions of Figs 9.1,
9.2 and 9.3, which were determined by simultaneous masking, are generally
similar in form to the basilar membrane and auditory nerve iso-intensity or
iso-response curves (tuning curves) of Figs 3.10A and B, and Figs 4.3 and
4.7. It is not currently doubted that, in broad terms at least, these functions
are all reflections of the basic frequency resolving power of the auditory

system. However, while the psychophysical and physiological data seem to be in general qualitative agreement, there is evidence that the agreement does not hold up in detail, and that the physiological and psychophysical functions are affected by different processes. There are three lines of evidence on this point:

1. The above psychophysical resolution bandwidths were measured by simultaneous masking techniques; that is, the response to one stimulus was measured in the presence of another. However, it is known that the cochlea behaves nonlinearly, so that the response to one stimulus is affected by the presence of another. Therefore, frequency resolution determined by methods where the signal and masker are present simultaneously may not give the same results as methods where only one stimulus is present at a time, as was the case in the physiological experiments of Figs 3.10, 4.3 and 4.7.

2. It was suggested above that non-simultaneous masking techniques, where the masker and signal are not present simultaneously, might provide the more accurate picture of the neural representation of auditory stimuli. This suggestion arose because non-simultaneous masking revealed lateral inhibitory or suppressive sidebands. As is indicated by Fig. 9.4, non-simultaneous masking techniques give bandwidths of frequency resolution that are narrower than those given by simultaneous masking. Simultaneous masking may therefore give bandwidths that are wider than neural tuning curves.

3. When psychophysical resolution bandwidths are measured behaviourally by simultaneous masking in the cat, the bandwidths are rather larger than the excitatory bandwidths of single auditory nerve fibres in the same animals (Pickles, 1975, 1979a, 1980). On the other hand, bandwidths determined by forward masking are similar to the bandwidths of single auditory nerve fibres (Pickles, 1980).

It is therefore suggested that simultaneous masking patterns produce wider resolution bandwidths than do basilar membrane and neural tuning curves, and that this occurs because of the operation of cochlear nonlinearity. Careful analysis is necessary here, in order to see how this occurs.

First, it should be noted that cochlear nonlinearity and sharp frequency resolution are inextricably related, and that they are both related to the same mechanism, probably to active mechanical feedback in the cochlea, and the contribution of outer hair cells to the sharp peak of the mechanical travelling wave (see Chapters 3 and 5).

However, the nonlinearity has a separate and additional effect, in that it enhances the apparent sharpness of a tuning curve, in a way that results simply from the method of its construction. Note from Fig. 3.10C that basilar membrane responses grow only slowly with stimulus amplitude for frequencies above the characteristic frequency. For instance, suppose that we use data similar to those of Fig. 3.10C to construct a tuning curve for the 1-nm response criterion, as in Fig. 9.5. This amplitude of response is

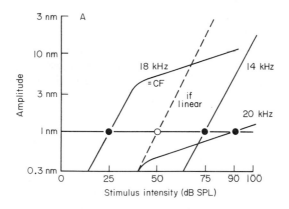

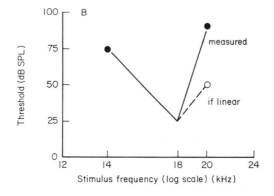

Fig. 9.5   The influence of nonlinearity on the measured slopes of a tuning curve. (A) Basilar membrane amplitude functions, similar to those in Fig. 3.10(C) (B) Tuning curve at the 1-nm vibration criterion derived from the data in part A (solid line). If the basilar membrane amplitude function at 20 kHz behaved linearly above 0.3 nm (dotted line, A), the high-frequency slope of the tuning curve would be shallower (dotted line, B). Note logarithmic scales in part A.

produced at 25 dB SPL for a stimulus at the characteristic frequency (18 kHz), and at 90 dB SPL for a stimulus at 20 kHz (Fig. 9.5A). However, if the response at 20 kHz grew linearly above 0.3 nm (dotted line), then a 1-nm vibration would have been produced at 50 dB SPL instead of 90 dB SPL. In other words, the effect of the nonlinearity has been to increase the steepness of the slope of the tuning curve for frequencies above the characteristic frequency (Fig. 9.5 B). The nonlinearity will therefore narrow the apparent bandwidth of the filtering function.

As suggested by Houtgast (1972, 1977), simultaneous masking methods reduce the influence of nonlinearity on the measured tuning curves. He argued that if nonlinearity changes the response to signal and masker to an equal extent, the signal-to-masker ratio will not be affected, and masked thresholds will not be altered by the nonlinearity (see Section C.1.d). This means that the steepening of the tuning curve due to the factor described in the last paragraph will not be seen with simultaneous masking methods. Simultaneous masking methods will therefore produce wider tuning curves, particularly on the high-frequency side, following the dotted line in Fig. 9.5B (see also Duifhuis, 1976; Pickles, 1984).

The logic of this analysis can be checked by measuring the responses of auditory nerve fibres to the stimuli used in the different types of masking task. In such experiments, a pure-tone tuning curve is first determined. Then tone pips are presented continuously at a fixed (e.g. 10 dB) intensity above the fibre's best threshold at the characteristic frequency. As in the conventional psychophysical tuning curve, the response is masked by a tone that is varied in frequency, and adjusted in intensity, so that the constant tone pips produce a fixed, criterion increment in firing above that produced

Fig. 9.6   (A) "Psychophysical tuning curve" determined by forward masking in a cat auditory nerve fibre (○) is similar in shape to the conventional tuning curve (thick line). Signal: 8 kHz and 24 dB SPL. From Bauer (1978, Fig. 5). (B) "Psychophysical tuning curve" determined by simultaneous masking in a single auditory nerve fibre (EPTC: thin line) is broader than the normal tuning curve (thick line), particularly on the high-frequency side. (▲), signal. For the figure, the EPTC was shifted down by 9 dB so that it coincided with the tuning curve at the tip, in order to permit comparison of the shapes. Guinea-pig. From Pickles (1986b, Fig. 5). (C) Psychophysical tuning curves in man, determined by forward masking (□) and simultaneous masking (△). The abscissa shows the frequency as fractional deviation from the probe frequency. Note the similarity to the curves in part (B) From Moore *et al.* (1984, Fig. 1).

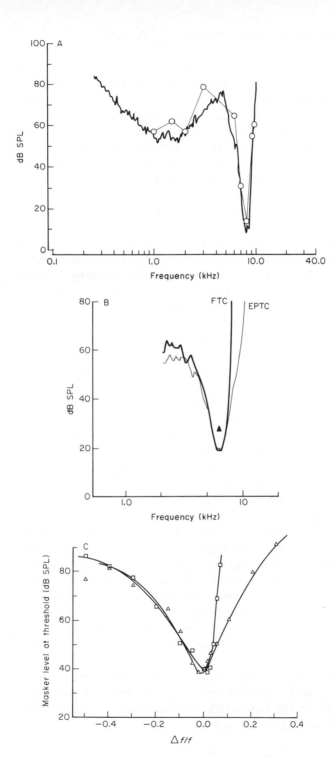

by the variable masker. If forward masking is used, the resulting electrophysiological "psychophysical tuning curves" are indeed similar in shape to the pure-tone tuning curve (Fig. 9.6A; Bauer, 1978; Harris and Dallos, 1979). On the other hand, if simultaneous masking is used, the single fibre "psychophysical tuning curve" is wider than the pure-tone tuning curve, particularly on the high-frequency side (Fig. 9.6B; Pickles, 1984). Figure 9.6C shows how closely these results relate to psychophysical data in man; tuning curves measured psychophysically by forward and simultaneous masking bear the same relation to one another as do the standard tuning curve and the "psychophysical tuning curve" measured electrophysiologically by simultaneous masking, even down to fine details.

This explanation of the influence of nonlinearity on simultaneous masking follows the logic initially proposed by Houtgast, described above. However, there is another possibility, which is that in determining the psychophysical tuning curve by simultaneous masking, the masker was *suppressing* the signal down to the signal's absolute threshold. In this case, the psychophysical tuning curves in Figs 9.6B and C would have been following the outer edges of the two-tone suppression areas (Fig. 4. 16). However, this was shown not to be the case for most experimental paradigms; it was shown psychophysically by Moore and Glasberg (1982b), and electrophysiologically by Pickles (1984), that in most cases suppression was not powerful enough. It was only in the special case of signals just above the absolute threshold, and maskers just above the characteristic frequency, that the masker was able to suppress the signal down to the absolute threshold.

Non-simultaneous masking techniques are not only able to produce an accurate measure of the excitatory part of the neural tuning curve. They can also be used to plot out suppression areas, and these turn out to be similar to the two-tone suppression areas of auditory nerve fibres (e.g. Fig. 4.16). Houtgast (1972, 1973) performed the psychophysical analogue of measuring neural two-tone suppression. With non-simultaneous masking, he used not one, but two, tones during the masking periods. One tone provided the masking, and the other suppressed the first tone. The net amount of excitation was assessed from the threshold of a third, or probe, tone, presented non-simultaneously. He was able to plot the contours of intensity and frequency within which the suppressing tone, or unmasker, provided 3 dB or more of release from masking. The resulting suppression areas appear very similar to the two-tone suppression areas of auditory nerve fibres (Fig. 9.7).

The conclusion is that non-simultaneous masking techniques can provide a reasonable measure of the neural representation of auditory stimuli. On the other hand, the bandwidths as determined by simultaneous masking techniques are rather too wide.

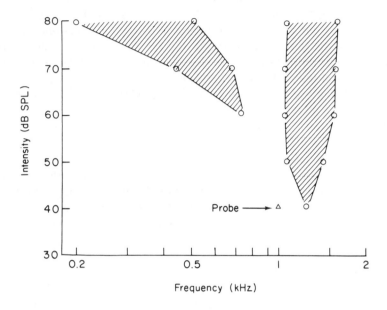

Fig 9.7   Two-tone suppression areas (shaded) can be revealed by psychophysical non-simultaneous masking techniques. From Houtgast (1974, Fig. 5.3).

3. Influence of Lateral Inhibition on Frequency Resolution

The mechanical frequency resolving power of the cochlea, as originally shown by von Békésy (1960), was rather poor, and he suggested that the greater frequency resolution of the whole organism was produced by neurally-mediated lateral inhibition. He suggested that the pattern of excitation in the cochlea was sharpened up by neural inhibitory networks in the cochlea and elsewhere. We now know that this view was wrong; neural lateral inhibition does not occur in the cochlea. Nevertheless, von Békésy's ideas both about the existence of neural lateral inhibition in the cochlea, and the role of inhibition in sharpening frequency resolution, often continue to be reproduced.

As was pointed out in Chapter 4, there is no evidence that there is any lateral inhibition in the cochlea mediated by inhibitory synapses (this is discounting effects produced by the olivocochlear fibres, but these act too slowly to account for the psychophysical results). On the other hand, could it be said that two-tone suppression, which produces "inhibitory" sidebands around a tuning curve, increases frequency resolution? Two-tone suppression depends on the nonlinearity of the cochlea, and the mechanism underlying the nonlinearity is inextricably linked with the mechanism producing sharp tuning. It is only to this extent, where two-tone suppression and sharp tuning are both products of the same mechanism, and also by the mechanism illustrated in Fig. 9.5, where nonlinearity steepens the measured slopes of tuning curves, that two-tone suppression can be said to be related to the increase of frequency resolution in the cochlea. It is important to emphasise here that two-tone suppression does not serve to increase frequency resolution *per se*; stimuli which activate both the excitatory and suppressive areas produce the same degree of tuning as do single tones. This is shown by the agreement between the tuning curves produced with single tones, and those produced with broadband noise as a stimulus, calculated with the reversed correlation technique (Fig. 4.13).

Nor does neural lateral inhibition in, say, the dorsal cochlear nucleus serve to sharpen neural tuning curves, at least around the tip. Cells of the dorsal cochlear nucleus have strong inhibitory sidebands (Chapter 6), but the sharpness of their tuning curves, as shown by the 10-dB bandwidths, is comparable to that of auditory nerve fibres (Goldberg and Brownell, 1973). This seems to be the position in the higher auditory nuclei as well, even though the degree of inhibition tends to increase higher up the auditory system (Aitkin and Webster, 1972). It is only well above threshold that lateral inhibition in these nuclei affects frequency resolution, by stopping the tuning curves from widening with intensity as much as they would otherwise have done.

So far the evidence has been presented as though the tips of tuning curves do not become any sharper at later stages of the auditory pathway, and that any sharpening produced by inhibition is confined to the region well above threshold. In one nucleus, however, this may not be the case. Aitkin *et al.* (1975) reported unusually sharp tuning curves in the inferior colliculus. They found 7 out of 92 neurones in the central nucleus which had tuning curve tips a little narrower than auditory nerve fibres, and 4 out of the 92 with tips that were substantially narrower. One had a Q_{10} with the extraordinarily high value of 38, as against a maximum of 10 for auditory nerve fibres. This very sharp tuning has so far been seen only for a very small proportion of the neurones in only one nucleus. At the moment, we do not know its significance for psychophysical tasks.

D. Frequency Discrimination

1. Introduction

Frequency discrimination refers to our ability to distinguish two tones on the basis of frequency, when they are separated in time. Frequency difference limens are very much smaller than critical bands. Two mechanisms are possible. For instance, the subject may detect shifts in the place of excitation in the cochlea. This is called the "place theory". Or he may use temporal information. We know that the firing in the auditory nerve is phase-locked to the stimulus waveform up to about 5 kHz. On this theory, called the "temporal" theory, the subject discriminates the two tones by using the time intervals between the neural firings. It is not certain which of the two mechanisms is used. Indeed the controversy has been active for more than 100 years, and the fact that it is not yet settled shows that we still do not have adequate evidence. Auditory physiologists divide into three groups, namely those that think only temporal information is used, those that think only place information is used, and an eclectic group who suppose that temporal information is used at low frequencies, and place information at high. In one version of the theories favoured by the eclectic group, both temporal and place information are used at low frequencies, and only place information at high.

In one of the most explicit formulations of the place theory, Zwicker (1970) suggested that frequency discrimination depended on detecting shifts in the place of excitation of the nerve fibre array (Fig. 9.8). The most extreme form of the temporal theory was put forward by Rutherford (1886) who said that each hair cell in the cochlea responded to every tone, and that frequency information was carried only in the frequency of the nerve impulses. Auditory nerve fibres do not fire continuously faster than about 300/s, and Wever (1949) formulated the principle of volleying by supposing that different fibres were activated on different cycles, so that the summed response was able to follow each cycle of the stimulus waveform up to much higher frequencies. A modern formulation would be, that in the frequency range below 5 kHz in which phase-locking is possible, it is the timing of the nerve action potentials that conveys frequency information.

There are several lines of evidence for and against these two theories, none of which is conclusive.

2. Evidence on Place Versus Time Coding of Frequency

(a) Small size of limits

Protagonists of both place and time theories point out how small the detectable limits are when translated into the terms of the other theory.

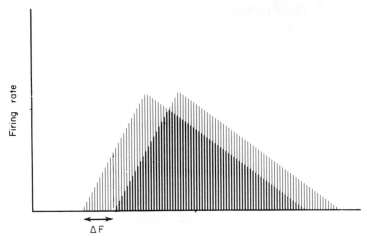

Fig. 9.8 The place theory of frequency discrimination. The pattern of activity in the nerve fibre array is represented schematically by a series of vertical lines, each line representing the firing rate of one neurone to the stimulus. The neurones are arranged in order of ascending characteristic frequency. Shifts in the place of excitation are detected.

Temporal theorists point out that a frequency discrimination limen of 3 Hz at 1 kHz corresponds to a shift in the pattern of excitation on the basilar membrane of 18 μm, or the width of two hair cells. Place theorists point out that the same limen corresponds to a time discrimination of 3 μs, as against some 1000 μs for the width of the nerve action potential, and a variability of some hundreds of μs in its initiation. In both cases averaging over many neurones would be able to reduce the limens below those set by the neurones considered individually.

(b) Differences above and below 5 kHz

Because phase-locking is lost in the auditory nerve above 5 kHz, only place theories must operate above that limit. If auditory perception is qualitatively different above and below that limit, the implication is that temporal information might be used. It seems that there are some important differences in the perception of tones above and below this limit, so that for instance tunes are hard to recognize when all the notes are above 5 kHz.

(c) Frequency discrimination with short stimuli

In a quasi-linear spectral analyser such as the cochlea the physical limits of frequency resolution are limited by the duration of the stimulus, as a result of spectral splatter: stimulus duration × spectral line width = 1. In other words, with place cues, very short stimuli can only be poorly resolved. Temporal theories are not so limited; indeed, given a sufficiently good signal-to-noise ratio, good discrimination is possible by measuring between two similar points only one cycle apart. On the hypothesis that place and not temporal cues are used, we can calculate a lower limit for the frequency difference limen as a function of the length of the stimulus. Moore (1973) showed that below 5 kHz frequency discrimination for short stimuli was up to an order of magnitude better than expected on a place basis. This again suggests that temporal information might be used below 5 kHz.

(d) The pitch of complex tones ("the missing fundamental")

If tones such as 1000 Hz, 1200 Hz and 1400 Hz are sounded together, a pitch corresponding to the fundamental of 200 Hz, which is not present in the stimulus, is also heard. The phenomenon was for a long time used as an argument for temporal processes in perception, because the resulting complex waveform has a periodicity of 200 Hz. Indeed, the phenomenon was once called "periodicity pitch". Alternative explanations are known as "pattern" theories. They suppose that the auditory system, by recognizing that the tones sounded are the upper harmonics of a low tone, supplies the missing fundamental that would have generated them. This is again an area which is controversial, and over the years opinions have swayed in favour of one hypothesis or the other. At the moment pattern theories are dominant, for the following reason: suppose high harmonics generate the low pitch. They will be relatively closely spaced and will not be resolved by the auditory system. Recognition of the spectral pattern will not therefore be possible, but the harmonics will be able to interact in the nervous system to produce periodically varying activity. Temporal theories are therefore supported. On the other hand, low harmonics will be resolved spectrally, and if they generate the low pitch, pattern models are possible. The harmonics will not be able to interact to produce a periodically varying waveform and temporal models become unlikely. Plomp (1967) and Ritsma (1967) showed that the low, resolved harmonics were dominant in generating the low pitch. Pattern models are therefore favoured, and the phenomenon is not now used to support temporal coding as much as it was before.

It is, however, recognized that some sensation of pitch can be conveyed by pure temporal information, for instance by the unresolved high harmonics or by amplitude-modulated noise (see Moore and Glasberg, 1986). However, the sensation of tonality is often much weaker than that evoked by pure tones, and in those cases may well reflect a different mechanism.

(e) Electrical stimulation of the cochlea

Electrodes have been placed within or near the cochlea, in an attempt to restore some hearing in otherwise deaf patients. Because of current spread, only the crudest of place information should be available. On the other hand, the temporal relations of the nerve firings should be unaffected, so that if correct timing is sufficient, discrimination should be normal. Different reports of the tonality of the sensation emerge: in some cases the stimuli appeared to be clearly tonal, in others completely atonal, like a noise. Similarly, the discrimination of frequency is in many cases poor or absent. The lack of success in many patients has therefore been taken as indicating that temporal information is by itself not sufficient for frequency discrimination or for the sensation of pitch. However, the basis of the good discrimination shown by some patients is not known; until more details become available, the results cannot conclusively be used to support the utility or otherwise of temporal coding. The position is discussed further in Chapter 10.

(f) Harmonic consonance

It might be thought that the pleasant consonance of simple musical intervals depends on the simple relations between their periods, resulting in synchronous nerve firing. However, once it is realized that most musical notes are rich in overtones, and that consonance might depend on a lack of beats between the harmonics, the argument cannot be used to support the importance of time information.

(g) Stimulus coding at high intensity

As will be discussed below, it is difficult to see how purely mean rate cues in auditory nerve firing can code details of a stimulus spectrum at high sound intensities. It is therefore suggested that timing information in the nerve, which preserves spectral information over a wide range of intensities, is used instead. The evidence will be discussed below in the sections dealing with intensity and the coding of speech. If timing information can indeed be used in this way, it is also possible that timing information can be used for frequency discrimination.

(h) A model for the analysis of temporal information

One of the greatest problems facing temporal theories of frequency discrimination is that of finding a realistic mechanism for analysing the information. When we discriminate a tone of 1000 Hz from one of 1010 Hz, we can activate our muscles to one stimulus and not to the other. We do not transmit nerve impulses to our muscles at 1000 Hz in one case and at 1010 Hz in the other! Temporal information must therefore be transformed into place information at some point in the nervous system. In fact, since timing information is degraded by synapses, the earlier the information is transformed the better.

A possible mechanism of transformation, described by Shamma (1985a,b), forms a simple and physiologically realistic way of extracting temporal information from the spike train at the level of the cochlear nucleus. The theory depends on an interaction of timing and place information, in a way that is dependent on the delay of the travelling wave as it travels up the cochlear duct. While the theory is currently a plausible one, it should be recognized that there is as yet no direct evidence that it is in fact true.

Shamma pointed out that the relative timing of nerve action potentials in the different fibres of the auditory nerve would be affected by the place of innervation along the cochlea. Since the travelling wave moves rapidly over the portion of the cochlea basal to the peak of the travelling wave, auditory nerve fibres innervating this region will tend to be activated together, and all in the same phase of the waveform (Fig. 9.9). However, the travelling wave moves more and more slowly as it passes through the peak region. Here the nerve fibres will tend to be activated at different phases of the input stimulus. Cells in the cochlear nucleus show lateral inhibition; that is, they are excited by nerve fibres innervating one region of the cochlear duct, and inhibited by nerve fibres from adjacent regions. In the case where nerve fibres are activated in the tail of the travelling wave, excitation and inhibition will tend to arrive in the same phase at cells of the cochlear nucleus, and the net response will be small. However, in the peak of the travelling wave, the phases of arrival of excitation and inhibition will be different. If the spatial separation of the excitatory and inhibitory inputs from different parts of the cochlea is suitable, it is possible that excitation will be transmitted in one phase of the stimulus, while inhibition is transmitted in the opposite phase. Therefore, there will be large fluctuations of excitation and inhibition in the cell during a cycle of the stimulus, and phase-locked responses in the cell will be enhanced (Fig. 9.9). If the mean firing rate responds nonlinearly to changes in membrane potential, the fluctuations will be reflected in the mean rate.

If the spread of lateral inhibition is small, temporal information will be most strongly extracted from the regions of the cochlea where the travelling wave is travelling most slowly – that is, from regions near the peak. Note that for this reason the extracted temporal information automatically maps onto the correct cells in the cochlear nucleus, that is, onto cells with

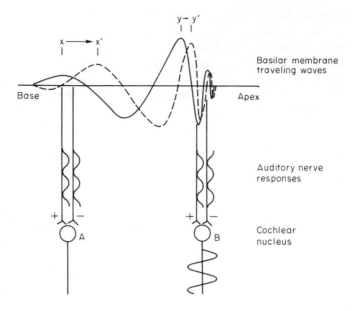

Fig. 9.9   Phase shifts in the travelling wave, combined with lateral inhibitory interactions in the cochlear nucleus, could decode timing information into place information, according to the theory presented by Shamma. Excitation and inhibition arrive at neurone A in the same phase, and so their effects cancel. However, excitation arrives at neurone B in one phase, and inhibition arrives in the other, so the responses are enhanced. Modified from Shamma (1985a, Fig. 4c).

characteristic frequencies corresponding to the time-intervals extracted. Thus, one of the conceptual problems in the re-coding of temporal information into a place code, of how the nervous system "knows" which time interval corresponds to which place, is solved by the mechanics of the cochlea. Note also that the theory requires the timing and place information in auditory nerve fibres to bear the correct relation to each other. Therefore, it is not surprising that when the relation is abnormal, as with electrical stimulation through the prosthesis, the analysis of temporal information should be poor.

   The theory requires that the excitatory and inhibitory post-synaptic potential changes should be able to follow individual cycles of the stimulus. They will be affected not only by the time-constants for opening and closing of membrane channels, but also by the time for neurotransmitter release and removal, and by the electrical time-constant of the cell membrane. The data of Møller (1976a) for cells of the cochlear nucleus suggest that they are able to follow fluctuations up to at least 200 Hz, after which the response would gradually decline. If the cells' electrical time constants are similar to those of hair cells, the sensitivity to phase-locked information should fall off with frequency as does phase-locking in the auditory nerve itself. The two factors in combination would double the rate at which phase-locking declines with increases in frequency. Therefore, we expect a severe high-frequency limitation for the model. However, a similar limitation applies to any model in which the temporal information is extracted by a mechanism which resides post-synaptically.

(i) Conclusions

On current evidence, it is not possible to decide between the temporal and place theories of frequency discrimination. One form of the eclectic view, which is that temporal information is used at low frequencies and spectral information at high, does not conflict with most of the evidence. A physiologically reasonable version of the eclectic view, which is one supported by the author, is that both place and timing information contribute at low frequencies, but for the timing information to be analysed, it needs to bear the correct relation to the place information (Shamma, 1985a,b). In any case, the best support for the eclectic view is the rather negative one that the evidence in favour of either of the other two theories is not conclusive, and this may be a function of the quality of the evidence available, rather than of the actual operation of the auditory system.

E. Intensity

1. Stimulus Coding as a Function of Intensity

As the stimulus intensity is raised, the tuning curves of auditory nerve fibres become wider and wider (e.g. Fig. 4.3). This in itself does not necessarily mean that frequency resolution deteriorates, because iso-rate functions determined at different rates above threshold show that resolution is practically unchanged as long as the firing is not saturated (Fig. 4.7A). When the firing is saturated, however, frequency resolution deteriorates sharply. This is indicated by the diagrams of the nerve fibre array in Figs 9.10A and B. Below saturation, the elements of a complex stimulus are resolved. Once the fibres are driven into saturation, we would expect resolution to be no longer possible if only rate information is used. But in cells with strong inhibitory sidebands, as in cells of the dorsal cochlear nucleus, frequency resolution is preserved (e.g. Palmer and Evans, 1982). We can understand how. The strong inhibitory sidebands will serve to keep the neural firing

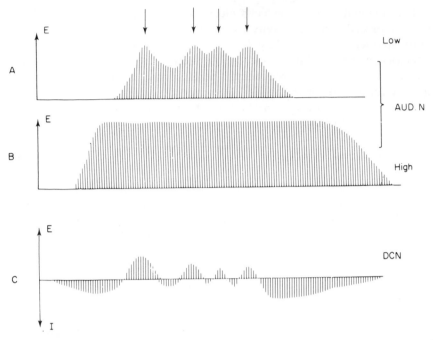

Fig. 9.10   The firing rates in the auditory nerve fibre array are shown to a complex acoustic stimulus at low (A) and high (B) intensities. At high intensities resolution is apparently lost in the auditory nerve (B), but not in the dorsal cochlear nucleus (C). Arrows mark peaks of energy in the stimulus. E, Excitation; I, inhibition.

rate near the middle of the range irrespective of overall stimulus intensity, so that the details of the stimulus pattern can be transmitted (Fig. 9.10C). Put in other terms, the inhibitory sidebands stop the tuning curve from becoming as wide as it would otherwise have been. The picture will also hold for cells with similar response characteristics at higher levels of the auditory system. Such responses can explain the great dynamic range, extending up to some 100 dB SPL, over which psychophysical frequency resolution can be preserved practically unchanged (e.g. Hawkins and Stevens, 1950; Scharf and Meiselman, 1977).

What is unexplained is how the information is transmitted by the auditory nerve in the first place, when the intensity is so high as to saturate the firing of the great majority of its fibres. The psychophysical dynamic range may be contrasted with a dynamic range of only 30–50 dB in the firing of single auditory nerve fibres. The great majority of auditory nerve fibres have thresholds in the bottom 10–15 dB of the range, and we would therefore expect nearly all fibres to be saturated by 50–65 dB above threshold.

Several hypotheses have been advanced to account for the wide psycho-physical range of hearing, compared to the restricted dynamic range of auditory nerve fibres.

(a) Range of thresholds

As described in Chapter 4, the great majority of auditory nerve fibres have thresholds in the bottom 10–15 dB of the range. Liberman and Kiang (1978) have shown that there is also a significant proportion of fibres with higher thresholds (Fig. 4.4; Geisler *et al.*, 1985). These fibres can signal the details of complex stimuli, when the fibres of lower threshold are saturated (Sachs and Young, 1979; Costalupes, 1985). But how is it possible for the compar-atively few fibres of high threshold to convey information as accurately at high intensities, as do the much larger number of low-threshold fibres at medium intensities? It is difficult to escape the conclusion that the low-threshold fibres must also have been contributing to some extent.

(b) Sloping saturation of rate-intensity functions

Sachs and Abbas (1974) showed that the firing of some auditory nerve fibres did not saturate in a sharply defined manner, but went on increasing gradually (Fig. 9.11). These fibres also tend to have higher thresholds than

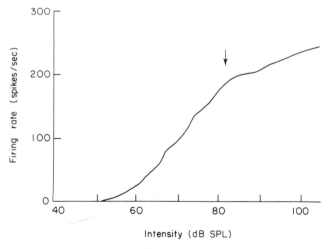

Fig. 9.11 In auditory nerve fibres exhibiting sloping saturation, the rate-intensity function shows a knee (arrow), and then goes on increasing. Such neurones may partly explain the wide dynamic range of hearing. From Sachs and Abbas (1974, Fig. 6).

the others. However, the proportion of fibres still unsaturated at high intensities is only a few per cent (Palmer and Evans, 1980). Again we have a problem in explaining why psychophysical abilities do not fall markedly at high intensities. We would, for instance, expect the detection thresholds for tones in wideband noise to deteriorate substantially. Not only would fewer fibres be able to convey information, but those fibres still unsaturated would be stimulated in a flatter part of their rate-intensity functions.

(c) Edge detection

The rate-intensity functions of Fig. 4.6 show that although a fibre may be saturated by stimuli at the characteristic frequency, it may not be saturated by stimuli of other frequencies. Figure 9.12 shows how at high intensities

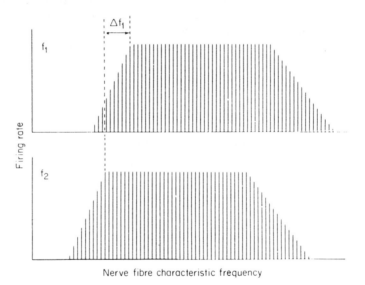

Fig. 9.12   The place theory of frequency discrimination for stimuli of high intensity. Changes on the edge of the active nerve fibre array are detected.

information can be conveyed by fibres at the edges of the active array of nerve fibres. Such mechanisms might explain how, for instance, frequency discrimination is possible at high intensities. The mechanisms will only operate for narrow band stimuli. The explanation will not work with wideband stimuli, where fibres of the whole frequency range are saturated.

(d) Two-tone suppression

Other sense organs, such as the retina, preserve a wide dynamic range by lateral inhibition, so that the response to a pattern is preserved relatively unchanged over a wide range of overall light intensities. In the auditory system, cells with strong inhibitory sidebands in, for instance, the dorsal cochlear nucleus behave in the same way. Could two-tone suppression in the auditory nerve function similarly? Two-tone suppression is after all maintained in saturation. Costalupes *et al.* (1984) recorded the responses of auditory nerve fibres to tones in background noise, and showed that the background noise could shift the rate-intensity functions to the tones upwards by as much as 30 dB. That is, intense background noise could alter the intensity range over which auditory nerve fibres signal the details of the stimulus. They presented evidence that this was due to two-tone suppression. For this type of stimulus at least, two-tone suppression does seem to increase the responsiveness of the system at high intensities. On the other hand, such a mechanism does not seem to help with all types of stimuli. Sachs and Young (1979) showed that for vowel sounds, two-tone suppression seems to *reduce* the spectral contrast of the formants at high intensity. That point will be discussed further, when the coding of speech is discussed.

(e) Temporal information

Auditory nerve fibres show phase-locking to stimuli below 5 kHz. The phase-locking is preserved at high intensities in spite of a saturation of the firing rate. Indeed, in spite of a loss of frequency selectivity in mean firing rate terms, the frequency selectivity indicated by temporal information deteriorates only slightly or not at all at high intensities. This is shown by the tuning curves determined from the time-pattern of the nerve firings by the reverse correlation technique (Fig. 4.13). Consider the following example showing how frequency resolution may be preserved. At low intensities a fibre may fire phase-locked to a tone at its characteristic frequency of 1.0 kHz, but not fire at all to one of 0.5-kHz. At high intensities, either one of the two tones will saturate the fibre when presented alone. The two tones will therefore activate the fibre to the same extent if presented separately. This occurs because the 0.5-kHz tone will drive the fibre as fast as it can fire, and the 1.0-kHz tone alone cannot produce more activity. But if the two tones are presented together, the 1.0-kHz tone will again show itself to be more effective in driving the fibre, and the nerve firings will be predominantly phase-locked to 1.0 kHz. The same point has been made particularly clearly for speech sounds, and those results will be discussed in detail below, when the coding of speech sounds is considered (Sachs and Young, 1979; Young and Sachs, 1979).

Psychophysical frequency resolution, as shown by the wideband masking of tones, deteriorates more with increasing intensity above than below the 5-kHz limit for phase-locking (Scharf and Meiselman, 1977; Moore, 1975).

The implication is that below 5 kHz temporal information had been used at high intensities, when the nerve firing was saturated. A possible mechanism for the extraction of temporal information was proposed by Shamma (1985a,b). As described above, it could transform the timing information into a place code in cells of the cochlear nucleus with inhibitory sidebands, i.e. in the types of cell which are known to have a wide dynamic range to auditory stimuli (Gibson *et al.*, 1985).

(f) The middle ear muscle reflex

Sound-elicited contractions of the middle ear muscles may attenuate the sound input by as much as 20 dB. Such powerful control will only be expected to occur for sound frequencies in the range where transmission is strongly affected by the middle ear muscle reflex, well below 1 kHz (Chapter 2). It will not explain the wide dynamic range of hearing for frequencies of 1 kHz and above.

(g) The olivocochlear bundle

The olivocochlear bundle attenuates the responses of auditory nerve fibres, probably by reducing the size of the travelling wave in the cochlea. Winslow and Sachs (1987) showed that stimulation of the olivocochlear bundle could enhance the contrast in a spectral pattern for certain types of stimuli at high intensity, although in itself it produced only a small increase in the upper limit of the dynamic range (Fig. 8.6B). The mechanism was discussed in Chapter 8.

(h) Conclusions

No one mechanism seems entirely satisfactory to explain the transmission of spectral information by the auditory nerve at high intensities. Plausible hypotheses suggest that information in the time-pattern of neural firings is used, together with the information from small variations in mean firing rates (based on a range of thresholds in auditory nerve fibres, sloping saturations of the rate-intensity functions, and two-tone suppression). The middle ear muscle reflex and the olivocochlear bundle may also help.

2. Loudness

It has been suggested that the sensation of loudness depends on the total sum of activity transmitted in the auditory nerve (e.g. Wever, 1949). When some auditory nerve fibres are destroyed unilaterally, as for instance by a tumour of the brainstem, the loudness of a stimulus can grow more slowly than normal. This can be measured by requiring the subject to match the loudness of a stimulus in the abnormal ear to one in the normal ear (Citron *et al.*, 1963). In a different type of experiment, Zwicker *et al.* (1957) asked normal subjects to match the loudness of tone complexes with the loudness of a standard stimulus. The bandwidth of the test stimulus was varied but its total power was kept constant. As the bandwidth of the stimulus increased up to and beyond the critical band, the loudness at first stayed constant and then, beyond the critical bandwidth, increased (Fig. 9.13A). The loudness of stimuli such as in Fig. 9.13A, and the loudness of many everyday sounds, can be explained very successfully by calculating the summed activity over all the critical band auditory filters stimulated (Zwicker and Scharf, 1965).

On the hypothesis that critical band filters can be identified with auditory nerve filters, the model of Zwicker and Scharf can be translated into the hypothesis described here, that loudness depends on the summed total of activity in the auditory nerve. This hypothesis was tested by Pickles (1983). Responses of auditory nerve fibres were measured to noise bands of different widths, but total constant intensity. However, instead of summing the responses of many fibres to noise bands centred on one frequency, the equivalent but complementary paradigm was used, that of summing the responses of *one* fibre to noise bands centred on *many* frequencies. The advantage of performing the experiment this way is that the variability of the data, due to variation in response properties from fibre to fibre, was reduced. The results agreed with the psychophysical data, in that the summed activity increased with stimulus bandwidth, for wider stimulus bandwidths (Fig. 9.13B). However, there was no clear sign of a flat portion in the function at narrow bandwidths. The reason for this is not known; the hypothesis that loudness is related to the summed total activity in auditory nerve fibres therefore does not seem to hold up at narrow bandwidths.

F. Sound Localization

A real sound source in space will stimulate both ears. Experiments with headphones have suggested that the side on which a source is heard depends on timing and intensity differences at the two ears. The intensity differences are contributed to by the pinna, which for high-frequency sounds produces shadows and reflections, and for some species a selective amplification for sound sources along a narrow axis (Phillips *et al.*, 1982).

The responses of neurones to stimuli differing in timing and intensity at

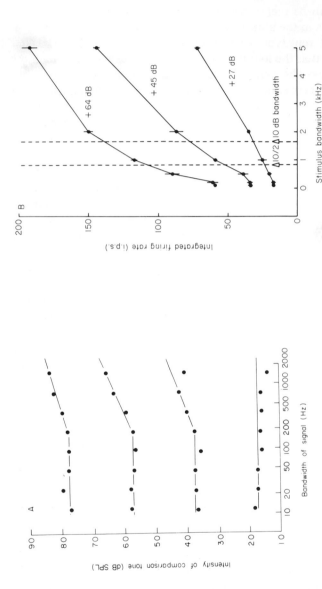

Fig. 9.13 (A) The loudness of a four-tone complex is plotted as a function of overall tone spacing. Stimuli centred on 1 kHz. From Zwicker et al. (1957, Fig. 3). (B) Integrated firing rate of an auditory nerve fibre, as a function of stimulus bandwidth, for noise bands of constant total intensity. The integrated firing rate was calculated as described in the text. If the function had a flat portion up to a bandwidth corresponding to the neural or psychophysical integration bandwidth, we might have expected to find the corner in the region of the dotted lines. However, the function grows steadily with stimulus bandwidth. The numbers marked on the curve show the stimulus intensity, with respect to the fibre absolute threshold. The dotted lines are marked at the 10-dB bandwidth and half the 10-dB bandwidth of the fibre. Fibre CF: 9.7 kHz. From Pickles (1983, Fig. 1, slightly modified).

the two ears have been investigated extensively. As was pointed out in Chapters 6 and 7, a large proportion of the neurones at and beyond the superior olivary complex are responsive to timing or intensity differences at the two ears. This is good *a priori* evidence that the cells are involved in sound localization; in some cases the results seem to agree with the results obtained with real sound sources in space. If a speaker is moved round the head while neurones are recorded electrophysiologically, the greatest responses are usually found with the speaker on the contralateral side (for neurones beyond the superior olive). In some cells the response area covers the whole contralateral hemifield, and in others it is sharply confined to certain specific directions (e.g. Middlebrooks and Pettigrew, 1981). Cells with narrow response areas tend to be high-frequency cells, and tend to have their directional selectivity aligned with the pinna axis (cat: Semple *et al.*, 1983; Middlebrooks *et al.*, 1981). In such cells, the directional sensitivity can be altered by manipulating the pinna nearest to the source (Aitkin *et al.*, 1984; Middlebrooks and Knudsen, 1987). In this type of cell, therefore, the fundamental directional selectivity is probably derived from the directionally-selective amplification produced by the pinna, enhanced and preserved by the sensitivity to interaural intensity differences. On the other hand, the selectivity of low-frequency neurones may be a result of the lower directional selectivity of the pinna in this frequency range (Middlebrooks and Pettigrew, 1981). It is also possible that timing cues are involved here (Aitkin *et al.*, 1985).

Extensive searches have been made for spatial "maps" in audition, analogous to the spatial maps known in vision. A clear and detailed map is known in the owl, in its homologue of the inferior colliculus, the lateral dorsal mesencephalic nucleus (Knudsen and Konishi, 1978; see Chapter 6). A less detailed map has been shown in the mammalian superior colliculus, in registration with the visual map in the nucleus (King and Palmer, 1983). On the other hand, spatial mapping in the specific auditory centres of mammals appears to be relatively poor (Middlebrooks and Pettigrew, 1981; Aitkin *et al.*, 1985).

One of the problems in understanding the basis of sound localization, is that many cells seem optimally responsive to timing and intensity disparities that are greater than could be produced by any real sound source in space. For instance, in the medial superior olive of the kangaroo rat the optimal or characteristic delays of 88% of neurones are greater than the maximum delay calculated from the separation of the ears (Crow *et al.*, 1978). The same point has been made for the cortex by Benson and Teas (1976). Benson and Teas also showed that the same was true for intensity differences: direct measurements have shown that the maximum interaural intensity difference expected in the chinchilla for stimuli below 2 kHz is 4–5 dB, yet cortical neurones of these characteristic frequencies were optimally sensitive to intensity differences as great as 20 dB. Such neurones may have a role, not in representing the direction of a sound source, but in *discriminating between* the directions of sound sources. The cyclic functions of Figs 6.13 and 7.10, obtained when firing rates were measured for different interaural timing disparities, indicate how. Note that we can best discriminate changes in time disparity by looking not at the peaks of the functions, but at the points of greatest slope. These generally occur for far smaller interaural disparities than the peaks, and will be in the range for real stimuli.

The role of binaural interaction in the brainstem in sound localization has been strongly supported by behavioural experiments. Moore *et al.* (1974) trained cats to discriminate the direction of sounds. The various tracts by which binaural interaction might occur were cut in different animals. Sound localization was affected only by cutting the crossing fibres in the trapezoid body. These are the fibres by which binaural interactions occur in the superior olivary complex. Transections of the fibres joining the inferior colliculi of both sides, or of the corpus callosum joining the two cerebral hemispheres, were without effect. Jenkins and Masterton (1982) later showed that, if unilateral lesions were made at levels above the superior olivary complex, there were severe deficits in sound localization, but only with sound sources on the contralateral side. So if binaural interaction had already occurred, information travelling up only one side of the brainstem was able to subserve sound localization, but only for sources on the contra-lateral side.

Brainstem mechanisms of sound localization are also dealt with in Chapter 6, Sections C, E and F. Cortical mechanisms are dealt with in Chapter 7, Sections A.3, B.2 and C.3.

G. Speech

The coding of steady vowel sounds was studied by Sachs and Young (1979), who presented anaesthetized cats with steady-state synthesized vowel sounds, consisting of three formants, or spectral peaks. They used a limited set of vowels, and each was presented at several intensities. The activity of a large number of auditory nerve fibres was sampled in each animal in response to the same stimulus set. In this way, a picture was built up of the response of the whole nerve fibre array to each of the stimuli.

When mean firing rates were plotted as a function of fibre characteristic frequency, clear peaks at the formant frequencies could be seen at low stimulus intensities (Fig. 9.14). However, as the stimulus intensity was raised to 68 dB SPL and above, the separate peaks disappeared, leaving a broad band of activation. Such a loss of spectral detail at medium and high intensities corresponds to that seen with other types of stimulus. It was only in a small population of fibres with high thresholds that spectral detail was preserved in the mean firing rates, even up to the highest intensities used (not shown in Fig. 9.14).

Sachs and Young investigated the reasons for the loss in spectral detail as the intensity was raised. One factor was due to the broadening of the tuning curves and iso-intensity functions at higher intensity, particularly apparent in the broad tails stretching to low frequencies (Figs 4.3 and 4.7C and D). Therefore, intense stimulus components, and particularly the first formant, will be able to activate fibres of a wide range of characteristic frequencies. A second factor was the flattening of the rate-intensity function at high intensities, so that the firing rate was no longer responsive to small variations in stimulus intensity (Fig. 4.6). A third factor was due to two-tone suppression. Sachs and Young (1980) showed that the formant F_1 suppressed the response to the formants F_2 and F_3, reducing their amplitude. However, fibres with characteristic frequencies in the gap between F_1 and F_2 were activated by F_1 itself, and so could not be suppressed by F_1. The result was that the peaks in the pattern were suppressed, but the troughs were not, so that the contrast in the pattern was reduced. Note that with this type of stimulus, two-tone suppression reduces the contrast in the representation of a spectral pattern. In this, it has the opposite effect to that described above with tones in wideband noise, where two-tone suppression enhances the spectral pattern (see Section E.1.d).

The reduction in contrast in the mean-rate pattern at what are really quite moderate intensities, has suggested that other cues might be used. In particular, it has prompted the search for temporal cues in the firing patterns.

Young and Sachs (1979) made Fourier transforms of the firing patterns of temporal auditory nerve fibres, so as to measure the degree of phase-locking to the individual component frequencies of the stimulus. In order to plot the results, they only counted phase-locking at stimulus frequencies within half of an octave of the fibre characteristic frequency. This method produced what was called the Averaged Localized Synchronized Rate (ALSR). The results showed that the representation of the stimulus spectrum was preserved without degradation at all stimulus levels (Fig. 9.15).

While this shows that temporal information is capable of coding the spectrum of a vowel, the results cannot be taken as proof that the activity is actually represented in this way. Note that the method of calculation only takes into account phase-locking at frequencies near the fibre characteristic frequency. Therefore, the main problem with mean-rate coding, namely the activation of fibres by frequency components remote from the characteristic

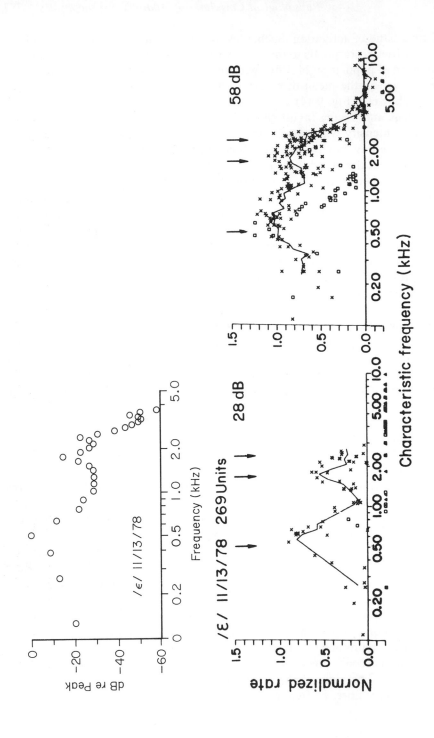

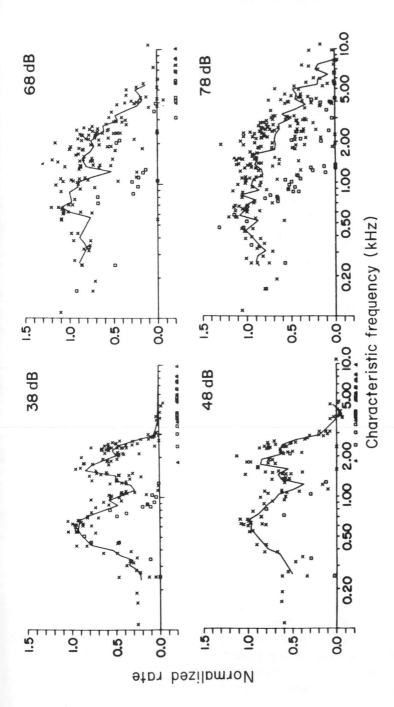

Fig. 9.14 Neurograms of auditory nerve fibre activity to the vowel sound /ɛ/ (spectrum top left). The activity of a large number of auditory nerve fibres was sampled in response to the one stimulus, and the activity of each fibre plotted against its characteristic frequency. Solid lines show the running means. Arrows indicate the frequencies of the formants. From Sachs and Young (1979, Figs 1 and 6).

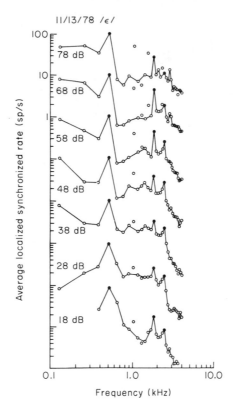

Fig. 9.15   The magnitude of phase-locking as a function of position in the nerve fibre array shows that the stimulus spectrum is preserved in the timing of the action potentials, over the whole range of stimulus intensities. Same stimulus as in Fig. 9.14. From Young and Sachs (1979, Fig. 6).

frequency, is avoided merely by the method of presenting the results. The plots of Fig. 9.15 would be similar to the patterns conveyed to the central nervous system only if the nervous system were able to do a similar calculation; in other words, if the nervous system were able to extract phase-locking to frequencies around the fibre characteristic frequency, while rejecting phase-locking to other frequencies.

According to the theory presented by Shamma (1985a,b; Fig. 9.9), inhibitory neural networks might be able to do just this. As described above (Section D.2.h), lateral inhibition in the cochlear nucleus would tend to enhance neural activity arising from points in the cochlea at which the phase of the travelling wave is changing particularly rapidly with distance along the cochlear duct, and attenuate it from points at which the phase is changing more slowly. This will tend to enhance activity phase-locked to frequencies near the characteristic frequency of the fibre, because the phase of a travelling wave changes particularly rapidly around its characteristic place in the cochlea. Shamma's simulations showed that the mechanism could theoretically retrieve the spectrum of the stimulus when the mean rate profile, as in Fig. 9.14, had become flat. However, as described above, we would expect a severe high-frequency limitation in the model. It is as yet uncertain whether it would be able to extract the upper formants of vowel sounds.

The above results were obtained with steady vowels. For other types of stimuli, a temporal analysis seems less appropriate. For instance, voiceless fricative consonants (e.g. /f/ or /s/) have their upper formants at frequencies too high to show phase-locking, and a Fourier analysis of the spike train does not detect them (Delgutte and Kiang, 1984). On the other hand, mean-rate profiles do give peaks at the formant frequencies. While at high intensities the peaks become less apparent in response to the steady-state component of the stimulus, they are still visible in response to the stimulus onset. Such enhanced contrast of spectral peaks is also seen in response to formants which are changing in time (as in the syllables /ba/ and /da/). These findings appear related to the slightly greater dynamic range that auditory nerve fibres show in response to temporally-varying stimuli (Smith and Brachman, 1980).

Neuronal responses to speech sounds have been investigated in much less detail in the central auditory nervous system. Based on the analysis with simpler stimuli, we would expect neurones in the anteroventral cochlear nucleus to have responses like auditory nerve fibres, while those with inhibitory sidebands, as in the dorsal cochlear nucleus, would be able to represent the spectrum of the stimulus over a wide range of stimulus intensities. Here, for instance, we would expect the formants to be represented by discrete peaks of activity, separated by bands of inhibition. We would expect this pattern to be preserved at many later stages of the auditory system. But even in the cochlear nucleus we expect other processing, for instance the selective responses to frequency and intensity transitions, to influence the results. Other complexities are also known. For instance, Moore and Cashin (1974) showed that energy in inhibitory sidebands could increase neurones' responses to the transients in speech sounds. Even at this early stage, therefore, the neurones will be responding to some of the higher-level aspects of the stimulus. Such an analysis is continued to the cortex, where the responses of cells to speech sounds are complex and cannot be predicted from the responses to simple stimuli (Chapter 7).

In the cortex the region primarily involved in the analysis of speech seems to be Wernicke's area, part of the auditory association cortex on the posterior

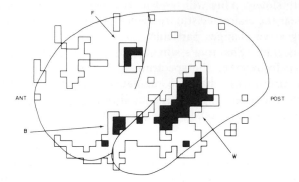

Fig. 9.16   When a conscious subject passively listens to speech, the greatest increases in blood flow are produced in Wernicke's area (W) in the dominant cerebral hemisphere. Increases were also produced in Broca's area (B) and in the frontal eye fields (F).

Radioactive Xe^{123} was dissolved in saline and injected into the arteries supplying the brain. The local concentration of Xe^{123} was measured by an array of scintillation counters placed over the skull. Areas of high relative blood flow are indicated by the contour lines. Data from Lassen *et al.* (1978).

part of the temporal lobe on the dominant, generally left, hemisphere of the brain. The information originally came from the effects of localized brain lesions. Lesions here led to a disability known as sensory or Wernicke's aphasia, a disorder in the production of speech which was considered to have a perceptual basis because it was associated with disorders of comprehension and errors in word selection. There is also strong evidence that the processing of speech occurs in the dominant hemisphere, because lesions in the left cerebral hemisphere, but rarely in the right, affect the perception of speech. Striking confirmation has come from studies in which the blood flow was measured by a radioactive tracer technique simultaneously in many regions of the brain while the subject was conscious (e.g. Nishizawa *et al.*, 1982). It seems that the local blood flow increases when neural activity is increased. When the subject passively listened to speech the greatest changes in blood flow were produced in Wernicke's area (Fig. 9.16). In this experiment, activity in the primary auditory cortex was presumaoly underrepresented because in man the cortex lies in the lateral fissure. There was also a small activation of Broca's area, an area generally thought to be concerned with the *production* of speech and situated on the lower posterior part of the frontal lobe. Anatomical studies show that the division of the cortex known as area 22, which contains Wernicke's area, is larger on the left side, as are the primary auditory areas. The areas are differentiated histologically, with larger and more widely separated cell columns, and with more widely ramifying dendrites, than the corresponding areas on the opposite side (Seldon, 1981a,b).

How are speech sounds analysed in the auditory cortex? Such electrophysiological studies as have been undertaken in animals suggest, as might be expected, that speech sounds undergo complex processing (e.g. Steinschneider *et al.*, 1982). But there is no clear evidence from such experiments that speech is processed in any way that is different from other complex, non-speech stimuli. We would not of course expect human speech to have any particular significance for non-human subjects, and some experimenters have tried to circumvent this by using the vocalizations of the species under investigation (e.g. Glass and Wollberg, 1983). However, such investigations have not yet shown any special processing for vocalizations. Another approach is based on inter-hemispheric differences. In behavioural experiments, mice and monkeys have shown left-hemispheric specializations for the detection of the vocalizations of their own species (Ehret, 1987; Heffner and Heffner, 1986b). Moreover, several species show anatomical left–right asymmetries in their auditory cortical areas (e.g. Sherman *et al.*, 1982). The results suggest that an approach linking anatomical asymmetries, species-specific vocalizations, and electrophysiological analysis might indeed be productive in future.

In man, attempts have been made to use psychophysical evidence to determine whether speech sounds are processed differently from non-speech sounds. For instance, Liberman *et al.* (1967) used psychoacoustic evidence to show that speech was processed in a special way, one that they called the "speech mode". As pointed out by Schouten (1980), many of the results on which they based their conclusions were determined by the basic psychoacoustic properties of the stimuli used, and many stimuli, particularly those containing many rapid transitions, also appear to be processed in the speech mode.

A further method of analysis in man relates to ear advantage. Each cortex seems to be particularly responsive to sounds on the contralateral side, and the right ear, contralateral to the generally dominant cerebral hemisphere, has often been found to have a small but significant advantage in the perception of speech (Kimura, 1967a). A right ear advantage is particularly seen for the abrupt onset of consonants (Dwyer *et al.*, 1982). The advantage of the right ear is not confined to speech: some tasks involving the detection of temporal transitions can also show a right ear advantage. For instance, Papçun *et al.* (1974) showed such an advantage for the recognition of Morse code. They suggested that the dominant cerebral hemisphere was important for the analysis of stimuli into its sequential components. It may well be that the complexity of the temporal pattern in the stimulus determines the extent of the right ear advantage along a continuum. Halperin *et al.* (1973) asked subjects to repeat the order of stimuli which varied either in frequency, or in duration. Their result was that the more frequency or duration transitions the stimulus contained, the more the ear advantage shifted from left to right. This suggests that one of the special functions of the dominant cerebral hemisphere, and hence possibly of Wernicke's area, is the analysis of complex auditory stimuli in terms of sequential patterns, and that this is one of the reasons for its special role in speech. The implication of the time dimension in cortical function is reminiscent of some of the effects of cortical lesions in animals (Chapter 7).

Our understanding of the way in which the auditory system processes speech is in its infancy and unravelling the neuronal mechanisms involved is one of the supreme challenges of auditory physiology.

H. Summary

1. The behavioural absolute threshold is related to the minimum thresholds of single auditory nerve fibres, determined from the mean firing rate.

2. Frequency resolution describes the ability to filter out, on the basis of frequency, one stimulus component from another in a complex stimulus. The psychophysical resolving power of the auditory system seems to approximately match its neural resolving power. However, closer analysis shows that it is necessary to specify the method of measurement when determining the correspondence.

3. A common way of measuring frequency resolution is to determine the masked threshold of a signal. Because the auditory system behaves nonlinearly, different results are obtained if the signal is presented simultaneously, or non-simultaneously, with the masker. The detailed mechanism behind the difference is discussed above in the text. The psychophysical resolution bandwidths with non-simultaneous masking are narrower than with simultaneous masking. The bandwidths determined by non-simultaneous masking match neural resolution bandwidths, and moreover the psychophysical filters are often surrounded by inhibitory sidebands. This suggests that non-simultaneous masking techniques, rather than simultaneous masking techniques, provide a more accurate picture of the neural representation of auditory stimuli.

4. In frequency discrimination, two tones are presented successively, and we have to tell whether there is a difference between them on the basis of frequency. This could be done by detecting a shift in the place of

excitation along the cochlea, or it could be done by detecting differences in the timing of nerve impulses. It is not yet possible to decide conclusively between these two theories. Many would support the view that temporal information is used at low frequencies, perhaps in conjunction with place information. However, only place information can be used at high frequencies, above the frequency range for phase-locking.

5. The psychophysical frequency resolving power of the auditory system is maintained practically unchanged in the intensity range for which auditory nerve firing is apparently saturated. It is not known how the effective frequency resolving power is preserved in this intensity range. It is difficult to escape the conclusion that for a variety of causes the firing of all fibres is not quite saturated. It is possible that the phase-locking in neural firing, which does not deteriorate to the same extent with intensity, is also used.

6. The sensation of loudness may well depend on the total quantity of activity in the auditory nerve, at least for wide stimulus bandwidths.

7. Sound localization seems to depend on binaural interactions of timing and intensity information in the superior olivary complex. Direction-specific changes in intensity produced by the pinna seem to play a role in some cells. In the owl, neurones in the homologue of the mammalian inferior colliculus form a spatial map of the environment. No clear maps have been found in mammals, except in the superior colliculus. In mammals, many of the neurones investigated seem to be involved, not in representing the directions of sound sources, but in discriminating *between* the directions of sound sources.

8. Speech sounds are transmitted by the auditory nerve in a way that can be understood from its response to simpler stimuli. Each fibre responds to a limited spectral analysis of the stimulus. The firings follow the temporal envelope of the stimulus, and at low frequencies, are phase-locked to the individual cycles of the stimulus. At medium and high intensities, when auditory nerve response areas broaden, the spectrum of the stimulus is not clearly represented in the mean firing rates of the nerve fibres. Temporal information may be used here. Speech sounds undergo complex analyses in the higher centres of the auditory system, depending for instance on transients of intensity and frequency. No single-cell electrophysiological analysis has so far shown ways in which speech sounds are treated differently from complex non-speech sounds.

9. The cortical analysis of speech occurs primarily in Wernicke's area in the dominant cerebral hemisphere. The dominant cerebral hemisphere seems particularly important for analysing sounds in terms of their sequential patterns.

I. Further Reading

The psychophysics of hearing is dealt with by Plomp (1976), Moore (1982), and several chapters in *Frequency Selectivity in Hearing*, edited by B. C. J. Moore (1986a), Academic Press, London. Frequency discrimination and resolution, and their relation to physiology, are discussed by Pickles (1986a). Mechanisms of sound localization are dealt with by Masterton and Imig (1984) and Aitkin (1986). The coding of speech in the auditory nerve is dealt with by Sachs (1984a,b), and cortical relations are discussed by Seldon (1985) and Efron (1985).

10. *Sensorineural Hearing Loss*

Some forms of cochlear pathology can be induced experimentally, and changes found in the responses of auditory nerve fibres. The correlation of these changes with the changes that can be shown morphologically and psychophysically will be described in this chapter. In addition, attempts to restore lost hearing by means of a cochlear prosthesis will be described. This chapter requires knowledge of Chapter 3 and the first part of Chapter 4 on the auditory nerve.

A. Types of Hearing Loss

Hearing loss arising in the auditory periphery is divided into two types, known as conductive and sensorineural loss. Hearing loss due to an abnormality *before* the cochlea is known as conductive loss. It may arise because the impedance transformation in the middle ear is disrupted, because for instance the ossicles are immobilized, or because the differential transfer of pressure to the oval and round windows is otherwise impaired. Conductive loss produces a simple though frequency-dependent attenuation of the stimulus, and can be compensated for by hearing aids. In some cases, the causes are amenable to treatment by antibiotics, and in severe cases the loss may be amenable to surgical intervention, for instance by a prosthesis replacing an ossified stapes in the oval window.

Hearing loss arising in the cochlea or auditory nerve is known as sensorineural hearing loss. That arising in the nerve often results from a tumour. However, the most common form of sensorineural impairment arises in the cochlea, when it is known by the rather cumbersome name of "sensorineural hearing loss of cochlear origin". In for instance middle and old age senile changes can produce a cochlear impairment known as presbyacusis, that can be severe, progressive, and is completely without cure. Cochlear hearing loss can also be caused by acoustic trauma, drugs, infections, or may be congenital. In addition, many cases are encountered where no cause can be assigned. Particularly vulnerable sites are the sensitive transducer cells, namely the hair cells of the cochlea, and the stria vascularis. If the hair cells are destroyed they cannot be replaced. Because treatment is so often inadequate, because hearing aids prove to be of limited use, and because the condition is widespread, the importance of sensorineural hearing loss of cochlear origin cannot be overestimated. Physiological studies have recently revealed some of the physiological changes associated with cochlear damage, and it is these changes with which the present chapter will be mostly concerned, as well as attempts to restore lost hearing by means of a cochlear prosthesis.

B. Sensorineural Hearing Loss of Cochlear Origin

1. Morphological and Biochemical Changes in Hair Cells

(a) Ototoxicity

A number of drugs in current clinical use are ototoxic. Perhaps the most notorious of these are the aminoglycoside antibiotics, which include drugs such as kanamycin, neomycin and streptomycin. While they produce changes in the stria vascularis, they also have a severe and apparently preferential effect on the outer hair cells.

In the early stages of intoxication, the changes in outer hair cells involve loss of links between stereocilia, fusion of their membranes, and loss of the stereocilia (Figs 10.1 and 10.2). Changes are also seen in the internal organelles. In the more severely affected cells the cell body degenerates, to be replaced by a scar as the adjacent supporting cells expand to fill the gap in the reticular lamina. Current hypotheses to explain aminoglycoside ototoxicity in hair cells divide the process into three stages (e.g. Lim, 1986; Schacht, 1986). First, the positively-charged aminoglycoside molecules may interact with the negatively-charged glycoconjugate coating on the cell membrane. Where this happens around the stereocilia, it allows the membranes to fuse, with subsequent joining of the stereocilia. Secondly, it is suggested that the aminoglycoside may be taken up actively by the cell membranes to bind to phosphatidylinositol biphosphate in the membrane. This would lead to structural changes in the membrane, making it leaky, and interfere with the phosphoinositide system which is known from other cells to act as a second messenger system controlling intracellular metabolism. Once inside the cell, the aminoglycosides can interfere with a variety of intracellular reactions, as known from biochemical analyses in other systems, such as RNA metabolism, protein synthesis, and

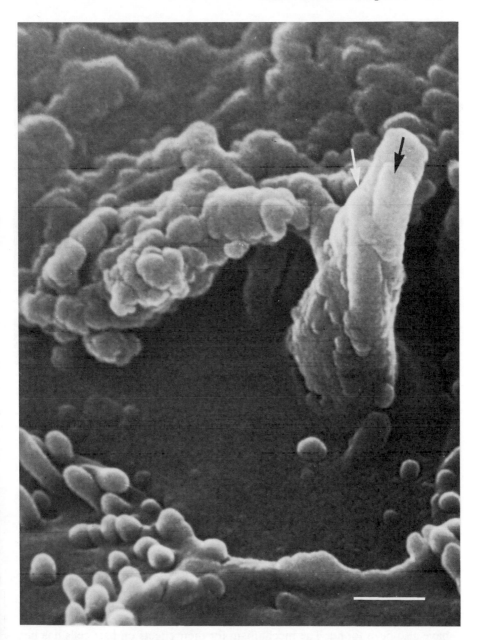

Fig. 10.1 Fused stereocilia remaining on an outer hair cell of the guinea-pig cochlea, after treatment with kanamycin. Arrows: remains of two individual stereocilia. Scale bar: 500 nm. From Pickles *et al.* (1987b, Fig. 3).

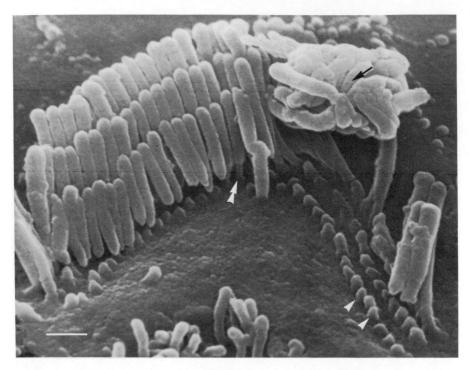

Fig. 10.2   In less severely affected outer hair cells, the stereocilia may be detached or fused (arrow), and the tip links lost. Single arrowheads: rootlets of stereocilia remaining in the cuticular plate. Double arrowhead: constricted rootlet of stereocilium. Guinea-pig. Scale bar: 500 nm. From Pickles *et al.* (1987b, Fig. 4).

carbohydrate metabolism (Schacht, 1986; Lim 1986; Beard *et al.*, 1969). Under these hypotheses, it is suggested that outer hair cells are particularly susceptible to aminoglycoside damage because they have a high metabolic rate which allows the greater active uptake of the drug, and because they have high levels of phosphoinositide metabolism. It is possible that these properties are ultimately related to the role of outer hair cells in the active amplification of the mechanical travelling wave (see Chapters 3 and 5).

A further group of ototoxic drugs are diuretics such as ethacrynic acid and furosemide, which reduce the absorption of fluid in the kidney by inhibiting the active absorption of Cl^- in the loop of Henle. They have their effects primarily on the stria vascularis, although effects on hair cells have also been found. The mechanism for their effects on hair cells has not been reported in such detail as with the aminoglycosides. Other ototoxic drugs include salicylates (aspirin), quinine, and drugs containing arsenic (see Brown and Daigneault, 1981, for a review).

(b) Acoustic trauma

Hair cells have been examined from cochleae either immediately after the induction of experimental acoustic trauma, or after a period in which the repair processes have had a chance to operate.

In the mildest cases of acoustic trauma, morphological changes are found only in the rootlets of the stereocilia, which become less dense in electron micrographs (Liberman and Dodds, 1987). It was suggested, from the size of the losses in the associated electrophysiological recordings, that this degree of change would have been reversible. In more severe cases, leading to permanent damage, the stereocilia kink or fracture at the rootlet, and the packed actin filaments which give the stereocilia their rigidity are depolymerized (Liberman, 1987; Tilney *et al.*, 1982). The links between the stereocilia break, and the stereocilia separate. The tip links, which may well be involved in transduction, also break. In inner hair cells, however, the tallest stereocilia in the bundle tend to break away from the others, the shorter stereocilia keeping their linkages intact (Fig. 10.3; Pickles *et al.*, 1987a).

In other cases the stereocilia become fused, bent, splayed apart or detached (Fig. 10.4). In more severe cases the whole cuticular plate can be ejected from the hair cell, and in still more severe cases, as with blast damage, the organ of Corti can be ripped apart (Hamernik *et al.*, 1986).

When cochleae are examined some weeks after the exposure, the stereocilia which remain on damaged hair cells may be fused into a single giant stereocilium. Where the whole hair cell has degenerated, the heads of the adjacent supporting cells expand to fill the gap in the reticular lamina (e.g. Slepecky, 1986).

The mechanisms of acoustic trauma have not been worked out in detail. Some of the changes may have been due to direct mechanical damage to the cytoskeleton. As with ototoxic effects, fusion of stereocilia may result from loss of negative charges from the membrane coat around the stereocilia (Lim, 1986). It is also possible that some of the changes in the cytoskeleton are secondary to changes in the enzymes which are necessary for its maintenance. In general, however, we do not know the extent to which contributions arise directly from mechanical damage, and indirectly from disordered metabolism (e.g. Slepecky, 1986).

2. Physiological Changes

(a) Damage to outer hair cells

As was discussed in Chapters 3 and 5, it is likely that outer hair cells help produce the sharp tuning and great sensitivity of the mechanical travelling

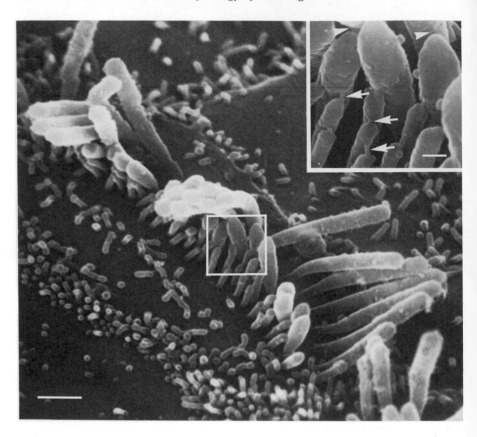

Fig. 10.3 The tallest stereocilia on an inner hair cell can show disarray after acoustic trauma, while the shorter stereocilia and their tip links remain intact. The insert shows the regions outlined on the main figure, and shows that tip links (arrows) remain on the shorter stereocilia. However, tip links connecting the shorter stereocilia to the tallest stereocilia are broken (arrowheads). It is possible that the shorter stereocilia may be moved by viscous drag of the surrounding fluid during acoustic stimulation. Scale bar: 1 μm on main figure, 200 nm on insert. From Pickles *et al.* (1987a, Figs 15 and 16).

wave in the cochlea, probably by means of an active mechanical process which feeds mechanical energy of biological origin back into the wave. Changes after the administration of ototoxic drugs such as the aminoglycoside antibiotics, which preferentially damage outer hair cells, show that neural tuning curves are raised in threshold and broadened in tuning, losing their sensitive, sharply-tuned tip (Fig. 10.5). The changes are reminiscent of those seen in the tuning of the basilar membrane itself as the preparation

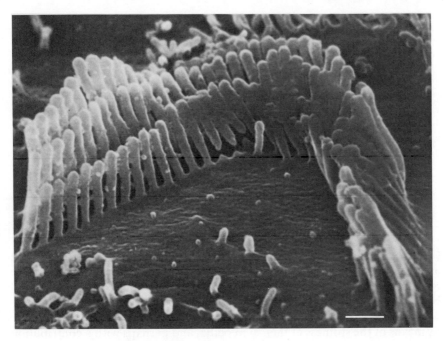

Fig. 10.4   An outer hair cell after acoustic trauma, showing fused and detached stereocilia, and loss of tip links. Scale bar: 500 nm. From Pickles *et al.* (1987a, Fig. 6).

deteriorates (Fig. 3.11 A). Since the very great majority of auditory nerve fibres innervate the *inner* hair cells, it is suggested that the changes in Fig. 10.5 occur because the loss of outer hair cells has altered the mechanical pattern of vibration on the basilar membrane.

In a series of careful experiments, Liberman and his colleagues recorded from auditory nerve fibres in acoustically damaged ears. The nerve fibres were filled with horseradish peroxidase after recording, so that the fibre could be traced back to the inner hair cell of origin. The cochlea was then analysed morphologically. It was found that if only the outer hair cell stereocilia had been damaged, the tuning curve was raised in threshold, and broadened in shape (Fig. 10.6A; Liberman and Dodds, 1984b). If, however, some outer hair cell stereocilia remained, a small, sharply-tuned tip could remain on the upper edge of the tuning curve (Fig. 10.6 B). These results confirm the role of the outer hair cells in increasing the sharp tuning and sensitivity of the travelling wave. As a second effect, the low-frequency tail of the tuning curve was *lowered* in threshold. This latter result suggests that at frequencies well below the characteristic frequency, the outer hair cells had been serving to *restrict* the movement of the basilar membrane, perhaps

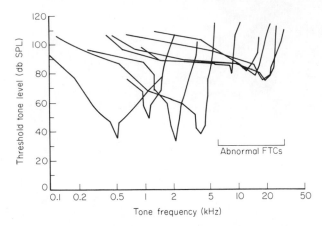

Fig. 10.5 After kanamycin, neural tuning curves are raised in threshold and broadened in shape. In this experiment, fibres from the high-frequency end of the cochlea were most affected ("abnormal FTCs"). Guinea-pig. From Evans and Harrison (1976, Fig. 1).

by means of the stiffness of their stereocilia, and perhaps with a contribution from the active mechanical process (Chapter 5, Sections D and E).

As shown in Fig. 3.11B, loss of the sharp tip of the basilar membrane tuning curve is associated with a steepening of the slopes of the intensity functions. In auditory nerve fibres, the change is reflected in a steepening of the rate-intensity function (Harrison, 1981). Nonlinear phenomena, such as two-tone suppression, are also affected.

(b) Damage to inner hair cells

The majority of experimental manipulations affect outer hair cells rather than inner hair cells. However, by tracing horseradish-peroxidase filled afferents to the inner hair cells after electrophysiological recordings, Liberman and Dodds (1984b) were able to show that where there was disarray

Fig. 10.6   (A) If the stereocilia of outer hair cells are lost, the auditory nerve fibres innervating the adjacent inner hair cells are raised in threshold and broadened in tuning (solid line in tuning curve *vs.* dotted line – normal). The low-frequency tail of the tuning curve becomes hypersensitive. (B) If some outer hair cell stereocilia remain, a short, sharply-tuned tip can be seen on the tuning curve. (C) If inner hair cell stereocilia are damaged while most outer hair cell stereocilia remain apparently normal, the tuning curve can have a nearly normal shape but be raised in threshold. From Liberman and Dodds (1984b, Fig. 14).

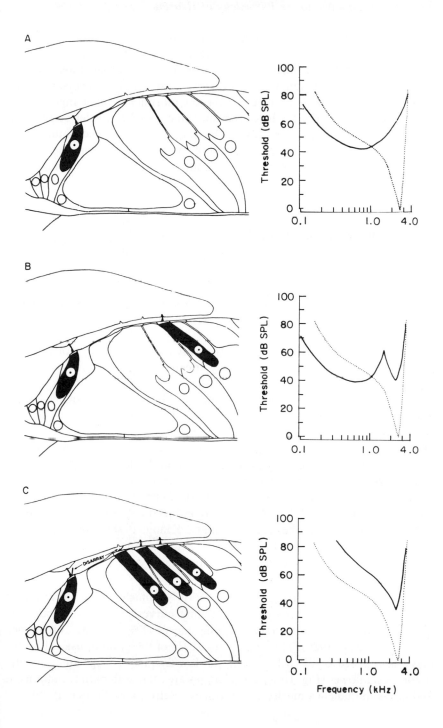

A

B

C

Threshold (dB SPL)

100
80
60
40
20
0

0.1   1.0   4.0

DISARRAY

Frequency (kHz)

mainly of inner hair cell stereocilia, the tuning curves were raised in threshold, but of nearly normal shape (Fig. 10.6C). These results suggest that inner hair cells are indeed only involved in detecting the movement of the basilar membrane, and not in producing sharp tuning. The disarray was usually only seen in the tallest stereocilia; the morphological changes shown in Fig. 10.3, where the tip links remain intact on the shorter stereocilia, suggest that such inner hair cells might indeed be able to continue transducing, but with reduced sensitivity. Auditory nerve fibres from these inner hair cells have much lowered rates of spontaneous activity (Liberman and Dodds, 1984a).

3. Psychophysical Correlates

(a) Sensitivity

The neural data should lead us to expect a loss of sensitivity in sensorineural deafness of cochlear origin, combined with a loss in frequency resolution. These are observed.

The loss of sensitivity is of course an easily recognizable sign of cochlear damage. High frequencies are generally affected first, for reasons which are not known. In man there is in addition a puzzling phenomenon, seen particularly after noise trauma, of a selective hearing loss around 4 kHz. This produces what is known as the "4-kHz notch" in the audiogram. Again, the reasons are not known. It may correspond to a particular vulnerability of the cochlea in this frequency region. A 4-kHz notch is also seen in the behavioural audiograms of apparently normal cats (Fig. 4.4), and is also seen in the thresholds of auditory nerve fibres with characteristic frequencies in that region. These small losses in apparently normal animals seem to be pathological, because they are reduced in cats raised from birth in a soundproofed chamber (Liberman and Kiang, 1978).

(b) Frequency resolution

The change in neural frequency resolution similarly has a correlate in psychophysical data. Psychophysical tuning curves, which are thought to give an approximation to neural tuning curves (Fig. 9.2), are broadened (Florentine *et al.*, 1980). The sharply tuned tip of the psychophysical tuning curve is reduced or abolished, and the low-frequency slope in particular becomes shallower (Fig. 10.7). The changes are seen with both simultaneous and non-simultaneous masking techniques (Nelson and Turner, 1980).

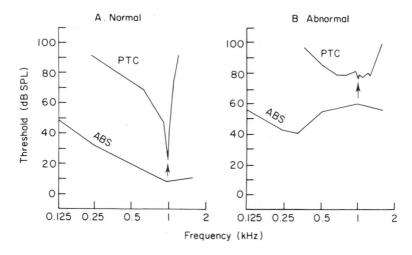

Fig. 10.7 Psychophysical tuning curves (PTC) determined with forward masking in a normal subject (A) and in a subject with sensorineural hearing loss (B). ABS, absolute threshold (audiogram). Arrow: frequency and intensity of probe. From Nelson and Turner (1980, Figs 1 and 5).

Critical bandwidths, being the bandwidths of the psychophysical filter (see p. 260), are also affected. For instance, estimates of the loudness of stimuli of different bandwidths can provide a measure of the critical bandwidth (Fig. 9.13A). Such loudness measures show the critical band to be wider than normal in patients with cochlear hearing loss (Bonding, 1979). In a different type of experiment, Pick *et al.* (1977) asked subjects to detect a probe tone in noise of rippled spectrum. They plotted the masked threshold as a function of the frequency spacing of the ripples and used the results to calculate the shape of the psychophysical filter. With hearing losses of up to 40 dB, the main change was in the length of the sharply tuned tip segment of the psychophysical filter. For greater losses, the tip segment increased in bandwidth as well. These patterns have a parallel in the single unit data; for small losses, the tips of the tuning curves are raised but do not become any broader; after a certain point, the tuning curves broaden substantially. There was also a large scatter in the psychophysical data, some patients with hearing losses of up to 70 dB showing normal frequency resolution bandwidths. Part of the scatter may arise because critical bandwidth tasks are comparatively difficult and complex anyway; but it may also be that the different patients were differently affected in the extent to which the sharply tuned tips of their tuning curves were affected. Some may have had tuning curves like those of Fig. 10.6A, with no sharply tuned tip segment, and others may have had ones like those of Fig. 10.6B, with a short remnant of the tip remaining.

The loss of frequency resolution may be one reason why speech perception is reduced in sensorineural hearing loss, and why even with amplification, speech perception cannot always be completely restored. Hearing aids may be able to restore sensitivity, and so the absolute threshold, but they cannot restore resolution.

(c) Loudness recruitment

Patients with sensorineural hearing loss show another characteristic feature, so specific that it can be used as a diagnostic tool – namely that of loudness recruitment. Loudness recruitment can also be explained in terms of the neural data. It can most readily be demonstrated if the hearing loss is unilateral, as can occur in Menière's disease. The subject is asked to match the loudness of a stimulus in his abnormal ear with one in his normal ear. Near threshold, the stimuli are matched in loudness when the stimulus in the abnormal ear is raised by the amount of hearing loss (Fig. 10.8). But as the stimulus intensity is raised, the match occurs nearer and nearer normal

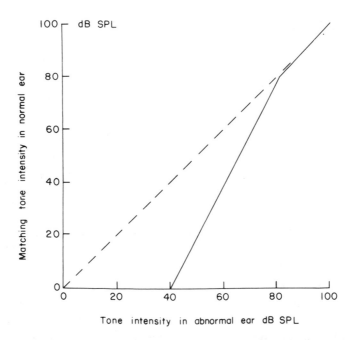

Fig. 10.8   Loudness recruitment with unilateral sensorineural hearing loss.

levels, so that often by 80 dB SPL the match is made with the two stimuli at the same intensity. Loudness in the affected ear therefore grows abnormally quickly with intensity.

Loudness recruitment can be understood from the changes in basilar membrane and auditory nerve intensity functions. When the low-threshold, sharply-tuned tip is lost, these functions steepen (Fig. 3.11B and Section B.2.a, this chapter). The evoked activity therefore grows more rapidly than normal with stimulus intensity. A second contribution to loudness recruitment can be understood from Fig. 10.9. In Fig. 10.9A, sample neural

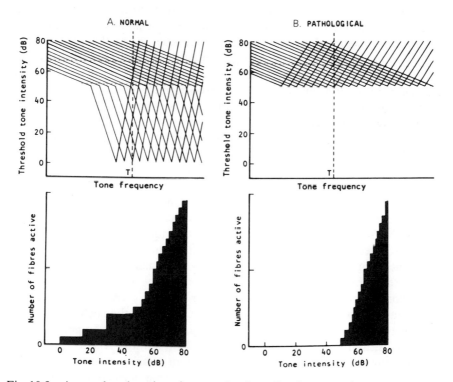

Fig. 10.9 A neural explanation of one mechanism of loudness recruitment. Loudness in the abnormal ear grows abnormally quickly with intensity once threshold is reached, because the tips of the tuning curves are missing. From Evans (1975b, Fig. 8).

tuning curves from a normal cochlea are shown. It seems likely, as was explained in the last chapter (p. 282), that the loudness of a stimulus depends on the total amount of activity in the auditory nerve. As the stimulus intensity is raised, the number of fibres activated at first rises slowly, while only the tip segments of the curves are activated, and then more abruptly, when the tails of the tuning curves are encountered. In the pathological ear (Fig. 10.9B), the tips of the tuning curves are missing, and as the stimulus intensity is raised the number of fibres activated increases rapidly, soon approaching normal levels.

(d) Tinnitus

One of the distressing accompaniments of sensorineural hearing loss is tinnitus, which can deprive the sufferer of even the dubious consolation of living in a silent world. Tinnitus is at the moment poorly controlled by drugs or surgery, and the only treatment of use is to mask it by external noise. This does not of course help if the patient is deaf anyway. Generally, sufferers just have to "live with it". Until recently, it was thought useful to divide tinnitus into "objective" and "subjective" tinnitus. Objective tinnitus was thought to arise peripheral to the cochlea, in say the musculature of the middle ear, and could be recorded objectively with microphones or heard by other listeners. Subjective tinnitus arose in the receptors, or more centrally, and therefore could not be measured objectively outside the ear. The discovery of the evoked cochlear mechanical response by Kemp (1978) showed that objectively measurable sound could be produced by the cochlea itself, so that the division of tinnitus into the objective and subjective does not have the obvious anatomical correlate that it once did.

Figure 5.20 shows the recording of the sound pressure in one of these cases of externally recordable tinnitus of cochlear origin. The origin of the evoked cochlear mechanical response is not certain, and some hypotheses were discussed in Chapter 5. It seems that when hair cells are activated some mechanical energy is fed back into the basilar membrane. If the spatial relations are right, mechanical energy may be reflected back and forth along the cochlea, further stimulating the hair cells, and leading to a self-sustaining oscillation, and so to tinnitus. Points in the cochlea at which its properties are changing abruptly seem particularly able to generate the oscillations, and it is possible that this type of tinnitus is associated with normal hair cells on the borders of regions of hearing loss (Clark *et al.*, 1984). Parenthetically, it also appears that some normal hair cells are necessary for the production of the evoked cochlear mechanical response (the cochlear echo), so this can be used as a method of objective audiometry (Bray and Kemp, 1987).

Although this is an intriguing explanation of one type of tinnitus, it is unlikely to explain the majority of cases encountered. We expect tinnitus associated with the evoked cochlear mechanical response to be continuous, and narrowband or tonal. Much tinnitus is discontinuous and atonal, like high pitched hissing or knocking noises. Moreover, the evoked cochlear mechanical response disappears in experimental cochlear deafness, and tinnitus can be particularly strong in patients with severe hearing loss.

It is very likely that there will be many causes of other types of tinnitus and a single explanation may therefore not be adequate. In some cases, it is possible that tinnitus can arise from an increase in the spontaneous activity in the auditory nerve. High doses of salicylate (aspirin) can produce acute tinnitus in man, and Evans *et al.* (1981) showed in animal experiments that salicylate could temporarily increase the spontaneous activity of auditory nerve fibres. However, it does seem that the tinnitus associated with some long-term sensorineural hearing loss does not arise from an increase of activity in auditory nerve fibres, because the spontaneous activity of auditory nerve fibres is reduced rather than increased in some cases of experimental cochlear pathology (Kiang *et al.*, 1970). One explanation of tinnitus which might apply in these cases is that it arises centrally after a disappearance of an input from the cochlea. It is possible therefore that with continued deprivation, and perhaps a lack of spontaneous activity in the auditory nerve, a hypersensitivity akin to denervation hypersensitivity may occur in the central nervous system. There is further evidence that some tinnitus arises centrally, because cutting the auditory nerve is often an ineffective treatment. Moreover, where it is possible to reverse the deprivation, for instance by stimulating the cochlea electrically by means of a prosthesis or with external skin electrodes, patients sometimes report an improvement in their tinnitus (e.g. Aran, 1981).

Gerken (1979) presented evidence that a denervation hypersensitivity did occur in the auditory system after cochlear damage. He implanted electrodes in a variety of auditory nuclei, and trained animals to respond behaviourally to electrical stimuli. Following cochlear damage, thresholds for the detection of electrical stimuli were lowered in a wide range of auditory nuclei, including the first nucleus of the auditory pathway, the cochlear nucleus.

C. Physiological Aspects of the Cochlear Prosthesis

1. Introduction

When the hair cells of the cochlea are lost they cannot be replaced. Therefore hearing cannot be restored. Under these circumstances the best hope for restoring some auditory function seems to be to bypass the transducer mechanism, and to attempt to stimulate what remains of the auditory nerve directly, by electrical means. In recent years, groups in several labratories have attempted to do this, by implanting electrodes within or near the cochlea. It is hoped that by stimulating early in the auditory pathway, the complex and specialized signal processing capability of the auditory system will be utilized. Well over 1000 patients have been implanted worldwide. Some of the results have been impressive; many of the others have not.

The aims of electrical stimulation have ranged from the very conservative to the very advanced. At the most basic level, patients with implanted prostheses report that one of the benefits is just being in some sort of auditory contact with the environment. Being able to hear alarm signals and approaching traffic is obviously valuable. At a slightly more advanced level, even a few auditory cues can be of use in lip reading, particularly if they help distinguish differences which do not appear in the lips, such as the difference between voiced and unvoiced sounds. Furthermore, some feedback from the patient's own voice is invaluable in helping him control it. Amelioration of tinnitus has also been reported after using the prosthesis. At a much more advanced level, some groups have the hope of conveying speech completely through the prosthesis.

2. Physiological Background

(a) Condition of the nerve

One question is: are there any auditory nerve fibres left when all hearing has been lost through cochlear deafness? Otte *et al.* (1978) counted cells of the spiral ganglion *post mortem* in 43 patients suffering from profound deafness. In 46% of the cases one-third or more of the spiral ganglion cells were present. They showed that in patients without total hearing loss, this number of nerve fibres is capable of subserving some degree of speech perception. The extent of the loss seemed to depend on the original cause of the deafness. There were large losses if the original deafness was due to bacterial labyrinthitis, but only small ones if due to ototoxic antibiotic administration. So it seems that there is a good survival in approximately 50% of patients which might justify an implant.

Before implanting a prosthesis, many laboratories test for the presence of auditory nerve fibres by stimulating the nerve extracochlearly and asking for patients' subjective reports or measuring brainstem evoked responses (e.g. Chouard *et al.*, 1985). However, it is difficult to tell from such tests how many fibres remain, or how they are distributed among the cochlear turns.

A second problem is that intracochlear insertion of the device or electrical stimulation through it may produce a further loss of nerve fibres. Schindler *et al.* (1977) showed that as long as there was no direct damage to the cochlear partitions when the device was inserted, the only degeneration was that of hair cells and not of nerve fibres. On the other hand, some reports suggest that electrical stimulation can stimulate bone growth, and cause morphological changes in surviving hair cells and auditory nerve fibres, and even produce changes in cells of the cochlear nucleus (Balkany *et al.*, 1985). In one report, morphological changes were found in hair cells of the *contralateral* cochlea as well (Dodson *et al.*, 1986). Therefore, it appears that electrical stimulation, while not necessarily causing loss of function in the surviving sensory elements, should be used with caution.

(b) Responses of auditory nerve fibres to electrical stimulation

In normal subjects, the auditory nerve transmits information by two principles of coding, namely (i) the place   principle, by which the place of maximum activation in the cochlear duct depends on the frequency of the stimulus, and (ii) the timing principle, by which individual action potentials tend to be locked to the temporal waveform of the stimulus. Furthermore (iii), at any one frequency, the firing rate increases sigmoidally with stimulus intensity. It is worth examining the extent to which these aspects can be reproduced by electrical stimulation.

(*i*) *The place principle.* Kiang and Moxon (1972) recorded the responses of cat auditory nerve fibres to electrical stimuli applied to the round window. The responses showed that auditory nerve fibres were not intrinsically tuned to electrical stimuli at all (Fig. 10.10). Irrespective of the fibres' place of origin along the cochlear duct, the lowest thresholds were produced with stimulus frequencies around 100 – 200 Hz, and there was a gradual rise in

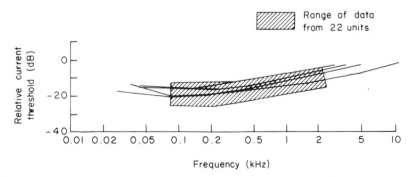

Fig. 10.10   Auditory nerve fibres show negligible tuning to electrical stimuli applied across the cochlea. These broad "tuning curves" should be compared to the sharp tuning curves to acoustic stimuli shown in Fig. 4.3. From Kiang and Moxon (1972, Fig. 1).

threshold towards higher frequencies of stimulation. This result has since been reproduced by many other investigators (e.g. Glass, 1983; Hartmann *et al.*, 1984). Therefore, if frequency information is to be transmitted as place information in the auditory nerve, ways must be found of restricting the stimulus to local regions of the cochlea. Current spread in the cochlea makes such localized stimulation difficult. Reports however show that it is possible to produce very restricted stimulation with bipolar electrode pairs, particularly if the pairs are oriented radially across the cochlear duct (van den Honert and Stypulkowski, 1987a). This suggests that localized activation of nerve fibres might indeed be possible.

Each ear in man gives rise to some 30,000 nerve fibres. However, the speech range can be covered by only 20 separate critical bands. Psychoacoustic studies suggest that even fewer channels might be required: experiments with vocoders suggest that only 8–10 are necessary. Therefore, perfect perception should be theoretically sustainable by stimulation at 8–10 appropriately chosen places along the cochlear duct. In practice, some have found that good perception is possible with many fewer channels (e.g. four; Eddington, 1983). Others, because of possible interaction between electrodes and because of possible patchy nerve survival, implant many more channels than this, sometimes as many as 16.

(*ii*) *The timing principle.* Temporal information is preserved with electrical stimulation. Indeed, firings tend to be restricted more closely to one point on the waveform than with acoustic stimulation (Kiang and Moxon, 1972). Although with electrical stimuli the maximum firing rate of auditory nerve fibres seems to be restricted to 500/s, a small degree of phase-locking has sometimes been reported at frequencies above 1 kHz (Hartmann *et al.*, 1984).

(*iii*) *Rate intensity functions.* Kiang and Moxon (1972) showed that rate-intensity functions were very steep with electrical stimulation. The dynamic range was often less than 10 dB (Fig. 10.11). This can be compared with the dynamic range of 20–50 dB to acoustic stimuli. Moreover, the spread of thresholds was very small, being less than 20 dB (van den Honert and Stypulkowski, 1987a). This means that addition of extra fibres can provide only a small extension of the dynamic range. The results mean that very reliable amplitude compression of the stimulus is necessary.

3. Results

(a) Intensity coding

As expected from the physiological experiments, the dynamic range, being the range between threshold and discomfort, is small, rarely more than

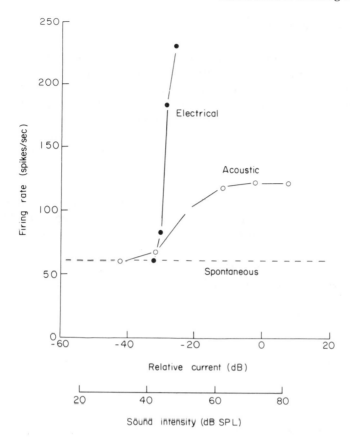

Fig. 10.11   The steep rate-intensity function seen with electrical stimuli, compared with the function found with acoustic stimuli. From Kiang and Moxon (1972, Fig. 3).

20 dB and often less than 10dB (e.g. Walsh *et al.*, 1980; Hochmair and Hochmair-Desoyer, 1985). It might be thought that if the stimulus were localized in the cochlea, increased current spread with intensity might increase the dynamic range. However, groups using bipolar stimulation and so a localized stimulus do not report wider dynamic ranges than others (Fugain *et al.*, 1985). The narrow dynamic range, presumably associated with a steep dependence of loudness on stimulus intensity, does not seem to have a correlate of particularly small difference limens for intensity. Differential intensity limens are about 1 dB, similar to those found with acoustic stimuli (e.g. Dillier *et al.*, 1980). Over the whole dynamic range, therefore, there may only be about 20 just noticeable differences for intensity, as compared with over 300 for acoustic stimuli in normal hearing.

(b) Frequency, pitch, and stimulus quality

If we believe the extreme position that at low frequencies information is carried purely by the temporal pattern of nerve impulses, then periodic electrical stimulation should produce faithful auditory sensations and good discrimination of frequencies. The results of electrical stimulation have on the whole been disappointing for such a prediction. In only a few cases do electrical stimuli seem to produce clear tonal sensations. A typical report is that tones sound like "comb and paper" (e.g. Fourcin *et al.*, 1979).

Where electrical stimuli of different frequencies can be discriminated or ranked, frequency discrimination limens are usually no better than 5% and sometimes as bad as 50%, one or two orders of magnitude worse than for acoustic stimuli. Discrimination limens fall markedly above 200–300 Hz. In some cases frequency discrimination is not possible at all at higher frequencies, although other patients can discriminate changes up to 1 kHz (Townshend *et al.*, 1987). When frequencies can be ranked or matched, the frequency associated with the perceived pitch does not necessarily increase in proportion to the stimulus frequency. In a typical report, Simmons *et al.* (1979) found that pitch increased linearly with stimulus frequency up to 150–200 Hz, and then at an increasingly accelerating rate up to a stimulus frequency of 400 Hz, after which no further pitch changes were produced. A further problem is that many patients show a strong interaction between stimulus intensity and perceived pitch.

The discussion so far has been concerned with the subjective correlate of the periodicity of the stimulus. But many patients report that the place of stimulation in the cochlea has a subjective frequency correlate as well. In these cases, and as might be expected from the physiology, stimulation near the base gives rise to sensations with a high-frequency correlate, described as a sharp timbre, and that nearer the apex to low-frequency sensations, described as a dull timbre. Accurate ranking according to relative electrode position in the cochlea is possible in some patients (e.g. Tong *et al.*, 1982). Where steps are taken to reduce the amount of current spread along the cochlear duct, the accuracy of ranking improves (Townshend *et al.*, 1987). There seems to be a periodicity component as well, because increasing the stimulus frequency at any one electrode raises the pitch of the sensation produced.

The generally poor quality of pitch sensations with nonlocalized stimulation suggests that timing information in nerve fibres is not *by itself* sufficient for the normal perception of auditory stimuli. One explanation for poor temporal coding is that the place of stimulation has to appropriate for the frequency of the stimulus used. Efforts have therefore been made to distribute electrodes more widely along the cochlear duct. Multiple electrodes have been inserted, so that for instance they cover a range of 8–25 mm from the round window (Patrick *et al.*, 1985). It can be calculated that the electrodes will stimulate places corresponding to characteristic frequencies of 400 Hz to several kHz. Electrophysiological experiments in animals show that electrical stimuli can produce phase-locking in the lower end of this frequency range. Therefore, it should theoretically be possible to provide some of the appropriate temporal and place cues together.

Theoretical analyses of temporal coding may help here. As was described in Chapter 9, a physiologically realistic mechanism for the analysis of temporal information in auditory nerve suggests that it might be accomplished by lateral inhibition in the cochlear nucleus. For this to work, the phase of activation has to change with place of stimulation in the appropriate way along the cochlear duct. The different cochlear places giving rise to excitatory and inhibitory inputs on single neurones of the cochlear nucleus have to be activated in phase opposition (Shamma, 1985a). The theory therefore suggests that optimal coding would be found with an electrode configuration producing this pattern of stimulation.

(c) Speech

The main goal in research on the auditory prosthesis has been to convey speech.

A small number of patients have been able to achieve good speech perception with single-channel stimulation (e.g. White, 1983; Hochmair-Desoyer *et al.*, 1985; Berliner and Eisenberg, 1987). The basis of good perception in these cases is not known. Patients with good difference limens for temporal discriminations tend to have better speech discrimination scores, suggesting that discrimination of stimulus envelope might sometimes be involved (Hochmair-Desoyer *et al.*, 1985). Moreover, they may be able to use temporal cues for frequency discrimination of individual cycles of the waveform at low frequencies, i.e. in the first-format range. In addition, these patients may be particularly good at making use of the minimal information remaining in a degraded input.

The above cases are exceptional. In the majority of cases, speech discrimination from "open" test sets has been poor or non-existent with single-channel electrodes. Here, the more advanced multi-electrode devices may be able to help. They can produce correct recognition rates of 20–30% in a substantial proportion of patients, and much higher recognition rates with closed test sets or with a single repeat of the stimulus (Schindler *et al.*, 1986; Clark, 1986). In multichannel devices, gains occur from electronic preprocessing of the stimulus, with the separate formants of the speech signal being presented to the appropriate electrodes in the cochlea. The fundamental voicing frequency can be transmitted by pulsing the current at each electrode (Clark, 1986; Blamey *et al.*, 1987; Holmes *et al.*, 1987).

Many simple devices seem able to assist lip reading. Even when the improvement in lip reading has not been dramatic, patients report at being delighted by the extra cues available (Fourcin *et al.*, 1979). Such assistance also has the benefit of reducing the considerable strain associated with lip reading, and also produces a rewarding sense of emotional contact with the speaker. Simple devices may be indicated in cases where patients would not be able to make use of the more sophisticated multi-electrode implants, for instance where nerve survival is poor, or in children where it is important not to compromise insertion of the more advanced devices which may become available at a later date.

(d) Conclusions

The auditory prosthesis is still in the experimental stage, although a large number of patients are already receiving substantial benefit. In order to convey speech sounds, many patients gain from multiple channels of stimulation, together with preprocessing of the stimulus. A further advantage of multichannel devices is that once implanted they can be driven in a single-channel configuration if that is later shown to be more appropriate. Multichannel prostheses, however, bring their own disadvantages of a greater necessary intrusion into the cochlea, and a concomitantly greater danger of damage to the remaining neural elements, together with a larger tissue reaction. They also require a larger surviving neural population. Although hopes for improvement direct attention to more sophistication in the auditory prosthesis, we should not lose sight of the benefits to be gained from simpler versions, including the detection of environmental warning signals, possible amelioration of tinnitus, assistance with lip reading, and the sense of emotional contact.

D. Summary

1. Hearing loss arises in the conductive apparatus before the oval window, where it is known as conductive loss, or in the cochlea, or more centrally. If it arises in the cochlea, it is known as sensorineural hearing loss of cochlear origin.

2. Sensorineural hearing loss of cochlear origin can be caused by ototoxic drugs. Aminoglycoside antibiotics are powerfully ototoxic and have been extensively investigated. In mild cases they cause damage to the

stereocilia, and in more severe cases loss of the hair cells. It has been postulated that the particular vulnerability of hair cells arises from the interaction of aminoglycosides with their external membranes as a preliminary to affecting intracellular biochemical reactions.

3. Noise trauma has been investigated extensively. With milder degrees of trauma, the changes are found in the stereocilia, particularly in the rootlets and in the cross-links. More severe effects include destruction of the hair cells.

4. Damage to outer hair cells produces a loss of the sharp tuning of the mechanical travelling wave. Auditory nerve fibres then lose the sharply-tuned tips of their tuning curves, with the result that they lose both sensitivity and frequency resolution. If inner hair cell stereocilia are selectively damaged, tuning curves keep their normal sharp tuning, but show reduced sensitivity.

5. The psychophysical results fit with the physiological results. In sensori-neural loss of cochlear origin, a loss in frequency resolution is often seen in addition to the loss in sensitivity. The loss in frequency resolution can be seen in widened psychophysical tuning curves and in widened critical bandwidths.

6. Loudness recruitment, also seen in sensorineural hearing loss, can simi-larly be explained in terms of the abnormal physiological responses. In pathological ears, the intensity functions for the basilar membrane and for auditory nerve fibres become steeper. Moreover, shallow tuning curves mean that activity spreads abnormally quickly along the nerve fibre array as the stimulus intensity is raised.

7. Some forms of tinnitus arising in the cochlea may give rise to objectively measurable sound emissions in the ear canal, as a result of the evoked cochlear mechanical response. However, in many cases of sensorineural loss, it seems that evoked and spontaneous activity in the auditory nerve both decrease. The tinnitus may well arise centrally, perhaps as a result of a hypersensitivity in the central nervous system when it is deprived of its normal input. There is some evidence that denervation hypersensi-tivity occurs in the central auditory nervous system.

8. Attempts have been made to restore hearing in cases of profound hearing loss by means of an auditory prosthesis, with electrical stimulation of the cochlea or auditory nerve. Problems arise because of poor dynamic

range of stimulus intensity, poor frequency range, and poor tonal quality. However, some patients can perceive speech with single channel devices, and others show good results with multichannel implants combined with preprocessing of the speech stimulus. Simple stimuli may be conveyed usefully by basic devices.

E. Further Reading

Mechanisms of ototoxicity are discussed by Brown and Daigneault (1981), Schacht (1986) and Lim (1986), who also discusses mechanisms of noise damage. Harrison (1985) discusses correlates of sensorineural hearing loss in the responses of auditory nerve fibres, and Phillips (1987) discusses the basis of loudness recruitment. Progress on the cochlear prosthesis has been reviewed in several symposium volumes, namely *Cochlear Prostheses*, edited by C.W. Parkins and S.W. Anderson (Proceedings of the National Academy of Sciences, Vol. **405**, 1983), *Cochlear Implants*, edited by R.A. Schindler and M.M. Merzenich (Raven Press, 1985), *The Cochlear Implant*, edited by T.J. Balkany (*Otolaryngol. Clin. North Am.* **19**, 215–449, 1986), *International Cochlear Implant Symposium and Workshop, Melbourne, 1985*, edited by G.M. Clark and P.A. Busby (*Ann. Otol., Rhinol. Laryngol.* **96** (Suppl. **128**), 1987), and *Am. J. Otol.* **8**, 189–268, 1987.

References

Abeles, M. and Goldstein, M. H. (1970). Functional architecture in cat primary auditory cortex: columnar organization and organization according to depth. *J. Neurophysiol.* **33**, 172–187.

Abeles, M. and Goldstein, M. H. (1972). Responses of single units in the primary auditory cortex of the cat to tones and to tone pairs. *Brain Res.* **42**, 337–352.

Abou-Madi, L., Pontarotti, P., Tramu, G., Cupo, A. and Eybalin, M. (1987). Co-existence of putative neuroactive substances in lateral olivocochlear neurons of rat and guinea pig. *Hearing Res.* **30**, 135–146.

Adams, J. C. (1979). Ascending projections to the inferior colliculus. *J. Comp. Neurol.* **183**, 519–538.

Adams, J. C. and Warr, W. B. (1976). Origins of axons in the cat's acoustic striae determined by injection of horseradish peroxidase into severed tracts. *J. Comp. Neurol.* **170**, 107–122.

Ades, H. W. and Engström, H. (1974). Anatomy of the inner ear. In *Handbook of Sensory Physiology* (eds W. D. Keidel and W. D. Neff), Vol. 5/1, pp. 125–158. Springer, Berlin.

Adrian, E. D. (1931). The microphonic action of the cochlea: an interpretation of Wever and Bray's experiments, *J. Physiol.* **71**, xxviii–xxix.

Aitkin, L. M. (1973). Medial geniculate body of the cat: responses to tonal stimuli of neurons in medial division. *J. Neurophysiol.* **36**, 275–283.

Aitkin, L. M. (1986). *The Auditory Midbrain*, 246 pp. Humana Press, Clifton.

Aitkin, L. M. and Phillips, S. C. (1984). Is the inferior colliculus an obligatory relay in the cat auditory system? *Neurosci. Lett.* **44**, 259–264.

Aitkin, L. M. and Prain, S. M. (1974). Medial geniculate body: unit responses in the awake cat. *J. Neurophysiol.* **37**, 512–521.

Aitkin, L. M. and Schuck, D. (1985). Low frequency neurons in the lateral central nucleus of the cat inferior colliculus receive their input predominantly from the medial superior olive. *Hearing Res.* **17**, 87–93.

Aitkin, L. M. and Webster, W. R. (1972). Medial geniculate body of the cat: organization and responses to tonal stimuli of neurons in ventral division. *J. Neurophysiol.* **35**, 365–380.

Aitkin, L. M., Webster, W. R., Veale, J. C. and Crosby, D. C. (1975). Inferior colliculus – I. Comparison of response properties of neurons in central, pericentral and external nuclei of adult cat. *J. Neurophysiol.* **38**, 1196–1207.

Aitkin L. M., Dickhaus, H., Schult, W. and Zimmerman, M. (1978). External nucleus of inferior colliculus: auditory and spinal somatosensory afferents and their interactions. J. *Neurophysiol.* **41**, 837–847.

Aitkin, L. M., Kenyon, C. E. and Philpott, P. (1981). The representation of the auditory and somatosensory systems in the external nucleus of the cat inferior colliculus. *J. Comp. Neurol.* **196**, 25–40.

Aitkin, L. M., Gates, G. R. and Phillips, S. C. (1984). Responses of neurons in inferior colliculus to variations in sound-source azimuth. *J. Neurophysiol.* **52**, 1–17.

Aitkin, L. M., Pettigrew, J. D., Calford, M. B., Phillips, S. C. and Wise, L. Z. (1985). Representation of stimulus azimuth by low-frequency neurons in inferior colliculus of the cat. *J. Neurophysiol.* **53**, 43–59.

Allen, W. F. (1945). Effect of destroying three localized cerebral cortical areas for sound on correct conditioned differential responses of the dog's foreleg. *Amer. J. Physiol.* **144**, 415–428.

Altman, J. A. and Kalmykova, I. V. (1986). Role of the dog's auditory cortex in discrimination of sound signals simulating sound source movement. *Hearing Res.* **24**, 243–253.

Altschuler, R. A. and Fex, J. (1986). Efferent neurotransmitters. In *Neurobiology of Hearing: The Cochlea* (eds R. A. Altschuler, R. P. Bobbin and D. W. Hoffman), pp. 383–396. Raven Press, New York.

Andersen, P., Junge, K. and Sveen, O. (1972). Corticofugal facilitation of thalamic transmission. *Brain Behav. Evol.* **6**, 170–184.

Andersen, R. A., Snyder, R. L. and Merzenich, M. M. (1980). The topographic organization of corticocollicular projections from physiologically defined loci in AI, AII and anterior cortical auditory fields of the cat. *J. Comp. Neurol.* **191**, 479–494.

Anderson, S. D. (1980). Some ECMR properties in relation to other signals from the auditory periphery. *Hearing Res.* **2**, 273–296.

Anniko, M. and Wroblewski, R. (1986). Ionic environment of cochlear hair cells. *Hearing Res.* **22**, 279–293.

Anniko, M., Lim, D. and Wroblewski, R. (1984). Elemental composition of individual cells and tissues in the cochlea. *Acta Otolaryngol.* **98**, 439–453.

Antoli-Candela, F. Jr. and Kiang, N. Y.-S. (1978). Unit activity underlying the N_1 potential. In *Evoked Electrical Activity in the Auditory Nervous System* (eds R. F. Naunton and C. Fernandez), pp. 165–189. Academic Press, London.

Aran, J. M. (1981). Electrical stimulation of the auditory system and tinnitus control. *J. Laryngol. Otol.* (Suppl. **4**), 153–162.

Art, J. J., Fettiplace, R. and Fuchs, P. A. (1984). Synaptic hyperpolarisation and inhibition of turtle cochlear hair cells. *J. Physiol.* **356**, 525–550.

Arthur, R. M., Pfeiffer, R. R. and Suga, N. (1971). Properties of "two-tone inhibition" in primary auditory neurones. *J. Physiol.* **212**, 593–609.

Ashmore, J. F. (1987). A fast motile response in guinea-pig outer hair cells: the cellular basis of the cochlear amplifier. *J. Physiol.* **388**, 323–347.

Balkany, T. J., Larsen, S. A., Reite, M., Rasmussen, K., Stypulkowski, P., Burgio, P., Asher, D., Rucker, N. C., Blanton, F. L. and Arenberg, I. K. (1985). Morphological and behavioral effects of chronic electrical stimulation of the primate cochlea. In *Cochlear Implants* (eds R. A. Schindler and M. M. Merzenich), pp. 83–92. Raven Press, New York.

Banks, W. F., Saunders, J. C. and Lowry, L. D. (1979). Olivocochlear bundle activity recorded in awake cats. *Otolaryngol. Head Neck Surg.* **87**, 463–471.

Baru, A. V. and Karaseva, T. A. (1972). *The Brain and Hearing*. Consultants Bureau, New York.

Bauer, J. W. (1978). Tuning curves and masking functions of auditory-nerve fibers in cat. *Sens. Proc.* **2**, 156–172.

Beard, N. S., Armentrout, S. A. and Weisberger, A. S. (1969). Inhibition of mammalian protein synthesis by antibiotics. *Pharmacol. Revs* **21**, 213–245.

von Békésy, G. (1952). DC resting potentials inside the cochlear partition. *J. Acoust. Soc. Am.* **24**, 72–76.

von Békésy, G. (1960). *Experiments in Hearing*. McGraw-Hill, New York.

Benson, D. A. and Teas, D. C. (1976). Single unit study of binaural interaction in the auditory cortex of the chinchilla. *Brain Res.* **103**, 313–338.

Berlin, C. I. and McNeil, M. R. (1976). Dichotic listening. In *Contemporary Issues in Experimental Phonetics* (ed. N. J. Lass), pp. 327–388. Academic Press, London.

Berliner, K. I. and Eisenberg, L. S. (1987). Our experience with cochlear implants: have we erred in our expectations? *Am. J. Otolaryngol.* **8**, 222–229.

Birt, D. and Olds, M. (1981). Associative response changes in lateral midbrain tegmentum and medial geniculate during differential appetitive conditioning. *J. Neurophysiol.* **46**, 1039–1055.

Birt, D., Nienhuis, R. and Olds, M. (1979). Separation of associative from non-associative short latency changes in medial geniculate and inferior colliculus during differential conditioning and reversal in rats. *Brain Res.* **167**, 129–138.

Blamey, P. J., Dowell, R. C., Clark, G. M. and Seligman, P. M. (1987). Acoustic parameters measured by a formant-estimating speech processor for a multiple-channel cochlear implant. *J. Acoust. Soc. Am.* **82**, 38–47.

Bock, G. R., Webster, W. R. and Aitkin, L. M. (1972). Discharge patterns of single units in inferior colliculus of the alert cat. *J. Neurophysiol.* **35**, 265–277.

de Boer, E. (1980). Physical principles in hearing theory I. *Physics Repts* **62**, 87–174.

de Boer, E. (1983). No sharpening? A challenge for cochlear mechanics. *J. Acoust. Soc. Am.* **73**, 567–573.

de Boer, E. and de Jongh, H. R. (1978). On cochlear encoding: potentialities and limitations of the reverse-correlation technique. *J. Acoust. Soc. Am.* **63**, 115–135.

de Boer, E. and Viergever, M. A. (1982). Validity of the Liouville-Green (or WKB) method for cochlear mechanics. *Hearing Res.* **8**, 131–155.

Bonding, P. (1979). Critical bandwidth in patients with a hearing loss induced by salicylates. *Audiology* **18**, 133–144.

Borg, E. (1973). On the neuronal organization of the acoustic middle ear reflex. A physiological and anatomical study. *Brain Res.* **49**, 101–123.

Bosher, S. K. and Warren, R. L. (1978). Very low calcium content of cochlear endolymph, an extracellular fluid. *Nature* **273**, 377–378.

Brawer, J. R., Morest, D. K. and Kane, E. C. (1974). The neuronal architecture of the cochlear nucleus of the cat. *J. Comp. Neurol.* **155**, 251–299.

Bray, P. and Kemp, D. T. (1987). An advanced cochlear echo technique suitable for infant screening. *Brit. J. Audiol.* **21**, 191–204.

Britt, R. and Starr, A. (1976a). Synaptic events and discharge patterns of cochlear nucleus cells. I. Steady-frequency tone bursts. *J. Neurophysiol.* **39**, 162–178.

Britt, R. and Starr, A. (1976b). Synaptic events and discharge patterns of cochlear nucleus cells. II. Frequency-modulated tones. *J. Neurophysiol.* **39**, 179–194.

Brown, A. M. and Kemp, D. T. (1984). Suppressibility of the $2f_1 - f_2$ stimulated acoustic emissions in gerbil and man. *Hearing Res.* **13**, 29–37.

Brown, A. M. and Pye, J. D. (1975). Auditory sensitivity at high frequencies in mammals. *Adv. Comp. Physiol. Biochem.* **6**, 1–73.

Brown, M. C. (1987a). Morphology of labeled afferent fibers in the guinea pig cochlea. *J. Comp. Neurol.* **260**, 591–604.

Brown, M. C. (1987b). Morphology of labeled efferent fibers in the guinea pig cochlea. *J. Comp. Neurol.* **260**, 605–618.

Brown, M. C., Nuttall, A. L., Masta, R. I. and Lawrence, M. (1983a). Cochlear inner hair cells: effects of transient asphyxia on intracellular potentials. *Hearing Res.* **9**, 131–144.

Brown, M. C., Nuttall, A. L. and Masta, R. I. (1983b). Intracellular recordings from cochlear inner hair cells: effects of stimulation of the crossed olivocochlear efferents. *Science* **222**, 69–72.

Brown, R. D. and Daigneault, E. A. (1981). *Pharmacology of Hearing*, 353 pp. Wiley, New York.

Brownell, W. E., Manis, P. B. and Ritz, L. A. (1979). Ipsilateral inhibitory responses in the cat lateral superior olive. *Brain Res.* **177**, 189–193.

Brownell, W. E., Bader, C. R., Bertrand, D. and Ribaupierre, Y. de (1985). Evoked mechanical responses of isolated cochlear outer hair cells. *Science* **227**, 194–196.

Brugge, J. F. and Geisler, C. D. (1978). Auditory mechanisms of the lower brainstem. *Ann. Rev. Neurosci.* **1**, 363–394.

Brugge, J. F. and Merzenich, M. M. (1973). Responses of neurons in auditory cortex of the macaque monkey to monaural and binaural stimulation. *J. Neurophysiol.* **36**, 1138–1158.

Brugge, J. F. and Reale, R. A. (1985). Auditory cortex. In *Cerebral Cortex, Vol. 4, Association and Auditory Cortices* (eds A. Peters and E. G. Jones), pp. 229–271. Plenum, New York.

Buño, W. (1978). Auditory nerve fiber activity influenced by contralateral ear sound stimulation. *Exp. Neurol.* **59**, 62–74.

Caird, D. and Klinke, R. (1983). Processing of binaural stimuli by cat superior olivary complex neurons. *Exp. Brain Res.* **52**, 385–399.

Caird, D. and Klinke, R. (1987). Processing of interaural time and intensity differences in the cat inferior colliculus. *Exp. Brain Res.* **68**, 379–392.

Calford, M. B. (1983). The parcellation of the medial geniculate body of the cat defined by the auditory response properties of single units. *J. Neurosci.* **3**, 2350–2364.

Calford, M. B. and Aitkin, L. M. (1983). Ascending projections to the medial geniculate body of the cat: evidence for multiple, parallel auditory pathways through the thalamus. *J. Neurosci.* **3**, 2365–2380.

Cant, N. B. and Gaston, K. C. (1982). Pathways connecting the right and left cochlear nuclei. *J. Comp. Neurol.* **212**, 313–326.

Cant, N. B. and Morest, D. K. (1978). Axons from non-cochlear sources in the anteroventral cochlear nucleus of the cat. A study with the rapid Golgi method. *Neuroscience* **3**, 1003–1029.

Cant, N. B. and Morest, D. K. (1984). The structural basis for stimulus coding in the cochlear nucleus of the cat. In *Hearing Science*, (ed. C. I. Berlin), pp. 371–421. College-Hill Press, San Diego.

Carmel, P. W. and Starr, A. (1963). Acoustic and non-acoustic factors modifying middle-ear muscle activity in waking cats. *J. Neurophysiol.* **26**, 598–616.

Caspary, D. M. (1986). Cochlear nuclei: functional neuropharmacology of the principal cell types. In *Neurobiology of Hearing: The Cochlea* (eds R. A. Altschuler, R. P. Bobbin and D. W. Hoffman), pp. 303–332. Raven Press, New York.

Cherry, E. C. (1953). Some experiments on the recognition of speech, with one and with two ears. *J. Acoust. Soc. Am.* **25**, 975–979.

Chouard, C.-H., Meyer, B. and Gegu, D. (1985). Pre- and per-operative electrical testing procedure. In *Cochlear Implants* (eds R. A. Schindler and M. M. Merzenich), pp. 365–374. Raven Press, New York.

Citron, L., Dix, M. R., Hallpike, C. S. and Hood, J. D. (1963). A recent clinico-pathological study of cochlear nerve degeneration resulting from tumor pressure and disseminated sclerosis, with particular reference to the finding of normal threshold sensitivity for pure tones. *Acta Otolaryngol.* **56**, 330–337.

Clark, G. M. (1986). The University of Melbourne/Cochlear Corporation (Nucleus) program. *Otolaryngol. Clin. North Am.* **19**, 329–354.

Clark, W. W., Kim, D. O., Zurek, P. M. and Bohne, B. A. (1984). Spontaneous otoacoustic emissions in chinchilla ear canals: correlation with histopathology and suppression by external tones. *Hearing Res.* **16**, 299–314.

Code R. A. and Winer J. A. (1986). Columnar organization and commissural connections in cat primary auditory cortex (AI). *Hearing Res.* **23**, 205–222.

Cody, A. R. and Johnstone, B. M. (1982). Temporary threshold shift modified by binaural acoustic stimulation. *Hearing Res.* **6**, 199–205.

Cody, A. R. and Russell, I. J. (1987). The responses of hair cells in the basal turn of the guinea pig cochlea to tones. *J. Physiol.* **383**, 551–569.

Colativa, F. B. (1972). Auditory cortical lesions and visual pattern discrimination in cat. *Brain Res.* **39**, 437–447.

Colativa, F. B. (1974). Insular-temporal lesions and vibrotactile temporal pattern discrimination in cats. *Psychol. Behav.* **12**, 215–218.

Colativa, F. B., Szeligo, F. V. and Zimmer, S. D. (1974). Temporal pattern discrimination in cats with insular-temporal lesions. *Brain Res.* **79**, 153–156.

Comis, S. D. (1970). Centrifugal inhibitory processes affecting neurones in the cat cochlear nucleus. *J. Physiol.* **210**, 751–760.

Comis, S. D. and Davies, W. E. (1969). Acetylcholine as a transmitter in the cat auditory system. *J. Neurochem.* **16**, 423–429.

Comis, S. D. and Whitfield, I. C. (1968). Influence of centrifugal pathways on unit activity in the cochlear nucleus. *J. Neurophysiol.* **31**, 62–68.

Comis, S. D., Pickles, J. O. and Osborne, M. P. (1985). Osmium tetroxide postfixation in relation to the cross linkage and spatial organization of stereocilia in the guinea-pig cochlea. *J. Neurocytol.* **14**, 113–130.

Corey, D. P. and Hudspeth, A. J. (1979a). Ionic basis of the receptor potential in a vertebrate hair cell. *Nature* **281**, 675–677.

Corey, D. P. and Hudspeth, A. J. (1979b). Response latency of vertebrate hair cells. *Biophys. J.* **26**, 499–506.

Corey, D. P. and Hudspeth, A. J. (1983). Kinetics of the receptor current in bullfrog saccular hair cells. *J. Neurosci.* **3**, 962–976.

Cornwell, P. (1967). Loss of auditory pattern discrimination following insular-temporal lesions in cats. *J. Comp. Physiol. Psychol.* **63**, 165–168.

Costalupes, J. A. (1985). Representation of tones in noise in the responses of auditory nerve fibers in cats. I. Comparison with detection thresholds. *J. Neurosci.* **5**, 3261–3269.

Costalupes. J. A., Young, E. D. and Gibson, D. J. (1984). Effects of continuous noise backgrounds on rate response of auditory nerve fibers in cat. *J. Neurophysiol*, **51**, 1326–1344.

Cowey, A. and Weiskrantz, L. (1976). Auditory sequence discrimination in *Macaca mulatta*: the role of the superior temporal cortex. *Neuropsychologia* **14**, 1–10.

Cranford, J. L. (1975). Role of neocortex in binaural hearing in the cat. I. Contralateral masking. *Brain Res.* **100**, 395–406.

Cranford, J. L. (1979a). Auditory cortex lesions and interaural intensity and phase-angle discrimination in cats. *J. Neurophysiol.* **42**, 1518–1526.

Cranford, J. L. (1979b). Detection versus discrimination of brief tones by cats with auditory cortex lesions. *J. Acoust. Soc. Am.* **65**, 1573–1575.

Cranford, J. L., Igarashi, M. and Stramler, J. H. (1976a). Effect of auditory neocortex ablation on pitch perception in the cat. *J. Neurophysiol.* **39**, 143–152.

Cranford, J. L., Igarashi, M. and Stramler, J. H. (1976b). Effect of auditory neo-cortex ablation on identification of click rates in cats. *Brain Res.* **116**, 69–81.

Crawford, A. C. and Fettiplace, R. (1981). An electrical tuning mechanism in turtle cochlear hair cells. *J. Physiol.* **312**, 377–422.

Crawford, A. C. and Fettiplace, R. (1985). The mechanical properties of ciliary bundles of turtle cochlear hair cells. *J. Physiol.* **364**, 359–379.

Crow, G., Rupert, A. L. and Moushegian, G. (1978). Phase-locking in monaural and binaural medullary neurons: implications for binaural phenomena. *J. Acoust. Soc. Am.* **64**, 493–501.

Dallos, P. (1973a). *The Auditory Periphery*. Academic Press, New York and London.

Dallos, P. (1973b). Cochlear potentials and cochlear mechanics. In *Basic Mechanisms in Hearing* (ed. A. Møller), pp. 335–372. Academic Press, New York and London.

Dallos, P. (1985). Response characteristics of mammalian cochlear hair cells. *J. Neurosci.* **5**, 1591–1608.

Dallos, P. (1986). Neurobiology of inner and outer hair cells: intracellular recordings. *Hearing Res.* **22**, 185–198.

Dallos, P. and Wang, C.-Y. (1974). Bioelectric correlates of kanamycin intoxication. *Audiology* **13**, 277–289.

Dallos, P., Schoeny, Z. G. and Cheatham, M. A. (1972). Cochlear summating potentials: descriptive aspects. *Acta Otolaryngol.* (Suppl. 302), 1–46.

Dallos, P., Santos-Sacchi, J. and Flock, A. (1982). Intracellular recordings from cochlear outer hair cells. *Science* **218**, 582–584.

David, E., Keidel, W. D., Mallert, S., Bechtereva, N. P. and Bundzen, P. V. (1977). Decoding processes in the auditory system and human speech analysis. In *Psychophysics and Physiology of Hearing* (eds E. F. Evans and J. P. Wilson), pp. 509–516. Academic Press, London.

Davis, H. (1958). Transmission and transduction in the cochlea. *Laryngoscope* **68**, 359–382.

Delgutte, B. and Kiang, N. Y.-S. (1984). Speech coding in the auditory nerve: III. Voiceless fricative consonants. *J. Acoust. Soc. Am.* **75**, 887–896.

Densert, O. and Flock, A. (1974). An electron-microscopic study of adrenergic innervation in the cochlea. *Acta Otolaryngol.* **77**, 185–197.

DeRosier, D. J., Tilney, L. G. and Egelman, E. (1980). Actin in the inner ear: the remarkable structure of the stereocilium. *Nature* **287**, 291–296.

Desmedt, J. E. (1975). Physiological studies of the efferent recurrent auditory system. In *Handbook of Sensory Physiology* (eds W. D. Keidel and W. D. Neff), Vol. 5/2, pp. 219–246. Springer, Berlin.

Desmedt, J. E. and Robertson, D. (1975). Ionic mechanism of the efferent olivo-cochlear inhibition studied by cochlear perfusion in the cat. *J. Physiol.* **247**, 407–428.

Dewson, J. H. (1964). Speech sound discrimination by cats. *Science* **144**, 555–556.

Dewson, J. H. (1968). Efferent olivocochlear bundle: some relationships to stimulus discrimination in noise. *J. Neurophysiol.* **31**, 122–130.

Dewson, J. H., Nobel, K. W. and Pribram, K. H. (1966). Corticofugal influence at cochlear nucleus of the cat: some effects of ablation of insular-temporal cortex. *Brain Res.* **2**, 151–159.

Dewson, J. H., Pribram, K. H. and Lynch, J. C. (1969). Effects of ablations of temporal cortex upon speech sound discrimination in the monkey. *Exp. Neurol.* **24**, 579–591.

Dewson, J. H., Cowey, A. and Weiskrantz, L. (1970). Disruptions of auditory sequence discrimination by unilateral and bilateral cortical ablations of superior temporal gyrus in the monkey. *Exp. Neurol.* **28**, 529–548.

Diamond, I. T. and Neff, W. D. (1957). Ablation of temporal cortex and discrimination of auditory patterns. *J. Neurophysiol.* **20**, 300–315.

Diamond, I. T., Jones, E. G. and Powell, T. P. S. (1969). The projection of the auditory cortex upon the diencephalon and the brain stem of the cat. *Brain Res.* **15**, 305–340.

Diependaal, R. J., de Boer, E. and Viergever, M. A. (1987). Cochlear power flux as an indicator of mechanical activity. *J. Acoust. Soc. Am.* **82**, 917–926.

Dillier, N., Spillmann, T., Fisch, U. P. and Leifer, L. (1980). Encoding and decoding of auditory signals in relation to human speech and its application to human cochlear implants. *Audiology* **19**, 146–163.

Dodson, H. C., Walliker, J. R., Frampton, S., Douek, E. E., Fourcin, A. J. and Bannister, L. H. (1986). Structural alteration of hair cells in the contralateral ear resulting from extracochlear electrical stimulation. *Nature* **320**, 65–67.

Drenckhahn, D., Kellner, J., Mannherz, H. G., Groschel-Stewart, U., Kendrick-Jones, J. and Scholey, J. (1982). Absence of myosin-like immunoreactivity in the stereocilia of cochlear hair cells. *Nature* **300**, 531–532.

Drenckhahn, D., Schafer, T. and Prinz, M. (1985). Actin, myosin and associated proteins in the auditory and vestibular organs: immunocytochemical and biochemical studies. In *Auditory Biochemistry* (ed. D. G. Drescher), pp. 317–335. Charles C. Thomas, Springfield.

Drescher, D. G. and Kerr, T. P. (1985). Na^+, K^+-activated adenosine triphosphatase and carbonic anhydrase: inner ear enzymes of ion transport. In *Auditory Biochemistry* (ed. D. G. Drescher), pp. 436–472. Charles C. Thomas, Springfield.

Duifhuis, H. (1976). Cochlear nonlinearity and second filter: possible mechanisms and implications. *J. Acoust. Soc. Am.* **59**, 408–423.

Durrant, J. D. and Lovrinic, J. H. (1977). *Bases of Hearing Science.* Williams and Wilkins, Baltimore.

Dwyer, J., Blumstein, S. E. and Ryalls, J. (1982). The role of duration and rapid temporal processing on the lateral perception of consonants and vowels. *Brain Lang.* **17**, 272–286.

Economo, C. von and Horn, L. (1930). Uber Windungsrelief, Masse und Rindenarchitektonik der Supratemporalfläche, ihre individuellen und ihre Seitenunterschiede. *Z. Ges. Neurol. Psychiat.* **130**, 678–757.

Eddington, D.K. (1983). Speech recognition in deaf subjects with multichannel intracochlear electrodes. *Ann. NY Acad. Sci.* **405**, 241–258.

Efron, R. (1985). The central auditory system and issues related to hemispheric development. In *Assessment of Central Auditory Dysfunction: Foundations and Clinical Correlates,* pp. 143–154. Williams and Wilkins, Baltimore.

Egan, J. P. and Hake, H. W. (1950). On the masking pattern of a simple auditory stimulus. *J. Acoust. Soc. Am.* **22**, 622–630.

Ehret, G. (1987). Left hemisphere advantage in the mouse brain for recognizing ultrasonic communication calls. *Nature* **325**, 249–251.

Eldredge D. H. (1974). Inner ear – cochlear mechanics and cochlear potentials. In *Handbook of Sensory Physiology* (eds W. D. Keidel and W. D. Neff), Vol. 5/1, pp. 549–584. Springer, Berlin.

Elliott, D. N. and Trahiotis, C. (1970). Cortical lesions and auditory discrimination. *Psychol. Bull.* **77**, 198–222.

Elliott, D. N., Stein, L. and Harrison, M. J. (1960). Discrimination of absolute-intensity thresholds and frequency-difference thresholds in cats. *J. Acoust. Soc. Am.* **32**, 380–384.

Elverland, H. H. (1977). Descending connections between the superior olivary and cochlear nucleus complexes in the cat studied by autoradiographic and horseradish peroxidase methods. *Exp. Brain Res.* **27**, 397–412.

Elverland, H. H. (1978). Ascending and intrinsic projections of the superior olivary complex in the cat. *Exp. Brain Res.* **32**, 117–134.

Engström, H. (1960). The cortilymph, the third lymph of the inner ear. *Acta Morphol. Neer. Scand.* **3**, 195–204.

Engström, H. and Engström, B. (1978). Structure of hairs on cochlear sensory cells. *Hearing Res.* **1**, 49–66.

Erulkar, S. D., Rose, J. E. and Davies, P. W. (1956). Single unit activity in the auditory cortex of the cat. *Bull. Johns Hopkins Hosp.* **99**, 55–86.

Evans, E. F. (1968). Cortical representation. In *Hearing Mechanisms in Vertebrates* (eds A. V. S. de Reuck and J. Knight), pp. 272–287. Churchill, London.

Evans, E. F. (1972). The frequency response and other properties of single fibres of the guinea-pig cochlear nerve. *J. Physiol.* **226**, 263–287.

Evans, E. F. (1975a). Cochlear nerve and cochlear nucleus. In *Handbook of Sensory Physiology* (eds W. D. Keidel and W. D. Neff), Vol. 5/2, pp. 1–108. Springer, Berlin.

Evans, E. F. (1975b). The sharpening of cochlear frequency selectivity in the normal and abnormal cochlea. *Audiology* **14**, 419–442.

Evans, E. F. (1977). Frequency selectivity at high signal levels of single units in cochlear nerve and nucleus. In *Psychophysics and Physiology of Hearing* (eds E. F. Evans and J. P. Wilson), pp. 185–192. Academic Press, London.

Evans, E. F. and Harrison, R. V. (1976). Correlation between outer hair cell damage and deterioration of cochlear nerve tuning properties in the guinea pig. *J. Physiol.* **256**, 43–44P.

Evans, E. F. and Nelson, P. G. (1973a). The responses of single neurones in the cochlear nucleus of the cat as a function of their location and anaesthetic state. *Exp. Brain Res.* **17**, 402–427.

Evans, E. F. and Nelson, P. G. (1973b). On the functional relationship between the dorsal and ventral divisions of the cochlear nucleus of the cat. *Exp. Brain Res.* **17**, 428–442.

Evans, E. F. and Whitfield, I. C. (1964). Classification of unit responses in the auditory cortex of the unanaesthetised cat. *J. Physiol.* **171**, 476–493.

Evans, E. F., Ross, H. F. and Whitfield, I. C. (1965). The spatial distribution of unit characteristic frequency in the primary auditory cortex of the cat. *J. Physiol.* **179**, 238–247.

Evans, E. F., Wilson, J. P. and Borerwe, T. A. (1981). Animal models of tinnitus. In *Tinnitus* (eds D. Evered and G. Lawrenson), pp. 108–129. CIBA Symposium No. **85**. Pitman Medical, London.

Eybalin, M. and Pujol, R. (1987). Choline acetyltransferase (ChAT) immunoelectron microscopy distinguishes at least three types of efferent synapses in the organ of Corti. *Exp. Brain Res.* **65**, 261–270.

Fawcett, D. W. (1986). *A Textbook of Histology.* W. B. Saunders, Philadelphia.

Feldman, A. M. (1981a). Cochlear biochemistry. In *Pharmacology of Hearing* (eds R. D. Brown and E. A. Daigneault), pp. 51–80. Wiley, New York.

Feldman, A. M. (1981b). Cochlear fluids: physiology, biochemistry and pharmacology. In *Pharmacology of Hearing,* (eds R. D. Brown and E. A. Daigneault), pp. 81–97. Wiley, New York.

Fernandez, C. (1951). The innervation of the cochlea (guinea pig). *Laryngoscope* **61**, 1152–1172.

Fex, J. (1959). Augmentation of the cochlear microphonics by stimulation of efferent fibres to cochlea. *Acta Otolaryngol.* **50**, 540–541.

Fex, J. (1962). Auditory activity in centrifugal and centripetal cochlear fibres in cat. *Acta Physiol. Scand.* **55** (Suppl. 189), 5–68.

Fex, J. and Altschuler, R. A. (1986). Neurotransmitter-related immunocytochemistry of the organ of Corti. *Hearing Res.* **22**, 249–263.

Fletcher, H. (1940). Auditory patterns. *Revs Modern Phys.* **12**, 47–65.

Flock, Å. (1977). Physiological properties of sensory hairs in the ear. In *Psychophysics and Physiology of Hearing* (eds E. F. Evans and J. P. Wilson), pp. 15–25. Academic Press, London.

Flock, Å. (1983). Hair cells: receptors with a motor capacity?. In *Hearing – Physiological Bases and Psychophysics* (eds R. Klinke and R. Hartmann), pp. 2–7. Springer, Berlin.

Flock, Å. and Cheung, H. C. (1977). Actin filaments in sensory hairs of inner ear receptor cells. *J. Cell Biol.* **75**, 339–343.

Flock, Å. and Strelioff, D. (1984). Studies on hair cells in isolated coils from the guinea pig cochlea. *Hearing Res.* **15**, 11–18.

Flock, Å., Flock, B. and Murray, E. (1977). Studies on the sensory hairs of receptor cells in the inner ear. *Acta Otolaryngol.* **83**, 85–91.

Flock, Å., Cheung, H. C., Flock, B. and Utter, G. (1981). Three sets of actin filaments in sensory cells of the inner ear. Identification and functional orientation determined by gel electrophoresis, immunofluorescence and electron microscopy. *J. Neurocytol.* **10**, 133–147.

Flock, Å., Bretscher, A. and Weber, K. (1982). Immunohistochemical localization of several cytoskeletal proteins in inner ear sensory and supporting cells. *Hearing Res.* **7**, 75–89.

Flock, Å., Flock, B. and Ulfendahl M. (1986). Mechanisms of movement in outer hair cells and a possible structural basis. *Arch. Otorhinolaryngol.* **243**, 83–90.

Florentine, M., Buus, S., Scharf, B. and Zwicker, E. (1980). Frequency selectivity in normally-hearing and hearing-impaired observers. *J. Speech Hear. Res.* **23**, 646–669.

Fourcin, A. J., Rosen, S. M., Moore, B. C. J., Douek, E. E., Clarke, G. P., Dodson, H. and Bannister, L. H. (1979). External electrical stimulation of the cochlea: clinical, psychophysical, speech-perceptual and histological findings. *Brit. J. Audiol.* **13**, 85–107.

Fox, J. E. (1979). Habituation and prestimulus inhibition of the auditory startle reflex in decerebrate rats. *Physiol. Behav.* **23**, 291–297.

Fugain, C., Meyer, B. and Chouard, C.-H. (1985). Speech processing strategies and clinical results of the French multichannel cochlear implant. In *Cochlear Implants* (eds R. A. Schindler and M. M. Merzenich), pp. 433–451. Raven Press, New York.

Funkenstein, H. H. and Winter, P. (1973). Responses to acoustic stimuli of units in the auditory cortex of awake squirrel monkeys. *Exp. Brain Res.* **18**, 464–488.

Galambos, R. (1956). Suppression of auditory nerve activity by stimulation of efferent fibers to cochlea. *J. Neurophysiol.* **19**, 424–437.

Gardner, M. B. and Gardner, R. S. (1973). Problem of localization in the median plane: effect of pinna cavity occlusion. *J. Acoust. Soc. Am.* **53**, 400–408.

Geisler, C. D., Deng, L. and Greenberg, S. R. (1985). Thresholds for primary auditory fibers using statistically defined criteria. *J. Acoust. Soc. Am.* **77**, 1102–1109.

Gerken, G. M. (1979). Central denervation hypersensitivity in the auditory system of the cat. *J. Acoust. Soc. Am.* **66**, 721–727.

Gershuni, G. V., Baru, A. V. and Karaseva, T. A. (1967). Role of auditory cortex in discrimination of acoustic stimuli. *Neural Sciences Trans.* **1**, 370–382.

Gibson, D. J., Young, E. D. and Costalupes, J. A. (1985). Similarity of dynamic range adjustment in auditory nerve and cochlear nuclei. *J. Neurophysiol.* **53**, 940–958.

Gifford, M. L. and Guinan, J. J. (1987). Effects of electrical stimulation of medial olivocochlear neurons on ipsilateral and contralateral cochlear responses. *Hearing Res.* **29**, 179–194.

Gilbert, A. G. and Pickles, J. O. (1980). Responses of auditory nerve fibres to noise bands of different widths. *Hearing Res.* **2**, 327–333.

Glaser, E. M., van der Loos, H. and Gissler, M. (1979). Tangential orientation and spatial order in dendrites of cat auditory cortex: a computer microscope study of Golgi-impregnated material. *Exp. Brain Res.* **36**, 411–431.

Glass, I. (1983). Tuning characteristics of cochlear nucleus units in response to electrical stimulation of the cochlea. *Hearing Res.* **12**, 223–237.

Glass, I. and Wollberg, Z. (1983). Responses of cells in the auditory cortex of awake squirrel monkeys to normal and reversed species-specific vocalizations. *Hearing Res.* **9**, 27–33.

Glendenning, K. K. and Masterton, R. B. (1983). Acoustic chiasm: efferent projections of the lateral superior olive. *J. Neurosci.* **3**, 1521–1537.

Glendenning, K. K., Hutson, K. A., Nudo, R. J. and Masterton, R. B. (1985). Acoustic chiasm. II: Anatomical basis of binaurality in lateral superior olive of cat. *J. Comp. Neurol.* **232**, 261–285.

Goblick, T. and Pfeiffer, R. R. (1969). Time domain measurements of cochlear nonlinearities using combination click stimuli. *J. Acoust. Soc. Am.* **46**, 924–938.

Godfrey, D. A., Kiang, N. Y.-S. and Norris, B. E. (1975a). Single unit activity in the posteroventral cochlear nucleus of the cat. *J. Comp. Neurol.* **162**, 247–268.

Godfrey, D. A., Kiang, N. Y.-S. and Norris, B. E. (1975b). Single unit activity in the dorsal cochlear nucleus of the cat. *J. Comp. Neurol.* **162**, 269–284.

Godfrey, D. A., Carter, J. A., Berger, S. J., Lowry, O. H. and Matchinsky, F. M. (1977). Quantitative histochemical mapping of candidate transmitter amino acids in cat cochlear nucleus. *J. Histochem. Cytochem.* **25**, 417–431.

Godfrey, D. A., Carter, J. A., Lowry, O. H. and Matchinsky, F. M. (1978). Distribution of gamma-amino butyric acid, glycine, glutamate and aspartate in the cochlear nucleus of the rat. *J. Histochem. Cytochem.* **26**, 118–126.

Godfrey, D. A., Bowers, M., Johnson, B. A. and Ross, C. D. (1984). Aspartate aminotransferase activity in fiber tracts of the rat brain. *J. Neurochem.* **42**, 1450–1456.

Godfrey, D. A., Park-Hellendall, J. L., Dunn, J. D. and Ross, C. D. (1987a). Effect of olivocochlear bundle transection on choline acetyltransferase activity in the rat cochlear nucleus. *Hearing Res.* **28**, 237–251.

Godfrey, D. A., Park-Hellendall, J. L., Dunn, J. D. and Ross, C. D. (1987b). Effects of trapezoid body and superior olive lesions on choline acetyltransferase activity in the rat cochlear nucleus. *Hearing Res.* **28**, 253–270.

Goldberg, J. M. and Brown, P. B. (1968). Functional organization of the dog superior olivary complex: an anatomical and electrophysiological study. *J. Neurophysiol.* **31**, 639–656.

Goldberg, J. M. and Brown, P. B. (1969). Response of binaural neurons of dog superior olivary complex to dichotic tonal stimuli: some physiological mechanisms of sound localization. *J. Neurophysiol.* **32**, 613–636.

Goldberg, J. M. and Brownell, W. E. (1973). Response characteristics of neurons in anteroventral and dorsal cochlear nuclei of cat. *Brain Res.* **64**, 35–54.

Goldberg, J. M., Adrian, H. O. and Smith, F. D. (1964). Response of neurons of the superior olivary complex of the cat to acoustic stimuli of long duration. *J. Neurophysiol.* **27**, 706–749.

Goldstein, J. L. (1967). Auditory nonlinearity. *J. Acoust. Soc. Am.* **41**, 676–689.

Goldstein, J. L. and Kiang, N. Y.-S. (1968). Neural correlates of the aural combination tone $2f_1 - f_2$. *Proc. IEEE* **56**, 981–992.

Goldstein, M. H. and Abeles, M. (1975). Single unit activity of the auditory cortex. In *Handbook of Sensory Physiology* (eds W. D. Keidel and W. D. Neff), Vol. 5/2, pp. 199–218. Springer, Berlin.

Goldstein, M. H., Hall, J. L. and Butterfield, B. O. (1968). Single-unit activity in the primary auditory cortex of unanesthetized cats. *J. Acoust. Soc. Am.* **43**, 444–455.

Goodman, D. A., Smith, R. L. and Chamberlain, S. C. (1982). Intracellular and extracellular responses in the organ of Corti of the gerbil. *Hearing Res.* **7**, 161–179.

Greenwood, D. D. and Goldberg, J. M. (1970). Response of neurons in the cochlear nuclei to variations in noise bandwidths and to tone–noise combinations. *J. Acoust. Soc. Am.* **47**, 1022–1040.

Groen, J. J. (1964). Super- and subliminal binaural beats. *Acta Otolaryngol.* **57**, 224–230.

Guinan, J. J. and Peake, W. T. (1967). Middle ear characteristics of anesthetized cats. *J. Acoust. Soc. Am.* **41**, 1237–1261.

Guinan, J. J., Guinan, S. S. and Norris, B. E. (1972). Single auditory units in the superior olivary complex. I. Responses to sounds and classification based on physiological properties. *Int. J. Neurosci.* **4**, 101–120.

Guinan, J. J., Warr, W. B. and Norris, B. E. (1983). Differential olivocochlear projections from lateral versus medial zones of the superior olivary complex. *J. Comp. Neurol.* **221**, 358–370.

Gummer, A. W., Johnstone, B. M. and Armstrong, N. J. (1981). Direct measurement of basilar membrane stiffness in the guinea pig. *J. Acoust. Soc. Am.* **70**, 1298–1309.

Hall, J. L. (1972). Auditory distortion products $f_2 - f_1$ and $2f_1 - f_2$. *J. Acoust. Soc. Am.* **51**, 1863–1871.

Halperin, Y., Nachson, I. and Carmon, A. (1973). Shift of ear superiority in dichotic listening to temporally-patterned nonverbal stimuli. *J. Acoust. Soc. Am.* **53**, 46–50.

Hamernik, R. P., Turrentine, G. and Roberto, M. (1986). Mechanically-induced morphological changes in the organ of Corti. In *Basic and Applied Aspects of*

Noise-induced Hearing Loss (eds R. J. Salvi, D. Henderson, R. P. Hamernik and Colletti, V.), pp. 69–83. Plenum, New York.

Harris, D. M. and Dallos, P. (1979). Forward masking of auditory nerve fiber responses. *J. Neurophysiol.* **42**, 1083–1107.

Harrison, J. M. and Howe, M. E. (1974a). Anatomy of the afferent auditory nervous system of mammals. In *Handbook of Sensory Physiology* (eds W. D. Keidel and W. D. Neff), Vol. 5/1, pp. 283–336. Springer, Berlin.

Harrison, J. M. and Howe, M. E. (1974b). Anatomy of the descending auditory system (mammalian). In *Handbook of Sensory Physiology* (eds W. D. Keidel and W. D. Neff), Vol. 5/1, pp. 363–388. Springer, Berlin.

Harrison, R. V. (1981). Rate-versus-intensity functions and related AP responses in normal and pathological guinea pig and human cochleas. *J. Acoust. Soc. Am.* **70**, 1036–1044.

Harrison, R. V. (1985). The physiology of the normal and pathological cochlear neurones – some recent advances. *J. Otolaryngol.* **14**, 345–356.

Hartmann, R., Topp, G. and Klinke, R. (1984). Discharge patterns of cat primary auditory fibers with electrical stimulation of the cochlea. *Hearing Res.* **13**, 47–62.

Hawkins, J. E. and Stevens, S. S. (1950). The masking of pure tones and of speech by white noise. *J. Acoust. Soc. Am.* **22**, 6–13.

Heffner, H. (1978). Effect of auditory cortex ablation on localization and discrimination of brief sounds. *J. Neurophysiol.* **41**, 963–976.

Heffner, H. E. and Heffner, R. S. (1986a). Hearing loss in Japanese macaques following bilateral auditory cortex lesions. *J. Neurophysiol.* **55**, 256–271.

Heffner, H. E. and Heffner, R. S. (1986b). Effect of unilateral and bilateral auditory cortex lesions on the discrimination of vocalizations by Japanese macaques. *J. Neurophysiol.* **56**, 683–701.

Held, G. (1893). Die centrale Gehörleitung. *Arch. Anat. Physiol. Anat. Abt.* **(1893)** 201–248.

Helmholtz, H. L. F. (1863). *Die Lehre von den Tonempfindungen als Physiologische Grundlage für die Theorie der Musik*. Eng. Trans. of 3rd. ed. by A. J. Ellis, *On the Sensations of Tone*, 1875. Longmans, Green, London.

Hillman, D. E. (1969). New ultrastructural findings regarding a vestibular ciliary apparatus and its possible functional significance. *Brain Res.* **13**, 407–412.

Hirokawa. N. and Tilney, L. G. (1982). Interactions between actin filaments and between actin filaments and membranes in quick-frozen and deeply etched hair cells of the chick ear. *J. Cell Biol.* **95**, 249–261.

Hochmair, E. S. and Hochmair-Desoyer, I. J. (1985). Aspects of sound signal processing using the Vienna intra- and extracochlear implants. In *Cochlear Implants* (eds R. A. Schindler and M. M. Merzenich), pp. 101–110. Raven Press, New York.

Hochmair-Desoyer, I. J., Hochmair, E. S. and Stiglbrunner, H.K. (1985). Psycho-acoustic temporal processing and speech understanding in cochlear implant patients. In *Cochlear Implants* (eds R. A. Schindler and M. M. Merzenich), pp. 291-304. Raven Press, New York.

Holley M.C. and Ashmore J. F. (1988). On the mechanism of a high-frequency force generator in outer hair cells isolated from the guinea pig cochlea. *Proc. Roy. Soc. Lond.* B **232**, 413-429.

Holmes, A. E., Kemker, F. J. and Merwin, G. E. (1987). The effects of varying the number of cochlear implant electrodes on speech perception. *Am. J. Otol.* **8**, 240–246.

Holton, T. and Hudspeth, A. J. (1986). The transduction channel of hair cells from the bull-frog characterized by noise analysis. *J. Physiol.* **375**, 195–227.

van den Honert, C. and Stypulkowski, P. H. (1987a). Single fiber mapping of spatial excitation patterns in the electrically stimulated auditory nerve. *Hearing Res.* **29**, 195–206.

van den Honert, C. and Stypulkowski, P. H. (1987b). Temporal response patterns of single auditory nerve fibers elicited by periodic electrical stimuli. *Hearing Res.* **29**, 207–222.

Honrubia, V. and Ward, P. H. (1969). Properties of the summating potential of the guinea pig's cochlea. *J. Acoust. Soc. Am.* **45**, 1443–1450.

Houtgast, T. (1972). Psychophysical evidence for lateral inhibition in hearing. *J. Acoust. Soc. Am.* **51**, 1885–1894.

Houtgast, T. (1973). Psychophysical experiments on "tuning curves" and "two-tone inhibition". *Acustica* **29**, 168–179.

Houtgast, T. (1974). *Lateral Suppression in Hearing*. Inst TNO, Soesterberg.

Houtgast, T. (1977). Auditory-filter characteristics derived from direct-masking data and pulsation-threshold data with a rippled-noise masker. *J. Acoust. Soc. Am.* **62**, 409–415.

Hubel, D. H. and Wiesel, T. N. (1962). Receptive fields, binocular interaction and functional architecture in the cat's visual cortex. *J. Physiol.* **160**, 106–154.

Hubel, D. H. and Wiesel, T. N. (1963). Shape and arrangement of columns in cat's striate cortex. *J. Physiol.* **165**, 559–568.

Hubel, D. H., Henson, C. O., Rupert, A. and Galambos, R. (1959). "Attention" units in the auditory cortex. *Science* **129**, 1279–1280.

Hudspeth, A. J. (1982). Extracellular current flow and the site of transduction by vertebrate hair cells. *J. Neurosci.* **2**, 1–10.

Hudspeth, A. J. (1985). The cellular basis of hearing: the biophysics of hair cells. *Science* **230**, 745–752.

Hudspeth, A. J. and Corey, D. P. (1977). Sensitivity, polarity, and conductance change in the response of vertebrate hair cells to controlled mechanical stimuli. *Proc. Natl Acad. Sci. USA* **74**, 2407–2411.

Hudspeth, A. J. and Jacobs, R. (1979). Stereocilia mediate transduction in vertebrate hair cells. *Proc. Natl Acad. Sci. USA* **76**, 1506–1509.

Imig, T. J. and Adrian, H. O. (1977). Binaural columns in the primary field (AI) of cat auditory cortex. *Brain Res.* **138**, 241–257.

Imig, T. J. and Brugge, J. F. (1978). Sources and terminations of callosal axons related to binaural and frequency maps in primary auditory cortex of the cat. *J. Comp. Neurol.* **182**, 637–660.

Imig, T. J., Ruggero, M. A., Kitzes, L. M., Javel, E. and Brugge, J. F. (1977). Organization of auditory cortex in the owl monkey *(Aotus trivirgatus)*. *J. Comp. Neurol.* **171**, 111–128.

Irvine, D. R. F. (1986). The auditory brainstem. *Prog. Sens. Physiol.* **7**, 1–279.

Itoh, M. (1982). Preservation and visualization of actin-containing filaments in the apical zone of cochlear sensory cells. *Hearing Res.* **6**, 277–289.

Jane, J. A., Masterton, R. B. and Diamond, I. T. (1965). The function of the tectum for attention to auditory stimuli in the cat. *J. Comp. Neurol.* **125**, 165–192.

Javel, E. (1981). Suppression of auditory nerve responses. I. Temporal analysis, intensity effects and suppression contours. *J. Acoust. Soc. Am.* **69**, 1735–1745.

Javel, E. (1986). Basic response properties of auditory nerve fibers. In *Neurobiology of Hearing: The Cochlea* (eds R. A. Altschuler, R. P. Bobbin and D. W. Hoffman), pp. 213–245. Raven Press, New York.

Javel, E., McGee, J., Walsh, E. J., Farley, G. R. and Gorga, M. P. (1983). Suppression of auditory nerve responses II: Suppression threshold and growth, iso-suppression contours. *J. Acoust. Soc. Am.* **74**, 801–813.

Jenkins, W. M. and Masterton, R. B. (1982). Sound localization: effects of unilateral lesions in central auditory system. *J. Neurophysiol.* **47**, 987–1016.

Jenkins, W. M. and Merzenich, M. M. (1984). Role of cat primary auditory cortex for sound-localization behavior. *J. Neurophysiol.* **52**, 819–847.

Jerger, J., Weikers, N. J., Sharbrough, F. W. and Jerger, S. (1969). Bilateral lesions of the temporal lobe. *Acta Otolaryngol.* (Suppl. **258**), 1–51.

Johnstone, B. M. and Boyle, A. J. F. (1967). Basilar membrane vibration examined with the Mössbauer technique. *Science* **158**, 389–390.

Johnstone, B. M. and Sellick, P. M. (1972). The peripheral auditory apparatus. *Quart. Revs Biophys.* **5**, 1–57.

Johnstone, B. M. and Yates, G. K. (1974). Basilar membrane tuning curves in the guinea pig. *J. Acoust. Soc. Am.* **55**, 584–587.

Johnstone, B. M., Johnstone, J. R. and Pugsley, I. D. (1966). Membrane resistance in endolymphatic walls of the first turn of the guinea pig cochlea. *J. Acoust. Soc. Am.* **40**, 1398–1404

Johnstone, B. M., Patuzzi, R. and Yates, G. K. (1986). Basilar membrane measurements and the travelling wave. *Hearing Res.* **22**, 147–153.

Kaas, J., Axelrod, S. and Diamond, I. T. (1967). An ablation study of the auditory cortex in the cat using binaural tonal patterns. *J. Neurophysiol.* **30**, 710–724.

Kachar B., Brownell, W. E., Altschuler, R. and Fex, J. (1986). Electrokinetic changes of cochlear outer hair cells. *Nature* **322**, 365–368.

Kane, E. C. (1973). Octopus cells in the cochlear nucleus of the cat: heterotypic synapses on homeotypic neurons. *Int. J. Neurosci.* **5**, 251–279.

Karaseva, T. A. (1972). The role of the temporal lobe in human auditory perception. *Neuropsychologia* **10**, 227–231.

Katsuki, Y., Watanabe, T. and Maruyama, N. (1959). Activity of auditory neurons in upper levels of brain of cat. *J. Neurophysiol.* **22**, 343–359.

Katsuki, Y., Suga, N. and Kanno, Y. (1962). Neural mechanism of the peripheral and central auditory system in monkeys. *J. Acoust. Soc. Am.* **34**, 1396–1410.

Kemp, D. T. (1978). Stimulated acoustic emissions from within the human auditory system. *J. Acoust. Soc. Am.* **64**, 1386–1391.

Kesner, R. P. (1966) Subcortical mechanisms of audiogenic seizure. *Exp. Neurol.* **15**, 192–205.

Kessel, R. G. and Kardon, R. H. (1979). *Tissues and Organs.* W. H. Freeman and Company, San Francisco.

Khanna, S. M. and Leonard, D. G. B. (1982). Basilar membrane tuning in the cat cochlea. *Science* **215**, 305–306.

Khanna, S. M. and Tonndorf, J. (1972). Tympanic membrane vibration in cats studied by time-averaged holography. *J. Acoust. Soc. Am.* **51**, 1904–1920.

Kiang, N. Y.-S. (1965). Stimulus coding in the auditory nerve and cochlear nucleus. *Acta Otolaryngol.* **59**, 186–200.

Kiang, N. Y.-S. (1968). A survey of recent developments in the study of auditory physiology. *Ann. Otol. Rhinol. Laryngol.* **77**, 656–675.

Kiang, N. Y.-S. (1980). Processing of speech by the auditory nervous system. *J. Acoust. Soc. Am.* **68**, 830–835.

Kiang, N. Y.-S. and Moxon, E. C. (1972). Physiological considerations in artificial stimulation of the inner ear. *Ann. Otol. Rhinol. Laryngol.* **81**, 714–730.

Kiang, N. Y.-S., Watanabe, T., Thomas, E. C. and Clark, L. F. (1965). Discharge Patterns of Single Fibers in the Cat's Auditory Nerve *(Res. Monogr. no. 35).* M. I. T. Press, Cambridge.

Kiang, N. Y.-S., Sachs, M. B. and Peake, W. T. (1967). Shapes of tuning curves for single auditory-nerve fibers. *J. Acoust. Soc. Am.* **42**, 1341–1342.

Kiang, N. Y.-S., Moxon, E. C. and Levine, R. A. (1970). Auditory-nerve activity in cats with normal and abnormal cochleas. In *Sensorineural Hearing Loss* (eds G. E. W. Wolstenholme and J. Knight), pp. 241–268. CIBA Foundation Symposium. Churchill, London.

Kim, D. O., Molnar, C. E. and Matthews, J. W. (1980). Cochlear mechanics: nonlinear behavior in two-tone responses as reflected in cochlear-nerve-fiber responses and in ear-canal sound pressure. *J. Acoust. Soc. Am.* **67**, 1704–1721.

Kimura, D. (1967a). Functional asymmetry of the brain in dichotic listening. *Cortex* **3**, 163–178.

Kimura, R. S. (1967b). Experimental blockage of the endolymphatic duct and sac and its effect on the inner ear of the guinea pig. *Ann. Otol. Rhinol. Laryngol.* **76**, 664–687.

King, A. J. and Palmer, A. R. (1983). Cells responsive to free-field auditory stimuli in guinea-pig superior colliculus: distribution and response properties. *J. Physiol.* **342**, 361–381.

Kitzes, L. M., Wrege, K. S. and Cassady, J. M. (1980). Patterns of responses of cortical cells to binaural stimulation. *J. Comp. Neurol.* **192**, 455–472.

Klinke, R. (1986). Neurotransmission in the inner ear. *Hearing Res.* **22**, 235–243.

Klinke, R. and Galley, N. (1974). Efferent innervation of vestibular and auditory receptors. *Physiol. Revs* **54**, 316–357.

Klinke, R., Boerger, G. and Gruber, J. (1969). Studies on the functional significance of efferent innervation in the auditory system: afferent neuronal activity as influenced by contralaterally-applied sound. *Pflüger's Arch. Ges. Physiol.* **306**, 165–175.

Knight, P. L. (1977). Representation of the cochlea within the anterior field (AAF) of the cat. *Brain Res.* **130**, 447–467.

Knudsen, E. I. and Konishi, M. (1978). A neural map of auditory space in the owl. *Science* **200**, 795–797.

Kohllöffel, L. U. E. (1972). A study of basilar membrane vibrations. II. The vibratory amplitude and phase pattern along the basilar membrane (post mortem). *Acustica* **27**, 66–81.

Konishi, T. and Yasuno, T. (1963). Summating potential of the cochlea in the guinea pig. *J. Acoust. Soc. Am.* **35**, 1448–1452.

Kronester-Frei, A. (1979). Localization of the marginal zone of the tectorial membrane *in situ,* unfixed and *in vivo*-like ionic milieu. *Arch. Otol. Rhinol. Laryngol.* **224**, 3–9.

Kryter, K. D. and Ades, H. W. (1943). Studies on the function of the higher acoustic centers in the cat. *Am. J. Psychol.* **56**, 501–536.

Kuijpers, W. and Bonting, S. L. (1969). Studies on the (Na^+-K^+)-activated ATPase. XXIV. Localization and properties of ATPase in the inner ear of the guinea pig. *Biochim. Biophys. Acta.* **173**, 477–485.

Kuijpers, W. and Bonting, S. L. (1970). The cochlear potentials. I. The effect of ouabain on the cochlear potentials of the guinea pig. *Pflüger's Arch. Ges. Physiol.* **320**, 348–358.

Lassen, N. A., Ingvar, D. H. and Skinhøj, E. (1978). Brain function and blood flow. *Scientific Am.* **239**, (4), 50–59.

Lauter, J. L., Herscovitch, P., Formby, C. and Raichle, M. E. (1985). Tonotopic organization in human auditory cortex revealed by positron emission tomography. *Hearing Res.* **20**, 199–205.

LePage, E. L. (1987). Frequency-dependent self-induced bias of the basilar membrane and its potential for controlling sensitivity and tuning in the mammalian cochlea. *J. Acoust. Soc. Am.* **82**, 139–154.

Liberman, A. M., Cooper, F. S., Shankweiler, D. P. and Studdert-Kennedy, M. (1967). Perception of the speech code. *Psychol. Rev.* **74**, 431–461.

Liberman, M. C. (1978). Auditory-nerve responses from cats raised in a low-noise chamber. *J. Acoust. Soc. Am.* **63**, 442–455.

Liberman, M. C. (1987). Chronic ultrastructural changes in acoustic trauma: serial-section reconstruction of stereocilia and cuticular plates. *Hearing Res.* **26**, 65–88.

Liberman, M. C. and Brown, M. C. (1986). Physiology and anatomy of single olivocochlear neurons in the cat. *Hearing Res.* **24**, 17–36.

Liberman, M. C. and Dodds, L. W. (1984a). Single neuron labeling and chronic cochlear pathology. II. Stereocilia damage and alterations of spontaneous discharge rates. *Hearing Res.* **16**, 43–53.

Liberman, M. C. and Dodds, L. W. (1984b). Single neuron labeling and chronic cochlear pathology. III. Stereocilia damage and alterations of threshold tuning curves. *Hearing Res.* **16**, 55–74.

Liberman, M. C. and Dodds, L. W. (1987). Acute ultrastructural changes in acoustic trauma: serial-section reconstruction of stereocilia and cuticular plates. *Hearing Res.* **26**, 45–64.

Liberman, M. C. and Kiang, N. Y.-S. (1978). Acoustic trauma in cats. *Acta Otolaryngol.* (Suppl. **358**), 1–63.

Liberman, M. C. and Kiang, N. Y.-S. (1984). Single neuron labeling and chronic cochlear pathology. IV. Stereocilia damage and alterations in rate- and phase-level functions. *Hearing Res.* **16**, 75–90.

Lim, D. J. (1980). Cochlear anatomy related to cochlear micromechanics. A review. *J. Acoust. Soc. Am.* **67**, 1686–1695.

Lim, D. J. (1986). Functional structure of the organ of Corti: a review. *Hearing Res.* **22**, 117–146.

Lowenstein, O. and Wersäll, J. (1959). A functional interpretation of the electron-microscopic structure of sensory hairs in the cristae of the elasmobranch *Raja clavata* in terms of directional sensitivity. *Nature* **184**, 1807–1808.

Lumer, G. (1987a). Computer model of cochlear preprocessing (steady state condition). I. Basics and results for one sinusoidal input signal. *Acustica* **62**, 282–290.

Lumer, G. (1987b). Computer model of cochlear preprocessing (steady state condition). II. Two-tone suppression. *Acustica* **63**, 17–25.

Lynch, T. J., Nedzelnitsky, V. and Peake, W. T. (1982). Input impedance of the cochlea in cat. *J. Acoust. Soc. Am.* **72**, 108–130.

Macartney, J. C., Comis, S. D. and Pickles, J. O. (1980). Is myosin in the cochlea a basis for active motility? *Nature* **288**, 491–492.

Maffi, C. L. and Aitkin, L. M. (1987). Differential neural projections to regions of the inferior colliculus of the cat responsive to high-frequency sounds. *J. Neurophysiol.* **26**, 211–219.

Majorossy, K. and Kiss, A. (1976). Specific patterns of neuron arrangement and of synaptic articulation in the medial geniculate body. *Exp. Brain Res.* **26**, 1–17.

Manley, G. (1983). Auditory nerve fibre activity in mammals. In *Bioacoustics* (ed. B. Lewis), pp. 207–232. Academic Press, London.

Manley, G. A. and Kronester-Frei, A. (1980). The electrophysiological profile of the organ of Corti. In *Psychophysical, Physiological and Behavioural Studies in Hearing* (eds G. van den Brink and F. A. Bilsen, pp. 24–31. Delft University Press, Delft.

Manley, J. A. and Müller-Preuss, P. (1978). Response variability of auditory cortex cells in the squirrel monkey to constant acoustic stimuli. *Exp. Brain Res.* **32**, 171–180.

Martin, M. R. (1985). The pharmacology of amino acid receptors and synaptic transmission in the cochlear nucleus. In *Auditory Biochemistry* (ed. D. G. Drescher), pp. 184–197. Charles C. Thomas, Springfield.

Massopust, L. C. and Ordy, J. M. (1962). Auditory organization of the inferior colliculi in the cat. *Exp. Neurol.* **6**, 465–477.

Mast, T. E. (1973). Dorsal cochlear nucleus of the chinchilla: excitation by contra-lateral sound. *Brain Res.* **62**, 61–70.

Masterton, R. B. and Diamond, I. T. (1964). Effects of auditory cortex ablation on discrimination of small binaural time differences. *J. Neurophysiol.* **27**, 15–36.

Masterton, R. B. and Diamond, I. T. (1967). The medial superior olive and sound localization. *Science* **155**, 1696–1697.

Masterton, R. B. and Imig, T. J. (1984). Neural mechanisms for sound localization. *Ann. Rev. Physiol.* **46**, 275–287.

Mees, K. (1983). Ultrastructural localization of K^+ dependent, ouabain-sensitive NPPase (Na-K-ATPase) in the guinea pig inner ear. *Acta Otolaryngol.* **95**, 277–289.

Melichar, I. and Syka, J. (1987). Electrophysiological measurements of the stria vascularis potentials in vivo. *Hearing Res.* **25**, 35–43.

Mendelson, J. R. and Cynader, M. S. (1985). Sensitivity of cat primary auditory cortex (AI) neurons to the direction and rate of frequency modulation. *Brain Res.* **327**, 331–335.

Merzenich, M. M. and Brugge, J. F. (1973). Representation of the cochlear partition on the superior temporal plane of the macaque monkey. *Brain Res.* **50**, 275–296.

Merzenich, M. M., Knight, P. L. and Roth, G. L. (1975). Representation of cochlea within primary auditory cortex in the cat. *J. Neurophysiol.* **38**, 231–249.

Merzenich, M. M., Roth, G. L., Andersen, R. A., Knight, P. L. and Colwell, S. A. (1977). Some basic features of organization of the central auditory nervous system. In *Psychophysics and Physiology of Hearing* (eds E. F. Evans and J. P. Wilson), pp. 485–497. Academic Press, London.

Meyer, D. R. and Woolsey, C. N. (1952). Effects of localized cortical destruction on auditory discriminative conditioning in cat. *J. Neurophysiol.* **15**, 149–162.

Middlebrooks, J. C. and Knudsen, E. I. (1987). Changes in external ear position modify the spatial tuning of auditory units in the cat's superior colliculus. *J. Neurophysiol.* **57**, 672–686.

Middlebrooks, J. C. and Pettigrew, J. D. (1981). Functional classes of neurons in primary auditory cortex of the cat distinguished by sensitivity to sound location. *J. Neurosci.* **1**, 107–120.

Middlebrooks, J. C., Dykes, R. W. and Merzenich, M. M. (1980). Binaural response-specific bands in primary auditory cortex (AI) of the cat: topographical organization orthogonal to isofrequency contours. *Brain Res.* **181**, 31–48.

Mitani, A., Shimokouchi, M., Itoh, K., Nomura, S., Kudo, M. and Mizuno, N.

(1985). Morphology and laminar organization of electrophysiologically-identified neurons in the primary auditory cortex of the cat. *J. Comp. Neurol.* **235**, 430–447.

Moiseff, A. and Konishi, M. (1983). Binaural characteristics of units in the owl's brainstem auditory pathway: precursors of restricted spatial receptive fields. *J. Neurosci.* **3**, 2553–2562.

Møller, A. R. (1965). An experimental study of the acoustic impedance of the middle ear and its transmission properties. *Acta Otolaryngol.* **60**, 129–149.

Møller, A. R. (1974). Function of the middle ear. In *Handbook of Sensory Physiology* (eds W. D. Keidel and W. D. Neff), Vol. 5/1, pp. 491–517. Springer, Berlin.

Møller, A. R. (1976a). Dynamic properties of primary auditory fibers compared with cells in the dorsal cochlear nucleus. *Acta Physiol. Scand.* **98**, 157–167.

Møller, A. R. (1976b). Dynamic properties of the responses of single neurones in the cochlear nucleus of the rat. *J. Physiol.* **259**, 63–82.

Møller, A. R. (1977). Frequency selectivity of single auditory nerve fibers in response to broadband noise stimuli. *J. Acoust. Soc. Am.* **62**, 135–142.

Møller, A. R. (1978). Coding of time-varying sounds in the cochlear nucleus. *Audiology* **17**, 446–468.

Moore, B. C. J. (1973). Frequency difference limens for short-duration tones. *J. Acoust. Soc. Am.* **54**, 610–619.

Moore, B. C. J. (1975). Mechanisms of masking. *J. Acoust. Soc. Am.* **57**, 391–399.

Moore, B. C. J. (1982). *An Introduction to the Psychology of Hearing.* Academic Press, London.

Moore, B. C. J. (ed.) (1986a). *Frequency Selectivity in Hearing.* Academic Press, London.

Moore, B. C. J. and Glasberg, B. R. (1982a). Contralateral and ipsilateral cueing in forward masking. *J. Acoust. Soc. Am.* **71**, 942–945.

Moore, B. C. J. and Glasberg, B. R. (1982b). Interpreting the role of suppression in psychophysical tuning curves. *J. Acoust. Soc. Am.* **72**, 1374–1379.

Moore, B. C. J. and Glasberg, B. R. (1986). The role of frequency selectivity in the perception of loudness, pitch and time. In *Frequency Selectivity in Hearing* (ed. B. C. J. Moore), pp. 251–308. Academic Press, London.

Moore, B. C. J., Glasberg, B. R. and Roberts, B. (1984). Refining the measurement of psychophysical tuning curves. *J. Acoust. Soc. Am.* **76**, 1057–1066.

Moore, C. N., Casseday, J. H. and Neff, W. D. (1974). Sound localization: the role of the commissural pathways of the auditory system of the cat. *Brain Res.* **82**, 13–26.

Moore, J. K. (1986b). Cochlear nuclei: relationship to the auditory nerve. In *Neurobiology of Hearing: The Cochlea* (eds R. A. Altschuler, R. Bobbin and D. W. Hoffman), pp. 283–301. Raven Press, New York.

Moore, J. K. (1987). The human auditory brain stem: a comparative view. *Hearing Res.* **29**, 1–32.

Moore, T. J. and Cashin, J. L. (1974). Response patterns of cochlear nucleus neurons to excerpts from sustained vowels. *J. Acoust. Soc. Am.* **56**, 1565–1576.

Morest, D. K. (1964). The neuronal architecture of the medial geniculate body of the cat. *J. Anat.* **98**, 611–630.

Morest, D. K. (1965). The laminar structure of the medial geniculate body of the cat. *J. Anat.* **99**, 143–160.

Morest, D. K. (1975). Synaptic relationships of Golgi type II cells in the medial geniculate body of the cat. *J. Comp. Neurol.* **162**, 157–194.

Morest, D. K. and Oliver, D. L. (1984) The neuronal architecture of the inferior colliculus in the cat: defining the functional anatomy of the auditory midbrain. *J. Comp. Neurol.* **222**, 209–236.

Morest, D. K., Kiang, N. Y.-S., Kane, E. C., Guinan, J. J. and Godfrey, D. A. (1973). Stimulus coding at caudal levels of the cat's auditory nervous system: II. Patterns of synaptic organization. In *Basic Mechanisms in Hearing* (ed. A. Møller), pp. 479–504. Academic Press, New York and London.

Morrison, D., Schindler R. A. and Wersäll, J. (1975). A quantitative analysis of the afferent innervation of the organ of Corti in guinea pig. *Acta Otolaryngol.* **79**, 11–23.

Mountain, D. C. (1980). Changes in endolymphatic potential and crossed olivocochlear bundle stimulation alter cochlear mechanics. *Science* **210**, 71–72.

Mountain, D. C. (1986). Electromechanical properties of hair cells. In *Neurobiology of Hearing: The Cochlea,* (eds R. A. Altschuler, R. P. Bobbin and D. W. Hoffman), pp. 77–90. Raven Press, New York.

Mountcastle, V. B. (1957). Modality and topographic properties of single neurons of cat's somatic sensory cortex. *J. Neurophysiol.* **20**, 408–434.

Moushegian, G., Rupert, A. L. and Gidda, J. S. (1975). Functional characteristics of superior olivary neurones to binaural stimuli. *J. Neurophysiol.* **38**, 1037–1048.

Nedzelnitsky, V. (1980). Sound pressures in the basal turn of the cat cochlea. *J. Acoust. Soc. Am.* **68**, 1676–1689.

Neely, S. T. (1981). Finite difference solution of a two-dimensional mathematical model of the cochlea. *J. Acoust. Soc. Am.* **69**, 1386–1393.

Neely, S. T. and Kim, D. O. (1983). An active cochlear model showing sharp tuning and high sensitivity. *Hearing Res.* **9**, 123–130.

Neely, S. T. and Kim, D. O. (1986). A model for active elements in cochlear biomechanics. *J. Acoust. Soc. Am.* **79**, 1472–1480.

Neff, W. D. (1961). Neural mechanisms of auditory discrimination. In *Sensory Communication* (ed. W. A. Rosenblith), pp. 259–278. Wiley, New York.

Neff, W. D. (1968). Localization and lateralization of sound in space. In *Hearing Mechanisms in Vertebrates* (eds A. V. S. de Reuck and J. Knight), pp. 207–231. CIBA Foundation Symposium. Churchill, London.

Neff, W. D., Diamond, I. T. and Casseday, J. H. (1975). Behavioral studies of auditory discrimination: central nervous system. In *Handbook of Sensory Physiology* (eds W. D. Keidel and W. D. Neff), Vol. 5/2, pp. 307–400. Springer, Berlin.

Nelson, D. A. and Turner, C. W. (1980). Decay of masking and frequency resolution in sensorineural hearing-impaired listeners. In *Psychophysical, Physiological and Behavioural Studies in Hearing* (eds G. van den Brink and F. A. Bilsen), pp. 175–182. Delft University Press, Delft.

Nelson, P. G., Erulkar, S. D. and Bryan, J. S. (1966). Responses of units of the inferior colliculus to time-varying acoustic stimuli. *J. Neurophysiol.* **29**, 834–860.

Nielsen, D. W. and Slepecky, N. (1986). Stereocilia. In *Neurobiology of Hearing: The Cochlea* (eds R. A. Altschuler, R. P. Bobbin and D. W. Hoffman), pp. 23–46. Raven Press, New York.

Niimi, K. and Matsuoka, H. (1979). Thalamocortical organization of the auditory system in the cat studied by retrograde axonal transport of horseradish peroxidase. *Adv. Anat. Embryol. Cell. Biol.* **57**, 1–56.

Nishizawa, Y., Olsen, T. S., Larsen, B. and Lassen, N. (1982). Left–right cortical asymmetries of regional cerebral blood flow during listening to words. *J. Neurophysiol.* **48**, 458–466.

Nomoto, M., Suga, N. and Katsuki, Y. (1964). Discharge pattern and inhibition of primary auditory nerve fibers in the monkey. *J. Neurophysiol.* **27**, 768–787.

Nuttall, A. L. (1986). Physiology of hair cells. In *Neurobiology of Hearing: The Cochlea* (eds R. A. Altschuler, R. P. Bobbin and D. W. Hoffman), pp. 47–75. Raven Press, New York.

Nuttall, A. L., Brown, M. C., Masta, R. I. and Lawrence, M. (1981). Inner hair cell responses to the velocity of basilar membrane motion in the guinea pig. *Brain Res.* **211**, 171–174.

Oatman, L. C. (1976). Effects of visual attention on the intensity of auditory evoked potentials. *Exp. Neurol.* **51**, 41–53.

Offner, D. L., Dallos, P. and Cheatham, M. A. (1987). Positive endocochlear potential: mechanism of production by marginal cells of the stria vascularis. *Hearing Res.* **29**, 117–124.

Ohmori, H, (1985). Mechano-electrical transduction currents in isolated vestibular hair cells of the chick. *J. Physiol.* **359**, 189–217.

Oliver, D. L. and Morest, D. K. (1984). The central nucleus of the inferior colliculus in the cat. *J. Comp. Neurol.* **222**, 237–264.

Oonishi, S. and Katsuki, Y. (1965). Functional organization and integrative mechanism on the auditory cortex of the cat. *Jap. J. Physiol.* **15**, 342–365.

Osborne, M. P., Comis, S. D. and Pickles, J. O. (1988). Further observations on the fine structure of tip links between stereocilia in the guinea pig cochlea. *Hearing Res.* (in press).

Osen, K. K. (1969). Cytoarchitecture of the cochlear nuclei in the cat. *J. Comp. Neurol.* **136**, 453–483.

Osen, K. K. and Roth, K. (1969). Histochemical localization of cholinesterases in the cochlear nuclei of the cat, with notes on the origin of acetylcholinesterase-positive afferents and the superior olive. *Brain Res.* **16**, 165–185.

Otte, J., Schuknecht, H. F. and Kerr, A. G. (1978). Ganglion cell populations in normal and pathological human cochleae. Implications for cochlear implantation. *Laryngoscope* **88**, 1231–1246.

Palmer, A. R. and Evans, E. F. (1980). Cochlear fibre rate-intensity functions: no evidence for basilar membrane nonlinearities. *Hearing Res.* **2**, 319–326.

Palmer, A. R. and Evans, E. F. (1982). Intensity coding in the auditory periphery of the cat: Response of cochlear nerve and cochlear nucleus neurons to signals in the presence of bandstop masking noise. *Hearing Res.* **7**, 305–323.

Palmer, A. R. and Russell, I. J. (1986). Phase-locking in the cochlear nerve of the guinea-pig and its relation to the receptor potential of inner hair-cells. *Hearing Res.* **24**, 1–15.

Pang, X. D. and Peake, W. T. (1986). How do contractions of the stapedius muscle alter the acoustic properties of the ear? In *Peripheral Auditory Mechanisms* (eds J. B. Allen, J. L. Hall, A. Hubbard, S. T. Neely and A. Tubis), pp. 36–43. Springer, Berlin.

Papçun, G., Krashen, S., Terbeek, D., Remington, R. and Harshman, R. (1974). Is the left hemisphere specialised for speech, language, and/or something else? *J. Acoust. Soc. Am.* **55**, 319–327.

Patrick, J. F., Crosby, P. A., Hirshorn, M. S., Kuzma, J. A., Money, D. K., Ridler, J. and Seligman, P. M. (1985). Australian multi-channel implantable hearing prosthesis. In *Cochlear Implants* (eds R. A. Schindler and M. M. Merzenich), pp. 93–100. Raven Press, New York.

Patterson, R. D. and Moore, B. C. J. (1986). Auditory filters and excitation patterns as representations of frequency resolution. In *Frequency Selectivity in Hearing* (ed. B. C. J. Moore), pp. 123–177. Academic Press, London.

Patterson, R. D., Nimmo-Smith, I., Weber, D. L. and Milroy, R. (1982). The deterioration of hearing with age: frequency selectivity, the critical ratio, the audiogram, and speech threshold. *J. Acoust. Soc. Am.* **72**, 1788–1803.

Patuzzi, R. and Sellick, P. M. (1984). The modulation of the sensitivity of the mammalian cochlea by low-frequency tones. II. Inner hair cell receptor potentials. *Hearing Res.* **13**, 9–18.

Patuzzi, R., Sellick, P. M. and Johnstone, B. M. (1984a). The modulation of the sensitivity of the mammalian cochlea by low-frequency tones. I. Primary afferent activity. *Hearing Res.* **13**, 1–8.

Patuzzi, R. Sellick, P. M. and Johnstone, B. M. (1984b). The modulation of the sensitivity of the mammalian cochlea by low-frequency tones. III. Basilar membrane motion. *Hearing Res.* **13**, 19–27.

Peake, W. T., Sohmer, H. S. and Weiss, T. F. (1969). Microelectrode recordings of intracochlear potentials. In *Quarterly Progress Report, M.I.T. Research Laboratory of Electronics* **94**, 293–304.

Peyret, D., Campistron, G., Geffard, M. and Aran, J. M. (1987). Glycine immunore-activity in the brainstem auditory and vestibular nuclei of the guinea pig. *Acta Otolaryngol.* **104**, 71–76.

Pfeiffer, R. R. (1966a). Classification of response patterns of spike discharges for units in the cochlear nucleus: tone-burst stimulation. *Exp. Brain Res.* **1**, 220–235.

Pfeiffer, R. R. (1966b). Anteroventral cochlear nucleus: wave forms of extracellularly recorded spike potentials. *Science* **154**, 667–668.

Pfeiffer, R. R. and Kim, D. O. (1973). Considerations of nonlinear response properties of single cochlear nerve fibers. In *Basic Mechanisms in Hearing* (ed. A. Møller), pp. 555–587. Academic Press, New York and London.

Pfingst, B. E., O'Connor, T. A. and Miller, J. M. (1977). Response plasticity of neurons in auditory cortex of the rhesus monkey. *Exp. Brain Res.* **29**, 393–404.

Phillips, D. P. (1987). Stimulus intensity and loudness recruitment: neural correlates. *J. Acoust. Soc. Am.* **82**, 1–12.

Phillips, D. P. and Brugge, J. F. (1985). Progress in neurophysiology of sound localization. *Ann. Rev. Psychol.* **36**, 245–274.

Phillips, D. P. and Irvine, D. R. F. (1983). Some features of binaural input to single neurons in physiologically-defined area AI of cat cerebral cortex. *J. Neurophysiol.* **49**, 383–395.

Phillips, D. P. and Orman, S. S. (1984). Responses of single neurons in posterior field of cat auditory cortex to tonal stimulation. *J. Neurophysiol.* **51**, 147–163.

Phillips, D. P., Calford, M. B., Pettigrew, J. D., Aitkin, L. M. and Semple, M. N. (1982). Directionality of sound pressure transformation at the cat's pinna. *Hearing Res.* **8**, 13–28.

Pick. G. F., Evans, E. F. and Wilson, J. P. (1977). Frequency resolution in patients with hearing loss of cochlear origin. In *Psychophysics and Physiology of Hearing* (eds E. F. Evans and J. P. Wilson), pp. 273–281. Academic Press, London.

Pickles, J. O. (1975). Normal critical bands in the cat. *Acta Otolaryngol.* **80**, 245–254.

Pickles, J. O. (1976a). Role of centrifugal pathways to cochlear nucleus in determin-ation of critical bandwidth. *J. Neurophysiol.* **39**, 394–400.

Pickles, J. O. (1976b). The noradrenaline-containing innervation of the cochlear nucleus and the detection of signals in noise. *Brain Res.* **105**, 591–596.

Pickles, J. O. (1979a). Psychophysical frequency resolution in the cat as determined by simultaneous masking and its relation to auditory-nerve resolution. *J. Acoust. Soc. Am.* **66**, 1725–1732.

Pickles, J. O. (1979b). An investigation of sympathetic effects on hearing. *Acta Otolaryngol.* **87**, 69–71.

Pickles, J. O. (1980). Psychophysical frequency resolution in the cat studied with forward masking. In *Psychophysical, Physiological and Behavioural Studies in Hearing* (eds G. van den Brink and F. A. Bilsen), pp. 118–126. Delft University Press, Delft.

Pickles, J. O. (1983). Auditory-nerve correlates of loudness summation with stimulus bandwidth, in normal and pathological cochleae. *Hearing Res.* **12**, 239–250.

Pickles, J. O. (1984). Frequency threshold curves and simultaneous masking functions in single fibres of the guinea pig auditory nerve. *Hearing Res.* **14**, 245–256.

Pickles, J. O. (1985a). Recent advances in cochlear physiology. *Prog. Neurobiol.* **24**, 1–42.

Pickles, J. O. (1985b). Hearing and listening. In *Scientific Basis of Clinical Neurology* (eds M. Swash and C. Kennard), pp. 188–200. Churchill, Edinburgh.

Pickles, J. O. (1985c). Physiology of the cerebral auditory system. In *Assessment of Central Auditory Dysfunction* (eds M. L. Pinheiro and F. E. Musiek), pp. 67–85. Williams and Wilkins, Baltimore.

Pickles, J. O. (1986a). The neurophysiological basis of frequency selectivity. In *Frequency Selectivity in Hearing* (ed. B. C. J. Moore), pp. 51–121. Academic Press, London.

Pickles, J. O. (1986b). Auditory-nerve fibre bandwidths determined by two different simultaneous masking procedures. In *Auditory Frequency Selectivity* (eds B. C. J. Moore and R. D. Patterson), pp. 171–178. Plenum, New York.

Pickles, J. O. (1987). The physiology of the ear. In *Scott-Brown's Otolaryngology 5th edn.* Vol. 1. Basic Sciences (ed. D. Wright), pp. 47–80. Butterworths Scientific, London.

Pickles, J. O. and Comis, S. D. (1973). Role of centrifugal pathways to cochlear nucleus in detection of signals in noise. *J. Neurophysiol.* **36**, 1131–1137.

Pickles, J. O. and Comis, S. D. (1976). Auditory-nerve fiber bandwidths and critical bandwidths in the cat. *J. Acoust. Soc. Am.* **60**, 1151–1156.

Pickles, J. O., Comis, S. D. and Osborne, M. P. (1984). Cross-links between stereocilia in the guinea pig organ of Corti, and their possible relation to sensory transduction. *Hearing Res.* **15**, 103–112.

Pickles, J. O., Osborne, M. P. and Comis, S. D. (1987a). The vulnerability of tip links between stereocilia to acoustic trauma in the guinea pig. *Hearing Res.* **25**, 173–183.

Pickles, J. O., Comis, S. D. and Osborne, M. P. (1987b). The effect of chronic application of kanamycin on stereocilia and their tip links in hair cells of the guinea pig cochlea. *Hearing Res.* **29**, 237–244.

Pickles, J. O., Brix, J., Gleich, O., Köppl, C., Manley, G. A., Osborne, M. P. and Comis, S. D. (1988). The fine structure and organization of tip links on hair cell stereovilli. In *Basic Issues In Hearing* (eds H. Duifhuis, J. W. Horst and H. P. Wit), pp. 56–63. Academic Press, London.

Pinheiro, M. L. and Musiek, F. E. (1985). *Assessment of Central Auditory Dysfunction: Foundations and Clinical Correlates.* Williams and Wilkins, Baltimore.

Plomp, R. (1967). Pitch of complex tones. *J. Acoust. Soc. Am.* **41**, 1526–1533.

Plomp, R. (1976). *Aspects of Tone Sensation.* Academic Press, London.

References   343

Rabie. A., Thomasset, M. and Legrand, Ch. (1983). Immunocytochemical detection of calcium-binding protein in the cochlear and vestibular hair cells of the rat. *Cell Tissue Res.* **232**, 691–696.

Rajan, R. and Johnstone, B. M. (1983). Crossed cochlear influences on monaural temporary threshold shifts. *Hearing Res.* **9**, 279–294.

Rasmussen, G. L. (1946). The olivary peduncle and other fiber projections to the superior olivary complex. *J. Comp. Neurol.* **84**, 141–220.

Rasmussen, G. L. (1964). Anatomic relationships of the ascending and descending auditory systems. In *Neurological Aspects of Auditory and Vestibular Disorders* (eds W. S. Fields and B. R. Alford), pp. 1–19. Thomas, Springfield.

Rassmusen, G. L. (1967). Efferent connections of the cochlear nucleus. In *Sensorineural Hearing Processes and Disorders* (ed. A. B. Graham), pp. 61–75. Little Brown, Boston.

Ravizza, R. J. and Belmore, S. M. (1978). Auditory forebrain: evidence from anatomical and behavioral experiments involving human and animal subjects. In *Handbook of Behavioral Neurobiology* (ed. R. B. Masterton), pp. 459–501. Plenum Press, New York.

Ravizza, R. J. and Masterton, R. B. (1972). Contribution of neocortex to sound localization in opossum (*Didelphis virginiana*). *J. Neurophysiol.* **35**, 344–356.

Reale, R. A. and Imig, T. J. (1980). Tonotopic organization in auditory cortex of the cat. *J. Comp. Neurol.* **192**, 265–291.

Reale, R. A. and Kettner, R. E. (1986). Topography of binaural organization in primary auditory cortex of the cat: effects of changing interaural intensity. *J. Neurophysiol.* **56**, 663–682.

Rhode, W. S. (1971). Observations of the vibration of the basilar membrane in squirrel monkeys using the Mössbauer technique. *J. Acoust. Soc. Am.* **49**, 1218–1231.

Rhode, W. S. (1978). Some observations on cochlear mechanics. *J. Acoust. Soc. Am.* **64**, 158–176.

Rhode, W. S. (1980). Cochlear partition vibration – recent views. *J. Acoust. Soc. Am.* **67**, 1696–1703.

Rhode, W. S. and Smith, P. H. (1986a). Encoding timing and intensity in the ventral cochlear nucleus of the cat. *J. Neurophysiol.* **56**, 261–286.

Rhode, W. S. and Smith, P. H. (1986b). Physiological studies on neurons in the dorsal cochlear nucleus of cat. *J. Neurophysiol.* **56**, 287–307.

Rhode, W. S., Geisler, C. D. and Kennedy, D. T. (1978). Auditory nerve fiber responses to wide-band noise and tone combinations. *J. Neurophysiol.* **41**, 692–704.

Rhode, W. S., Smith, P. H. and Oertel, D. (1983a). Physiological response properties of cells labelled intracellularly with horseradish peroxidase in cat dorsal cochlear nucleus. *J. Comp. Neurol.* **213**, 426–447.

Rhode, W. S., Oertel, D. and Smith, P. H. (1983b). Physiological response properties of cells labelled intracellularly with horseradish peroxidase in cat ventral cochlear nucleus. *J. Comp. Neurol.* **213**, 448–463.

Ritsma, R. J. (1967). Frequencies dominant in the perception of the pitch of complex sounds. *J. Acoust. Soc. Am.* **42**, 191–198.

Ritz, L. A. and Brownell, W. E. (1982). Single unit analysis of the posteroventral nucleus of the decerebrate cat. *Neuroscience* **7**, 1995–2010.

Robards, M. J. (1979). Somatic neurons in the brain-stem and neocortex projecting to the external nucleus of the inferior colliculus: an anatomical study in the opossum. *J. Comp. Neurol.* **184**, 547–566.

Robards, M. J., Watkins, D. W. and Masterton, R. B. (1976). An anatomical study of some somesthetic afferents to the intercollicular terminal zone of the midbrain of the opossum. *J. Comp. Neurol.* **170**, 499–524.

Robertson, D. (1984). Horseradish peroxidase injection of physiologically characterised afferent and efferent neurones in the guinea pig spiral ganglion. *Hearing Res.* **15**, 113–121.

Robertson, D. and Gummer, M. (1985). Physiological and morphological characterization of efferent neurones in the guinea pig cochlea. *Hearing Res.* **20**, 63–77.

Robertson, D. and Johnstone, B. M. (1979). Aberrant tonotopic organization in the inner ear damaged by kanamycin. *J. Acoust. Soc. Am.* **66**, 466–469.

Robertson, D. and Johnstone, B. M. (1981). Primary auditory neurons: Nonlinear responses altered without changes in sharp tuning. *J. Acoust. Soc. Am.* **69**, 1096–1098.

Robertson, D., Anderson, C.-J. and Cole, K. S. (1987). Segregation of efferent projections to different turns of the guinea pig cochlea. *Hearing Res.* **25**, 69–76.

Robles, L., Ruggero, M. A. and Rich, N. C. (1986a). Basilar membrane mechanics at the base of the chinchilla cochlea. I. Input–output functions, tuning curves, and response phases. *J. Acoust. Soc. Am.* **80**, 1364–1374.

Robles, L., Ruggero, M. A. and Rich, N. C. (1986b). Mössbauer measurements of the mechanical response to single-tone and two-tone stimuli at the base of the chinchilla cochlea. In *Peripheral Auditory Mechanisms* (eds J. B. Allen, J. L. Hall, A. Hubbard, S. T. Neely and A. Tubis), pp. 121–128. Springer, Berlin.

Rockel, A. J. and Jones, E. G. (1973a). The neuronal organization of the inferior colliculus of the adult cat. I. The central nucleus. *J. Comp. Neurol.* **147**, 11–60.

Rockel, A. J. and Jones, E. G. (1973b). The neuronal organization of the inferior colliculus of the adult cat. II. The pericentral nucleus. *J. Comp. Neurol.* **149**, 301–334.

Rose, J. E. (1949). The cellular structure of the auditory region of the cat. *J. Comp. Neurol.* **91**, 409–439.

Rose, J. E. and Woolsey, C. N. (1958). Cortical connections and functional organization of the thalamic auditory system of the cat. In *Biological and Biochemical Bases of Behavior* (eds H. F. Harlow and C. N. Woolsey), pp. 127–150. University of Wisconsin Press, Madison.

Rose, J. E., Galambos, R. and Hughes, J. R. (1959). Microelectrode studies of the cochlear nuclei of the cat. *Bull. Johns Hopkins Hosp.* **104**, 211–251.

Rose, J. E., Galambos, R. and Hughes, J. (1960). Organization of frequency sensitive neurons in the cochlear nuclear complex of the cat. In *Neural Mechanisms of the Auditory and Vestibular Systems* (eds G. L. Rasmussen and W. F. Windle), pp. 116–136. Thomas, Springfield.

Rose, J. E., Greenwood, D. D., Goldberg, J. M. and Hind, J. E. (1963). Some discharge characteristics of single neurons in the inferior colliculus of the cat. I. Tonotopic organization, relation of spike counts to tone intensity, and firing patterns of single elements. *J. Neurophysiol.* **26**, 294–320.

Rose, J. E., Gross, N. B., Geisler, C. D. and Hind, J. E. (1966). Some neural mechanisms in the inferior colliculus of the cat which may be relevant to the localization of a sound source. *J. Neurophysiol.* **29**, 288–314.

Rose, J. E., Hind, J. E., Anderson, D. J. and Brugge, J. F. (1971). Some effects of stimulus intensity on response of auditory nerve fibers in the squirrel monkey. *J. Neurophysiol.* **34**, 685–699.

Rosowski, J. J., Carney, L. H., Lynch, T. J. and Peake, W. T. (1986). The effectiveness of external and middle ears in coupling acoustic power into the cochlea. In *Peripheral Auditory Mechanisms* (eds J. B. Allen, J. L. Hall, A. Hubbard, S. T. Neely and A. Tubis), pp. 3–12. Springer, Berlin.

Roth, G. L., Aitkin, L. M., Andersen, R. A. and Merzenich, M. M. (1978). Some features of the spatial organization of the central nucleus of the inferior colliculus of the cat. *J. Comp. Neurol.* **182**, 661–680.

Rouiller, E.M. and Ryugo, D.K. (1984). Intracellular marking of physiologically characterized cells in the ventral cochlear nucleus of the cat. *J. Comp. Neurol.* **225**, 167–186.

Ruggero, M. A. and Rich, N. C. (1987). Timing of spike initiation in cochlear afferents: dependence on site of innervation. *J. Neurophysiol.* **58**, 379–403.

Ruggero, M. A., Santi, P. A. and Rich, N. C. (1982). Type II cochlear ganglion cells in the chinchilla. *Hearing Res.* **8**, 339–356.

Runhaar, G. and Manley, G. A. (1987). Potassium concentration in the inner sulcus is perilymph-like. *Hearing Res.* **29**, 93–103.

Russell, I. J. (1983). Origin of the receptor potential in inner hair cells of the mammalian cochlea – evidence for Davis' theory. *Nature* **301**, 334–336.

Russell, I. J. and Cowley, E. M. (1983). The influence of transient asphyxia on receptor potentials in inner hair cells of the guinea pig cochlea. *Hearing Res.* **11**, 373–384.

Russell, I. J. and Sellick, P. M. (1978). Intracellular studies of hair cells in the mammalian cochlea. *J. Physiol.* **284**, 261–290.

Russell, I. J. and Sellick, P. M. (1983). Low-frequency characteristics of intracellularly recorded receptor potentials in guinea-pig cochlear hair cells. *J. Physiol.* **338**, 179–206.

Russell, I. J., Cody, A. R. and Richardson, G. P. (1986a). The responses of inner and outer hair cells in the basal turn of the guinea-pig cochlea grown *in vitro*. *Hearing Res.* **22**, 199–216.

Russell, I. J., Richardson, G. P. and Cody, A. R. (1986b). Mechanosensitivity of mammalian auditory hair cells *in vitro*. *Nature* **321**, 517–519.

Rutherford, W. (1886). A new theory of hearing. *J. Anat. Physiol.* **21**, 166–168.

Ryan, A. and Miller, J. (1977). Effects of behavioral performance on single-unit firing patterns in inferior colliculus of rhesus monkey. *J. Neurophysiol.* **40**, 943–956.

Ryan, A. and Miller, J. (1978). Single unit responses in the inferior colliculus of the awake and performing rhesus monkey. *Exp. Brain Res.* **32**, 389–407.

Ryan, A. F. and Dallos, P. (1984). Physiology of the cochlea. In *Hearing Disorders*, (ed. J. L. Northern), pp. 253–266. Little Brown, Boston.

Ryan, A. F., Wickham, G. M. and Bone, R. C. (1980). Studies of ion distribution in the inner ear: scanning electron microscopy and X-ray microanalysis of freeze-dried cochlear specimens. *Hearing Res.* **2**, 1–20.

Ryugo, D. K. and Weinberger, N. M. (1976). Corticofugal modulation of the medial geniculate body. *Exp. Neurol.* **51**, 377–391.

Sachs, M. B. (1984a). Neural coding of complex sounds: speech. *Ann. Rev. Physiol.* **46**, 261–273.

Sachs, M. B. (1984b). Speech encoding in the auditory nerve. In *Hearing Science* (ed. C. I. Berlin), pp. 263–307. College-Hill Press, San Diego.

Sachs, M. B. and Abbas, P. J. (1974). Rate versus level functions for auditory-nerve fibers in cats: tone-burst stimuli. *J. Acoust. Soc. Am.* **56**, 1835–1847.

Sachs, M. B. and Kiang, N. Y.-S. (1968). Two-tone inhibition in auditory nerve fibers. *J. Acoust. Soc. Am.* **43**, 1120–1128.

Sachs, M. B. and Young, E. D. (1979). Encoding of steady-state vowels in the auditory nerve: representation in terms of discharge rate. *J. Acoust. Soc. Am.* **66**, 470–479.

Sachs, M. B. and Young, E. D. (1980). Effects of nonlinearities on speech encoding in the auditory nerve. *J. Acoust. Soc. Am.* **68**, 858–875.

Sand, O. (1975). Effects of different ionic environments on the mechano-sensitivity of lateral line organs in the mudpuppy. *J. Comp. Physiol. Ser. A* **102**, 27–42.

Santi, P. A. and Anderson, C. B. (1987). A newly identified surface coat on cochlear hair cells. *Hearing Res.* **27**, 47–65.

Saunders, J. C., Bock, G. R., James, R. and Chen, C. S. (1972). Effects of priming for audiogenic seizure on auditory evoked responses in the cochlear nucleus and inferior colliculus of BALB/C mice. *Exp. Neurol.* **37**, 388–394.

Schacht, J. (1986). Molecular mechanisms of drug-induced hearing loss. *Hearing Res.* **22**, 297–304.

Schacht, J. and Zenner, H.-P. (1987). Evidence that phosphoinositides mediate motility in cochlear outer hair cells. *Hearing Res.* **31**, 155–160.

Scharf, B. (1970). Critical bands. In *Foundations of Modern Auditory Theory* (ed. J. V. Tobias), Vol. 1, pp. 159–202. Academic Press, New York and London.

Scharf, B. and Meiselman, C. H. (1977). Critical bandwidth at high intensities. In *Psychophysics and Physiology of Hearing* (eds E. F. Evans and J. P. Wilson), pp. 221–232. Academic Press, London.

Scharlock. D. P., Neff, W. D. and Strominger, N. L. (1965). Discrimination of tone duration after bilateral ablation of cortical auditory areas. *J. Neurophysiol.* **28**, 673–681.

Scheibel, M. E. and Scheibel, A. B. (1974). Neurophil organization in the superior olive of the cat. *Exp. Neurol.* **43**, 339–348.

Schindler, R. A., Merzenich, M. M., White, M. W. and Bjorkroth, B. (1977). Multielectrode cochlear implants: nerve survival and stimulation patterns. *Arch. Otolarynyol.* **103**, 691–699.

Schindler, R. A., Kessler, D. K., Rebscher, S. J. and Yanda, J. (1986). The University of California, San Francisco/Storz cochlear implant program. *Otolaryngol. Clin. North Am.* **19**, 287–305.

Schouten, M. E. H. (1980). The case against a speech mode of perception. *Acta Psychologica* **44**, 71–98.

Schreiner, C. E. and Cynader, M. S. (1984). Basic functional organization of second auditory cortical field (AII) of the cat. *J. Neurophysiol.* **51**, 1284–1305.

Schreiner, C. E. and Urbas, J. V. (1986). Representation of amplitude modulation in the auditory cortex of the cat. I. The anterior auditory field (AAF). *Hearing Res.* **21**, 227–241.

Schubert, D. (1978). History of research on hearing. In *Handbook of Perception* (eds E. C. Carterette and M. P. Friedman), Vol. 4, pp. 41–80. Academic Press, New York and London.

Schuknecht, H. F. (1960). Neuroanatomical correlates of auditory sensitivity and pitch discrimination in the cat. In *Neural Mechanisms of the Auditory and Vestibular Systems* (eds G. L. Rasmussen and W. F. Windle), pp. 76–90. Thomas, Springfield.

Seldon, H. L. (1981a). Structure of human auditory cortex. I. Cytoarchitectonics and dendritic distributions. *Brain Res.* **229**, 277–294.

Seldon, H. L. (1981b). Structure of human auditory cortex. II. Axon distributions and morphological correlates of speech perception. *Brain Res.* **229**, 295–310.

Seldon, H. L. (1985). The anatomy of speech perception: human auditory cortex. In *Cerebral Cortex, Vol. 4, Association and Auditory Cortices* (eds A. Peters and E. G. Jones), pp. 273–327. Plenum, New York.

Sellick, P. M. and Johnstone, B. M. (1974). Differential effects of ouabain and ethacrynic acid on labyrinthine potentials. *Pflügers Arch.* **352**, 339–350.

Sellick, P. M. and Russell, I. J. (1979). Two-tone suppression in cochlear hair cells. *Hearing Res.* **1**, 227–236.

Sellick, P. M. and Russell, I. J. (1980). The responses of inner hair cells to basilar membrane velocity during low-frequency auditory stimulation in the guinea pig. *Hearing Res.* **2**, 439–445.

Sellick, P. M., Patuzzi, R. and Johnstone, B. M. (1982). Measurement of basilar membrane motion in the guinea pig using the Mössbauer technique. *J. Acoust. Soc. Am.* **72**, 131–141.

Sellick, P. M., Yates, G. K. and Patuzzi, R. (1983). The influence of Mössbauer source size and position on phase and amplitude measurements of the guinea pig basilar membrane. *Hearing Res.* **10**, 101–108.

Semple, M. N. and Aitkin, L. M. (1979). Representation of sound frequency and laterality by units in central nucleus of cat inferior colliculus. *J. Neurophysiol.* **42**, 1626–1639.

Semple, M. N., Aitkin, L. M., Calford, M. B., Pettigrew, J. D. and Phillips, D. P. (1983). Spatial receptive fields in the cat inferior colliculus. *Hearing Res.* **10**, 203–215.

Shamma, S. A. (1985a). Speech processing in the auditory system I: The representation of speech sounds in the responses of the auditory nerve. *J. Acoust. Soc. Am.* **78**, 1612–1621.

Shamma, S. A. (1985b). Speech processing in the auditory system II: Lateral inhibition and the central processing of speech evoked activity in the auditory nerve. *J. Acoust. Soc. Am.* **78**, 1622–1632.

Shamma, S. A. and Symmes, D. (1985). Patterns of inhibition in auditory cortical cells in awake squirrel monkeys. *Hearing Res.* **19**, 1–13.

Shaw, E. A. G. (1974). The external ear. In *Handbook of Sensory Physiology*, (eds W. D. Keidel and W. D. Neff), Vol. 5/1, pp. 455–490. Springer, Berlin.

Shaw, E. A. G. and Stinson, M. R. (1983). The human external and middle ear: models and concepts. In *Mechanics of Hearing* (eds E. de Boer and M. A. Viergever), pp. 3–18. Martinus Nijhoff, The Hague.

Sherman G. F., Galaburda, A. M. and Geschwind, N. (1982). Neuroanatomical asymmetries in non-human species. *Trends in Neurosci.* **5**, 429–431.

Shofner, W. P. and Young, E. D. (1985). Excitatory/inhibitory response types in the cochlear nucleus: relationships to discharge patterns and responses to electrical stimulation of the auditory nerve. *J. Neurophysiol.* **54**, 917–939.

Shofner, W. P. and Young, E. D. (1987). Inhibitory connections between AVCN and DCN: evidence from lidocaine injection in AVCN. *Hearing Res.* **29**, 45–53.

Siegel, J. H. and Dallos, P. (1986). Spike activity recorded from the organ of Corti. *Hearing Res.* **22**, 245–248.

Siegel, J. H. and Kim, D. O. (1982). Efferent neural control of cochlear mechanics? Olivocochlear bundle stimulation affects cochlear biomechanical nonlinearity. *Hearing Res.* **6**, 171–182.

Siegel, J. H., Kim, D. O. and Molnar, C. E. (1982). Effects of altering organ of

Corti on cochlear distortion products $f_2 - f_1$ and $2f_1 - f_2$. *J. Neurophysiol.* **47**, 303–328.

Simmons, F. B. (1964). Perceptual theories of middle ear muscle function. *Ann. Otol. Rhinol. Laryngol.* **73**, 724–740.

Simmons, F. B., Matthews, R. G., Walker, M. G. and White, R. L. (1979). A functioning multichannel auditory nerve stimulator. *Acta Otolaryngol.* **87**, 170–175.

Slepecky, N. (1986). Overview of mechanical damage to the inner ear: noise as a tool to probe cochlear function. *Hearing Res.* **22**, 307–321.

Slepecky, N. and Chamberlain, S. C, (1985a). Immunoelectron microscopic and immunofluorescent localization of cytoskeletal and muscle-like contractile proteins in inner ear sensory hair cells. *Hearing Res.* **20**, 245–260.

Slepecky, N. and Chamberlain, S. C. (1985b). The cell coat of inner ear sensory and supporting cells as demonstrated by ruthenium red. *Hearing Res.* **17**, 281–288.

Smith, C. A. (1968). Ultrastructure of the organ of Corti. *Adv. Sci.* **24**, 419–433.

Smith, C. A. (1978). Structure of the cochlear duct. In *Evoked Electrical Activity in the Auditory Nervous System* (eds R. F. Naunton and C. Fernandez), pp. 3–19. Academic Press, London.

Smith, C. A., Lowry, O. H. and Wu, M. L. (1954). The electrolytes of the labyrinthine fluids. *Laryngoscope* **64**, 141–153.

Smith, D. E. and Moskowitz, N. (1979). Ultrastructure of layer IV of the primary auditory cortex of the squirrel monkey. *Neuroscience* **4**, 349–359.

Smith, R. L. (1979). Adaptation, saturation and physiological masking in single auditory-nerve fibers. *J. Acoust. Soc. Am.* **65**, 166–178.

Smith, R. L. and Brachman, M. L. (1980). Response modulation of auditory-nerve fibers by AM stimuli: effects of average intensity. *Hearing Res.* 2, 123–133.

Smolders, J. W. T., Aertsen, A. M. H. J. and Johannesma, P. I. M. (1979). Neural representation of the acoustic biotope: a comparison of the response of auditory neurons to tonal and natural stimuli in the cat. *Biol. Cybern.* **35**, 11–20.

Sohmer, H. (1966). A comparison of the efferent effects of the homolateral and contralateral olivo-cochlear bundles. *Acta Otolaryngol.* **62**, 74–87.

Sousa-Pinto, A. (1973). The structure of the first auditory cortex in the cat. I. Light microscopic observations on its organization. *Arch. Ital. Biol.* **111**, 112–137.

Sovijärvi A. R. A. and Hyvärinen J. (1974). Auditory cortical neurons in the cat sensitive to the direction of sound source movement. *Brain Res*, **73**, 455–471.

Spangler, K. M., Cant, N. B., Henkel, C. K., Farley, G. R. and Warr, W. B. (1987). Descending projections from the superior olivary complex to the cochlear nucleus of the cat. *J. Comp. Neurol.* **259**, 452–465.

Spoendlin, H. (1968). Ultrastructure and peripheral innervation pattern of the receptor in relation to the first coding of the acoustic message. In *Hearing Mechanisms in Vertebrates*, (eds A. V. S. de Reuck and J. Knight), pp. 89–119. Churchill, London.

Spoendlin, H. (1972). Innervation densities of the cochlea. *Acta Otolaryngol.* **73**, 235–248.

Spoendlin, H. (1978). The afferent innervation of the cochlea. In *Evoked Electrical Activity in the Auditory Nervous System* (eds R. F. Naunton and C. Fernandez), pp. 21–39. Academic Press, London.

Spoendlin, H. and Lichtensteiger, W. (1966). The adrenergic innervation of the labyrinth. *Acta Otolaryngol.* **61**, 423–434.

Starr, A. and Wernick, J. S. (1968). Olivocochlear bundle stimulation: effects on spontaneous and tone-evoked activities of single units in cat cochlear nucleus. *J. Neurophysiol.* **31**, 549–564.

Steinschneider, M., Arezzo, J. and Vaughan, H. G. (1982). Speech evoked activity in the auditory radiations and cortex of the awake monkey. *Brain Res.* **252**, 353–365.

Strominger, N. L. (1969). Localization of sound in space after unilateral and bilateral ablation of auditory cortex. *Exp. Neurol.* **25**, 521–533.

Swanson, L. W. and Hartman, B. K. (1975). The central adrenergic system. An immunofluorescence study of the location of cell bodies and their efferent connections in the rat utilizing dopamine-β-hydroxylase as a marker. *J. Comp. Neurol.* **163**, 467–506.

Syka, J., Sykova, E., Patuzzi, R and Johnstone, B.M. (1987). Potassium concentration changes in the organ of Corti during loud sound stimulation. *Inner Ear Biology Abstracts, 24 Mtg, Nijmegen*, p. 56.

Tasaki. I. (1954). Nerve impulses in individual auditory nerve fibers of guinea pig. *J. Neurophysiol.* **17**, 97–122.

Tasaki, I. and Spyropoulos, C. S. (1959). Stria vascularis as a source of endocochlear potential. *J. Neurophysiol.* **22**, 149–155

Tasaki, I., Davis, H. and Legouix, J. P. (1952). The space–time pattern of the cochlear microphonics (guinea pig), as recorded by differential electrodes. *J. Acoust. Soc. Am.* **24**, 502–519.

Tasaki, I., Davis, H. and Eldredge, D. H. (1954). Exploration of cochlear potentials in guinea pig with a microelectrode. *J. Acoust. Soc. Am.* **26**, 765–773.

Thompson, G. C. and Masterton, R. B. (1978). Brain stem auditory pathways involved in reflexive head orientation to sound. *J. Neurophysiol.* **41**, 1183–1202.

Thompson, R. F. (1960). Function of auditory cortex of cat in frequency discrimination. *J. Neurophysiol.* **23**, 321–334.

Tilney, L. G., DeRosier, D. J. and Mulroy, M. J. (1980). The organization of actin filaments in the stereocilia of cochlear hair cells. *J. Cell Biol.* **86**, 244–259.

Tilney, L. G., Saunders, J. C., Egelman, E. and DeRosier, D. J. (1982). Changes in the organization of actin filaments in the stereocilia of noise-damaged lizard cochleae. *Hearing Res.* **7**, 181–197.

Tilney L. G., Egelman, E. H., DeRosier, D. J. and Saunders, J. C. (1983). Actin filaments, stereocilia, and hair cells of the bird cochlea. II. Packing of actin filaments in the stereocilia and in the cuticular plate and what happens to the organization when the stereocilia are bent. *J. Cell Biol.* **96**, 822–834.

Tong, Y., Clark, G., Blamey, P., Busby, P. and Dowell, R. (1982). Psychophysical studies for two multiple-channel cochlear implant patients. *J. Acoust. Soc. Am.* **71**, 153–160.

Townshend, B., Cotter, N., van Compernolle, D. and White, R. L. (1987). Pitch perception by cochlear implant subjects. *J. Acoust. Soc. Am.* **82**, 106–115.

Trahiotis, C. and Elliott, D. N. (1970). Behavioral investigation of some possible effects of sectioning the crossed olivocochlear bundle. *J. Acoust. Soc. Am.* **47**, 592–596.

Tsuchitani, C. (1977). Functional organization of lateral cell groups of cat superior olivary complex. *J. Neurophysiol.* **40**, 296–318.

Tsuchitani, C. and Boudreau, J. C. (1966). Single unit analysis of cat superior olive S-segment with tonal stimuli. *J. Neurophysiol.* **29**, 684–697.

Tunturi, A. R. (1952). A difference in the representation of auditory signals for the left and right ears in the iso-frequency contours of the right middle ectosylvian auditory cortex of the dog. *Am. J. Physiol.* **168**, 712–727.

Ulehlova, L., Voldrich, L. and Janisch, R. (1987). Correlative study of sensory cell density and cochlear length in humans. *Hearing Res.* **28**, 149–151.

Viergever, M. A. (1986). Cochlear macromechanics – a review. In *Peripheral Auditory Mechanisms* (eds J. B. Allen, J. L. Hall, A. Hubbard, S. T. Neely and A. Tubis), pp. 63–72. Springer, Berlin.

Viergever, M. A. and Diependaal, R. J. (1986). Quantitative validation of cochlear models using the Liouville-Green approximation. *Hearing Res.* **21**, 1–15.

Voigt, H. F. and Young, E. D. (1980). Evidence for inhibitory interactions between neurons in dorsal cochlear nucleus. *J. Neurophysiol.* **44**, 76–96.

Walsh, S. M., Merzenich, M. M., Schindler, R. A. and Leake-Jones, P. A. (1980). Some practical considerations in development of multichannel scala tympani prostheses. *Audiology* **19**, 164–175.

Warr, W. B. (1975). Olivocochlear and vestibular efferent neurons of the feline brain-stem: their location, morphology and number determined by retrograde axonal transport and acetylcholinesterase histochemistry. *J. Comp. Neurol.* **161**, 159–182.

Warr, W. B. (1978). The olivocochlear bundle: its origins and terminations in the cat. In *Evoked Electrical Activity in the Auditory Nervous System* (eds R. F. Naunton and C. Fernandez), pp. 43–63. Academic Press, New York and London.

Warr, W. B. and Guinan, J. J. (1979). Efferent innervation of the organ of Corti: two separate systems. *Brain Res.* **173**, 152–155.

Warr, W. B., Guinan, J. J. and White, J. S. (1986). Organization of efferent fibers: the lateral and medial olivocochlear systems. In *The Neurobiology of Hearing: The Cochlea* (eds R. A. Altschuler, R. P. Bobbin and D. W. Hoffman), pp. 333–348. Raven Press, New York.

Watanabe, T., Yanagisawa, K., Kanzaki, J. and Katsuki, Y. (1966). Cortical efferent flow influencing unit responses of medial geniculate body to sound stimulation. *Exp. Brain Res.* **2**, 302–317.

Wegener, J. G. (1973). The sound localizing behavior of normal and brain-damaged monkeys. *J. Aud. Res.* **13**, 191–219.

Weinberger, N. M. and Diamond D. M. (1987). Physiological plasticity in auditory cortex: rapid induction by learning. *Prog. Neurobiol.* **29**, 1–55.

Wenthold, R. J. (1987). Evidence for a glycinergic pathway connecting the two cochlear nuclei: an immunocytochemical and retrograde transport study. *Brain Res.* **415**, 183–187.

Wenthold, R. J. and Martin, M. R, (1984). Neurotransmitters of the auditory nerve and central nervous system. In *Hearing Science* (ed. C. I. Berlin), pp. 341–369. College-Hill Press, San Diego.

Wever, E. G. (1949). *Theory of Hearing.* Wiley, New York.

Wever, E. G. and Bray, C. W. (1930). Action currents in the auditory nerve in response to acoustical stimulation. *Proc. Natl Acad. Sci. USA* **16**, 344–350.

Wever, E. G. and Lawrence, M. (1954). *Physiological Acoustics.* Princeton University Press, Princeton.

Wever, E. G. and Vernon, J. A. (1955). The effects of the tympanic muscle reflexes upon sound transmission. *Acta Otolaryngol.* **45**, 433–439.

White, M. W. (1983). Formant frequency discrimination and recognition in subjects implanted with intracochlear stimulating electrodes. *Ann. NY Acad. Sci.* **405**, 348–359.

Whitfield, I. C. (1979). The object of the sensory cortex. *Brain Behav. Evol.* **16**, 129–154.

Whitfield, I. C. and Evans, E. F. (1965). Responses of auditory cortical neurons to stimuli of changing frequency. *J. Neurophysiol.* **28**, 655–672.

Whitfield, I. C. and Purser, D. (1972). Microelectrode study of the medial geniculate body in unanaesthetised, free-moving cats. *Brain Behav. Evol.* **6**, 311–322.

Wiederhold, M. L. (1970). Variations in the effects of electric stimulation of the crossed olivocochlear bundle of cat single auditory-nerve-fiber responses to tone bursts. *J. Acoust. Soc. Am.* **48**, 966–977.

Wiederhold, M. L. (1986). Physiology of the olivocochlear system. In *Neurobiology of Hearing: The Cochlea* (eds R. A. Altschuler, R. P. Bobbin and D. W. Hoffman), pp. 349–370. Raven Press, New York.

Wiener, F. M. and Ross, D. A. (1946). The pressure distribution in the auditory canal in a progressive sound field. *J. Acoust. Soc. Am.* **18**, 401–408.

Willott, J. F., Shnerson. A. and Urban, G. P. (1979). Sensitivity of the acoustic startle response and neurons in subnuclei of the mouse inferior colliculus to stimulus parameters. *Exp. Neurol.* **65**, 625–644.

Wilson, J. P. (1980a). Subthreshold mechanical activity within the cochlea. *J. Physiol.* **298**, 32–33P.

Wilson, J. P. (1980b). Evidence for a cochlear origin for acoustic re-emissions, threshold fine structure and tonal tinnitus. *Hearing Res.* **2**, 233–252.

Wilson, J. P. and Johnstone, J. R. (1975). Basilar membrane and middle ear vibration in guinea pig measured by capacitive probe. *J. Acoust. Soc. Am.* **57**, 705–723.

Winer, J. A. (1985a). The medial geniculate body of the cat. *Adv. Anat. Embryol. Cell Biol.* **86**, 1–98.

Winer, J. A. (1985b). Structure of layer II in cat primary auditory cortex (AI). *J. Comp. Neurol.* **238**, 10–37.

Winer, J. A., Diamond, I. T. and Raczkowski, D. (1977). Subdivisions of the auditory cortex of the cat: the retrograde transport of horseradish peroxidase to the medial geniculate body and posterior thalamic nuclei. *J. Comp. Neurol.* **176**, 387–417.

Winslow, R. L. and Sachs, M. B. (1987). Effect of electrical stimulation of the crossed olivocochlear bundle on auditory nerve response to tones in noise. *J. Neurophysiol.* **57**, 1002–1021.

Woolsey, C. N. (1960). Organization of cortical auditory system: a review and a synthesis. In *Neural Mechanisms of the Auditory and Vestibular Systems* (eds G. L. Rasmussen and W. F. Windle), pp. 165–180. Thomas, Springfield.

Wright, C. G. and Barnes, C. D. (1972). Audio-spinal reflex responses in decerebrate and chloralose-anaesthetised cats. *Brain Res.* **36**, 307–331.

Yates, G. K. (1986). Frequency selectivity in the auditory periphery. In *Frequency Selectivity in Hearing* (ed. B. C. J. Moore), pp. 1–50. Academic Press, London.

Yin, T. C. T. and Kuwada, S. (1983). Binaural interaction in low-frequency neurons in inferior colliculus of the cat. III. Effects of changing frequency. *J. Neurophysiol.* **50**, 1020–1042.

Yin, T. C. T., Chan, J. C. K. and Irvine, D. R. F. (1986). Effects of interaural time delays of noise stimuli on low-frequency cells in the cat's inferior colliculus. I. Responses to wideband noise. *J. Neurophysiol.* **55**, 280–300.

Young, E. D. (1984). Response characteristics of neurons of the cochlear nuclei. In *Hearing Science* (ed. C. I. Berlin), pp. 423–460. College-Hill Press, San Diego.

Young, E. D. and Brownell, W. E. (1976). Responses to tones and noise of single cells in dorsal cochlear nucleus of unanesthetized cats. *J. Neurophysiol.* **39**, 282–300.

Young, E. D. and Sachs, M. B. (1979). Representation of steady-state vowels in the temporal aspects of the discharge patterns of populations of auditory-nerve fibers. *J. Acoust. Soc. Am.* **66**, 1381–1403.

Young, E. D. and Voigt, H. F. (1982). Response properties of type II and type III units in dorsal cochlear nucleus. *Hearing Res.* **6**, 153–169.

Zakrisson, J.-E. and Borg, E. (1974). Stapedius reflex and auditory fatigue. *Audiology* **13**, 231–235.

Zenner, H.- P. (1981). Cytoskeletal and muscle-like elements in cochlear hair cells. *Arch. Otorhinolaryngol.* **230**, 81–92.

Zenner, H.-P. (1986). Motile responses in outer hair cells. *Hearing Res.* **22**, 83–90.

Zenner, H.-P., Zimmerman, U. and Schmitt, U. (1985). Reversible contraction of isolated mammalian cochlear hair cells. *Hearing Res.* **18**, 127–133.

Zwicker, E. (1970). Masking and psychological excitation as consequences of the ear's frequency analysis. In *Frequency Analysis and Periodicity Detection in Hearing* (eds R. Plomp and G. F. Smoorenberg), pp. 376–394. Sijthoff, Leiden.

Zwicker, E. (1974). On a psychoacoustical equivalent of tuning curves. In *Facts and Models in Hearing* (eds E. Zwicker and E. Terhardt), pp. 132–141. Springer, Berlin.

Zwicker, E. (1986a). A hardware cochlear nonlinear preprocessing model with active feedback. *J. Acoust. Soc. Am.* **80**, 146–153.

Zwicker, E. (1986b). "Otoacoustic" emissions in a nonlinear cochlear hardware model with feedback. *J. Acoust. Soc. Am.* **80**, 154–162.

Zwicker, E. (1986c). Suppression and $(2f_1 - f_2)$-difference tones in a nonlinear cochlear preprocessing model with active feedback. *J. Acoust. Soc. Am.* **80**, 163–176.

Zwicker, E. and Scharf, B. (1965). A model of loudness summation. *Psychol. Rev.* **72**, 3–26.

Zwicker, E., Flottorp, G. and Stevens, S. S. (1957). Critical band-width in loudness summation. *J. Acoust. Soc. Am.* **29**, 548–557.

Zwislocki, J. J. (1965). Analysis of some auditory characteristics. In *Handbook of Mathematical Psychology* (eds R. Luce, R. Bush and E. Galanter), Vol. 3, pp. 1–97. Wiley, New York.

Zwislocki, J. J. (1979). Tectorial membrane: a possible sharpening effect on the frequency analysis in the cochlea. *Acta Otolaryngol.* **87**, 267–269

Zwislocki J. J. (1986). Are nonlinearities observed in firing rates of auditory nerve afferents reflections of a nonlinear coupling between the tectorial membrane and the organ of Corti? *Hearing Res.* **22**, 217–221.

Index